Crop Pollination by Bees, Volume 1

Evolution, Ecology, Conservation, and Management

2nd Edition

Crop Pollination by Bees, Volume 1

Evolution, Ecology, Conservation, and Management

2nd Edition

Keith S. Delaplane

CABI

CABI is a trading name of CAB International

CABI
Nosworthy Way
Wallingford
Oxfordshire OX10 8DE
UK

CABI
WeWork
One Lincoln St
24th Floor
Boston, MA 02111
USA

Tel: +44 (0)1491 832111
Fax: +44 (0)1491 833508
E-mail: info@cabi.org
Website: www.cabi.org

Tel: +1 (617)682-9015
E-mail: cabi-nao@cabi.org

A catalogue record for this book is available from the British Library, London, UK.

Library of Congress Cataloging-in-Publication Data

Names: Delaplane, Keith S, author.
Title: Crop Pollination by Bees, 2nd Edition, Volume 1: Evolution, Ecology, Conservation, and Management / Keith S Delaplane.
Description: Second edition. | Boston, MA, USA : CAB International, [2021] | At head of title: Volume I. | Includes bibliographical references and index. | Contents: Angiosperms and Bees: The Evolutionary Bases of Crop Pollination -- Biology of Bees -- What Makes a Good Pollinator? -- Economic and Ecosystem Benefits of Bee Pollination -- State of the World's Bee Pollinators and its Consequences for Crop Pollination -- Applied Bee Conservation -- Honey Bees -- Bumble Bees -- Managed Solitary Bees -- Wild Bees -- The Stingless Bees, Tribe Meliponini. | Summary: "A practical guide to bees and how they pollinate essential crops. Provides simple, succinct advice on how to increase bee abundance and pollination. Very useful for farmers, horticulturalists, gardeners, and those interested in insect ecology and conservation, including students of entomology and crop protection"-- Provided by publisher.
Identifiers: LCCN 2020050170 (print) | LCCN 2020050171 (ebook) | ISBN 9781786393494 (v. 1 ; paperback) | ISBN 9781786393500 (v. 1 ; ebook) | ISBN 9781786393517 (v. 1 ; epub)
Subjects: LCSH: Pollination by bees. | Honeybee. | Bee culture. | Food crops--Breeding.
Classification: LCC QK926 .D35 2021 (print) | LCC QK926 (ebook) | DDC 571.8/642--dc23
LC record available at https://lccn.loc.gov/2020050170
LC ebook record available at https://lccn.loc.gov/2020050171

References to Internet websites (URLs) were accurate at the time of writing.

ISBN-13: 9781786393494 (Paperback)
        9781786393500 (ePDF)
        9781786393517 (ePub)

DOI: 10.1079/9781786393494.0000

Commissioning Editor: Ward Cooper
Editorial Assistant: Lauren Davies
Production Editor: Tim Kapp

Typeset by SPi, Pondicherry, India

To my beloved Pilar, who celebrated every word and wanted to see this day.

Contents

Author Biography

Dr Keith S. Delaplane is Professor of Entomology and Walter B. Hill Fellow at the University of Georgia, USA. He received his MS and PhD degrees from Louisiana State University and his BS degree from Purdue University. At the University of Georgia he pursues research interests in honeybee social evolution, bee health management and pollination. He was inducted into the Most Excellent Order of the British Empire as an Honorary Member for his work on beekeeping research and education in the UK. Professor Delaplane lives in Athens, Georgia, USA.

Preface to the 2000 Edition

Pollination is the most important contribution bees make to human economies. The value of honey and beeswax pales in comparison to the value of fruits, vegetables, seeds, oils and fibres whose yields are optimized by pollinating bees. There was a time when it was relatively easy to overlook this benefit, and it may still be possible in particular areas and cropping systems in which there are large and sustainable populations of bees, whether managed or naturally occurring. In such places the rich background of pollinators means that pollination is rarely a limiting factor in crop production. Many parts of North America prior to the 1980s fit this description. The cropping systems and pollinator demographics in many countries, however, are changing profoundly, and a let-alone approach to pollination will prove increasingly inadequate for meeting the demands for an abundant, high-quality food supply into the 21st century.

It is becoming manifestly clear that our bee pollinators are a valuable and limited natural resource that should be conserved and encouraged at all costs. This awareness stems in part from an apparent decline of the western honeybee *Apis mellifera*, that is occurring in many parts of the world. The decline of honeybees stems from more than one cause, but the most straightforward explanation is the rapid spread of parasitic *Varroa* spp. mites that occurred worldwide in the closing decades of the 20th century. *Varroa* spp. is relatively innocuous on its natural host, the eastern honeybee *A. cerana*, but on *A. mellifera* it is devastating. The parasite occurs now on every continent on which *A. mellifera* is kept, except Australia, and it is considered the most serious health threat to apiculture (Matheson, 1993, 1995). The perception of a 'pollination crisis' proceeds also from a general increase in the area of bee-pollinated crops. In some countries the demand for pollination is increasing at the very time that the supply of managed pollinators is decreasing.

The so-called pollination crisis has generated a renewed interest in the management, culture and conservation of pollinating bees. We believe that it also creates the need for an updated book on applied bee management and conservation for crop pollination.

We are heavily indebted to two authoritative texts, S.E. McGregor's (1976) *Insect Pollination of Cultivated Crop Plants* and J.B. Free's (1993) *Insect Pollination of Crops*, 2nd edition. These texts virtually define the state of the science of crop pollination and remain the first stop for academics looking for comprehensive research reviews. With this volume our goal was not to duplicate another comprehensive review, but rather to synthesize the latest scientific literature into principles and practices that are relevant to workers in crop pollination. This volume is primarily for agricultural consultants, extension specialists, plant and bee conservationists, crop growers, beekeepers, and others with an interest in applied pollination.

We concentrate on bee-pollinated crops of significant or emerging economic importance in the temperate developed world, crops for which there is a strong bee pollination story in the literature, and crops for which pollination is historically a limiting factor. Pollination is a multifaceted component of crop production and not easily reduced to formulaic recommendations. Nevertheless, some practical guidance should come out of a book like this if we hope to help crop growers and beekeepers. One example is a recommended density of bees. This information is difficult to synthesize because the literature is often scarce or incongruent. It is scarce because it is difficult and expensive to experimentally control large acreages for rigorous scientific studies or to separate out the contribution of any one bee species. It is incongruent because results vary among different regions and researchers do not always test the same hypotheses or measure the same parameters. Rather than weary readers with a review of this difficult literature, we present research and extension service guidelines in table format for most crops and give a literature average for recommended bee densities. Although other considerations must enter the decision making process, this approach gives growers and beekeepers a rational starting point.

In much of the developed world, the last 30 years have seen changes in the beekeeping industry that approach in magnitude the technological revolutions of the 19th century. Chemical controls aimed at parasitic *Varroa* spp. mites have transformed the industry from one that was relatively pesticide-free to one that is now virtually pesticide-dependent. Africanized honeybees, a highly defensive race of bee introduced to Brazil from Africa in the 1950s, spread through tropical and subtropical regions of the Americas, altering beekeeping practices, raising liability risks, disrupting crop pollination and competing with native pollinators. Faced with problems like these, many beekeepers have gone out of business, leaving behind a pollination vacuum.

One result is a renewed interest in species other than honeybees, some of which are very good pollinators. Called 'non-managed bees', pollen bees, wild bees or non-*Apis* bees, these are solitary or social bees that nest primarily in simple burrows in grass thatch, wood, plant stems or soil. Methods for mass-rearing most of them are impractical, and their management often translates to conserving and enhancing wild populations. Bee conservation is not a mature science; in Europe it is in its adolescence; in North America it is embryonic. In this volume we highlight the emerging principles and, where justified, give recommendations for enhancing populations of those species other than honeybees. This requires some discussion of bee ecology and conservation biology, but here again our goal is to make the science relevant in the context of crop pollination.

Finally, in this volume we hope to engender an appreciation for *all* bee pollinators – managed or non-managed, exotic or native – and an honest recognition of the assets and limitations of each. The western honeybee is an exotic species in much of its modern range. It is rarely the most efficient pollinator, but it is very manageable. Conversely, some native specialist bees are extremely efficient pollinators, but their numbers can be low and unpredictable. It is counterproductive to debate the comparative strengths and weaknesses of different bee pollinators or, even worse, to advocate only one pollinator or group of pollinators. The truth is, we need all the pollinators we can get. The goal of this volume is to promote a large, diverse, sustainable and dependable bee pollinator workforce that can meet the challenge for optimizing food production well into the 21st century.

K.S. Delaplane[1] and D.F. Mayer[2]
[1]Athens, Georgia, USA
[2]Prosser, Washington, USA
October 1999

Preface to the 2021 Edition

The first time I set eyes on the 2000 edition of this volume I was at a vendor's stall at a convention in London. Being from a UK publisher, copies had not yet crossed the Atlantic, so it was a pleasant surprise for me to see that it had been released. I was basking in that author's glow for maybe 30 seconds before a new thought, not altogether pleasant, took root and solidified: a scientific book is a static thing, long in the making and obsolete before the ink dries.

This is certainly true if one's science is a vigorous affair, generous with researchable problems; charismatic in the eyes of the public; attractive to students, young scientists and funding agencies; and attached to outputs that explain the evolution of plant and insect life on this planet while helping to feed its billions of human beings. Crop pollination is that kind of science.

Although the 2000 edition was condemned to the obsolescence natural to books in the sciences, it had a good run for its money and helped summarize the state of 20th century agricultural pollination. I could not shake the feeling that its obsolescence was on an unusually fast track, however. To get a perspective on things, I plotted the annual number of new scientific papers searchable in Google Scholar by the key words 'crop pollination bees' for each year going back to 1980. The result is the following graph.

The 2000 edition was in the vanguard of an explosion of new knowledge on crop pollination by bees that continues to this day. What is driving this? Which on-ground indicators? What philosophies are ascendant that compel universities to create and fill research positions in pollinator conservation and crop pollination and, equally importantly, motivate granting agencies to fund their research?

I think the first answer is an awakening to the essential fiction of an autonomous agriculture independent of the webs of connectivity that enliven and stabilize natural ecosystems, a stability in which agriculture, differently practised, could participate. Second, and deriving from this framing mindset, is an understanding of the magnitude and quality of the pollination performed by wild bees.

These animators follow on the heels of decades of pollination centred on managed honeybees which itself draws from a broader historic context. Industrial agriculture in the 20th century was, at its nadir, functionally ambivalent to nature, imagining itself more or less independent of strictures of ecology and geoscience. Instead of valuing the profit-giving, sustainable and free benefits of ecosystem services, there was an approach that first simplified the ecosystem to an extreme then reintroduced its necessary services in the form of caricatures of the real processes: synthetic fertilizers in place of nutrient cycling across trophic levels; groundwater irrigation in place of rainfall; and pesticides in place of the networks of competitors, predators, herbivores, pathogens and parasites that mark a stable ecosystem. The irony is, in such reduced landscapes the effect size of these inputs is huge, reinforcing the delusion that a farm can be hermetically sealed off from nature. By the middle of the 20th century, pollination was understood to be another of those ecosystem services that agriculture cannot do without, and the most obvious candidate for the job – lacking armies of human labourers wielding so many paint brushes of pollen – was the pollinator already in the domestic fold, the honeybee. Numbers of beehives in the USA were entering their post-World War II apogee of 5.9 million in 1947, and the convergence of need and means seemed obvious. The following decades were the hegemony of honeybee pollination, the presumptions of which were captured in Professor Roger Morse's (1991) triumphalist paean 'Honeybees forever' published in the journal *Trends in Ecology and Evolution*.

Even before 1991, however, the cracks in the honeybee monolith were beginning to show; it was a natural extension of the times when appreciation was reawakening for the interconnectedness between food production and natural systems. Today the situation is very different. Thanks to a new generation of entomologists and pollinator conservators, we can state with evidence that non-managed 'wild bees', this catch-all term that includes bees native and exotic, solitary and social, are the heavy lifters of agricultural pollination. The only

Fig. i. Scientific publications searchable in Google Scholar by the key words 'crop pollination bees'.

exceptions are those systems persisting in extreme intensification, where the autonomous agriculture paradigm is so profitable that departures from it are hard to imagine – contexts such as California almond and any kind of greenhouse crop.

I do not call this a revolution. Rather, in my opinion, it is an evolution of the best kind where the merits of diverse species are being recognized, appreciated and integrated. The process has not been without partisanship (see section 7.3, this volume), but what the data are beginning to show and experts are starting to promote is the overriding value of *large and taxonomically diverse local admixtures of pollinators*. When a farm's natural conditions permit a robust assemblage of wild pollinators, there is no need to import managed bees. Indeed, to do so is a waste of money (see section 4.4.2, this volume). Equally, when wild bees and honeybees are in the mix together there is a positive synergy that capitalizes on the pollen-freeing abilities of wild bees and the sheer numbers of honeybees to effect superior pollen movement (see section 7.3.1). In any case, we now know that the foundation of pollination management is the conservation and encouragement of wild bees.

It is my goal to synthesize this burgeoning literature in compact form to a general audience of science-minded readers, with generosity toward all pollinators and love for this beautiful world, in the interest of improving the lives of bees, sustainably and humanely managing their yield-enhancing powers, and justly sharing the fruits of their labours with the whole human family.

Keith S. Delaplane
Athens, Georgia, USA
October 2020

Acknowledgements

Many colleagues helped me with constructive comments, provided images, or answered questions about aspects of their research. I thank Paul Arnold (Young Harris College, USA), Ricardo Ayala (Universidad Nacional Autónoma de Mexico), John Bergstrom (University of Georgia, USA), James Cane (ARS Pollinating Insects Research Unit, Logan, Utah, USA), Sydney Cameron (University of Illinois, USA), Dewey Caron (University of Delaware, USA), Arnon Dag (Gilat Research Center, Israel), Selim Dedej (Toronto, Canada), Elaine Evans (University of Minnesota, USA), Conor Fair (University of Georgia), Christine Cairns Fortuin (University of Georgia), Josh Fuder (University of Georgia Cooperative Extension), Nicola Gallai (Université de Toulouse, France), Jack Garrison (University of Georgia), Jason Gibbs (University of Manitoba, Canada), Ernesto Guzman-Novoa (University of Guelph, Canada), Terry Houston (Western Australian Museum, Australia), Zachary Huang (Michigan State University, USA), Thomas Lawrence (University of Georgia), José Octavio Macías-Macías (Universidad de Guadalajara, Mexico), Lora Morandin (Pollinator Partnership Canada), David Onstad (Corteva Agriscience), Juliet Osborne (University of Exeter, UK), Theresa L. Pitts-Singer (ARS Pollinating Insects Research Unit, Logan, Utah), Francis Ratnieks, (University of Sussex, UK), David Roubik (Smithsonian Tropical Research Institute, Panama), Tim Smalley (University of Georgia), Doug Soltis (University of Florida, USA), James Strange (Ohio State University, USA), Amber Vinchesi-Vahl (University of California Cooperative Extension, USA) and Douglas Walsh (Washington State University, USA).

All images and illustrations by the author unless otherwise indicated.

1 Angiosperms and Bees: The Evolutionary Bases of Crop Pollination

This volume is about one of the most celebrated relationships between species in all of natural history – that relationship between the bees and the flowering plants, the *angiosperms*. To be precise, this volume explores the relationship between bees and those angiosperms that make up modern crop plants that depend on bee pollination.

1.1. Sex: Diversity with Stability

For the plants, it is all about sex – that most extravagantly successful (and arguably popular) invention of natural selection that set multicellular organisms on their path toward global dominance. 'Global dominance?' you ask, 'How's that?' That's a fair question when one considers the other successful life alternatives.

The *single-celled* life alternative is indeed amply represented in Earth's biota. Just consider the bacteria and archaebacteria that carpet the planet, colonizing virtually every terrestrial and aquatic niche, even penetrating kilometres deep into the planet's crust. It is these simplest representatives of the biological continuum that baffle us with their boundary bending tolerances to environmental extremes (Merino *et al.*, 2019), making them figure prominently in our discussions about the evolution of life on other planets (Sundarasami *et al.*, 2019).

At the opposite pole of biological organization we have those assemblies of multicellular organisms who have banded together so tightly that we have to consider the group, not the individuals who make it up, as a Darwinian unit of selection. These we call the *superorganisms* (Wilson and Hölldobler, 2009), most famously represented by the termites, ants, and the social wasps and bees (most wasps and bees are not social), although quirky representatives exist in the forms of a genus of shrimps (*Synalpheus* spp.) and the naked mole rats of Africa (Heterocephalidae). The ecological impact of the superorganisms is wildly out of proportion to their species count. As one example, the ants and termites make up only 2% of the estimated 900,000 known species of insects on Earth, yet together account for more than half of total insect biomass (Wilson and Kinne, 1990). These are nature's great recyclers and soil conditioners. Another example is those superorganisms represented by the social bees; these will figure prominently in this volume about bee pollinators of crop plants, although we will also see that their solitary bee cousins are the real workhorses of pollination. To be plain, it is 'beeness' that makes a good pollinator, not 'superorganismness'.

Superorganismality, however fascinating and ecologically important its representatives, is nevertheless a bit of an evolutionary oddball. As far as we can tell, it has independently evolved only 28 times in the history of Earth (Bourke, 2011); all but two of those independent events happening in the insects.

It is the *multicellular organisms* (hereafter simply 'organisms') who occupy the middle of our biological continuum, those bundles of cooperating eukaryotic cells (cells whose DNA is enclosed in a nucleus) who together form a contiguous entity; share a common genetic fate; specialize for the diverse functions of procuring nutrients, defending self and reproducing; and by one means or another resolve internal genetic conflicts. They are the protists, fungi, plants and animals. Together, they are called the Eukarya, one of life's three *domains*, or highest taxonomic ranks, standing alongside the Eubacteria and Archaea.

If there is a case to be made that organisms are dominant in the grand scheme of things, it is because they have resolved many of the impediments that hazard the single-celled or superorganismal options. The feverish diversity of body plans and life strategies expressed in organisms have let them approach a measure of the global niche penetration achieved by the more nimble single-celled forms. And owing to the genetic clonality of their

© Keith S. Delaplane 2021. *Crop Pollination by Bees,* 2nd Edition, Volume 1 (K.S. Delaplane)
DOI: 10.1079/9781786393494.0001

1

body cells, each organism is far more genetically stable than the superorganism, every member of which is an organism in their own right and always poised for mutiny.

Diversity with stability. It is sex that makes all this possible.

Beginning with the eukaryotic single cells and carrying on into the eukaryotic multicellular organism, sex permitted a fresh roll of the genetic dice with every generation, the repeated pairing of unprecedented gene combinations, providing raw fodder for natural selection to act upon. Gene combinations whose phenotypes favoured their transmittal to the next generation were, by logical extension, preserved; unsuccessful combinations were, with symmetrical extension, not. In this way a population's genes were winnowed and tried against all the extremes its habitat could throw at it. The result was a species optimally adapted to its habitat.

So much for diversity; what about stability?

An emergent outcome of sexual reproduction in organisms is the single-celled *zygote*, or *embryo* – that product of the female's ova fertilized with the male's sperm. In that one special cell reside all the genetic resources of the future individual. After fertilization and when growth conditions permit, the zygote divides, then divides again, then divides again (1, 2, 4, 8, 16, etc.) in exponential progression until the mature organism is in place. However, the critical point here is that at every division the entire genome is replicated in virtual perfection. The somatic cells of an organism are genetically identical. They are clones and by definition cannot be in conflict.

The sexually derived single-celled zygote is thus the genetic bottleneck that harmonizes genetic variation with clonal compatibility. It is the secret to organisms' morphological and behavioural diversity, structural complexity, and ecological success among Earth's biological experiments. It is no accident that it is organisms that come to mind for most of us when we think about life on Earth; it is organisms that Darwin (1859) considered when he wrote *On the Origin of Species*.

Sex is a big deal then, and it was taking place at the very beginning for the angiosperms and plants in general.

1.2. Sex in the Gymnosperms

Rather than begin with primitive plants, let us jump to the *gymnosperms*, the nearest older relatives to the angiosperms (Fig. 1.1). Gymnosperm ovules are 'naked' (hence the Greek name *gymnos*) and remain exposed on the surface of leaf-derived structures called *bracts*, which when tightly concentrated together are called *cones*. The sexual structures are segregated into male cones and female cones. Pollen is transferred from male to female cones by abiotic vectors such as water and wind, the first pollinating agents (Ollerton and Coulthard, 2009). The morphology of windborne pollen reflects its mode of transfer by wind. Under magnification, windborne pollen grains appear dry, smooth and small to moderate in size; moreover, the pollen is produced in huge quantities (Ackerman, 2000). Anyone who lives in pine regions where windborne pollen blankets the landscape every spring, can appreciate the vast scales in quantity and space possible with gymnosperm pollination. However impressive these seasonal surges, from a biological point of view they are indiscriminate in pollen's spread and deposition, profligate in their wastage of it, and ultimately limited in the efficiency by which they ensure plant mating, reproduction, and range expansion.

Among the surfaces indiscriminately dusted with pine pollen are female cones and their exposed ovules. Each ovule excretes a solution called a *pollination drop* that extends beyond the terminus of the *micropyle* – a small opening at the apex of each ovule. (Fig. 1.2). This pollination drop serves as a landing site for airborne pollen. Once pollen lands on it, the drop recedes back into the interior of the ovule, carrying the pollen with it, facilitating pollination and subsequent fertilization and maturation of the seed.

1.3. Flower Morphology and Fertilization

Now for some terms.

A *flower* is a plant organ unique to the angiosperms, evolved for increasing efficiency of sexual reproduction. An *inflorescence* is an arrangement of flowers on a stem. The main stem of an inflorescence is the *peduncle*, and the stem of any individual flower is the *pedicel*. The thickened end of the pedicel forming the base of the flower is the *receptacle*. The configuration of inflorescences are variations on the presence, arrangement and point of origin of pedicels relative to the peduncle; a sampling of their multiplicity of form is shown in Fig. 1.3. A *raceme* has a series of unbranching pedicels along a central axis and no terminal flower. A *spike*

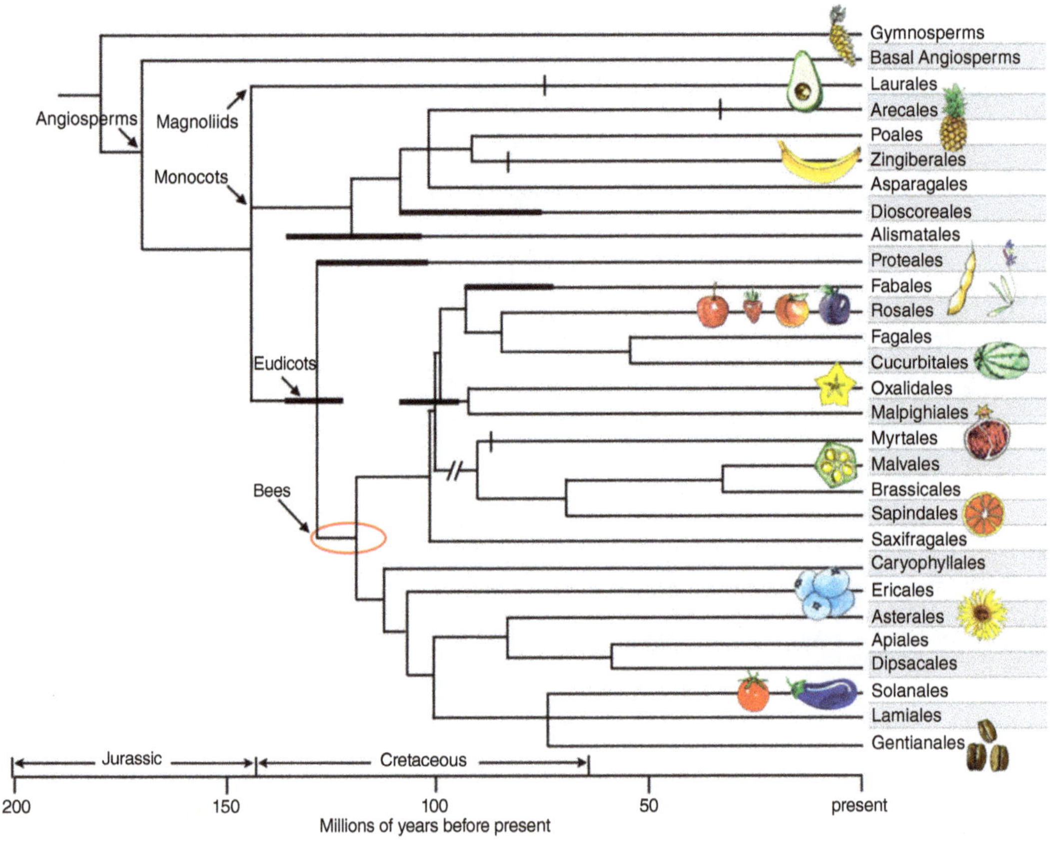

Fig. 1.1. Phylogeny showing chronology of angiosperm divergence and position of orders containing the major bee-pollinated crop plants listed in Table 3.1. Adapted from topology of Byng *et al.*, 2016, superimposed with geological divergence dates of Bell *et al.*, 2010. Gymnosperms are supported as a monophyletic sister group to the angiosperms from Bowe *et al.*, 2000. Bold lines indicate where topology is sustained with the confidence intervals of Bell *et al.*, 2010. Vertical tick marks indicate divergence chronology for the crown taxon. Divergence dates for bees from Cardinal and Danforth, 2013. Icons show representative crop members of each order.

is a kind of raceme without pedicels. A *head*, also called a composite flower, can be thought of as a concentrated raceme in which individual flowers are massed together onto one enlarged receptacle. A *corymb* is flat topped or convex, with long proximal pedicels becoming increasingly shorter as they move distally, and lacking a terminal flower. An *umbel* resembles a corymb, but all pedicels are of equal length and originate from one point of attachment. A *panicle*, or compound raceme, is irregularly branched with each branch possessing a terminal flower.

A flower's outer whorl of petals is called the *corolla* and functions to protect the interior sexual parts, to exclude ineffective pollinators, to attract

effective pollinators, and to direct effective pollinators toward the inside of the flower (Fig. 1.4). In legume-type flowers, two anterior petals are modified to form a *keel* inside which are housed the sexual parts of the flower. At the base of the corolla are the *calyx* or *sepals*; typically green and leaf-like, sepals protect the flower in bud and provide structural support in bloom. Collectively, the non-sexual parts of the flower – the sepals and petals – are called the *perianth*. In some of the basal angiosperms it is difficult to distinguish sepals from petals, in which case the structures are called *tepals* (Endress, 2008). The aggregate structure comprising the bases of the sepals, corolla and stamens is called the *hypanthium* and often contains the *nectaries*.

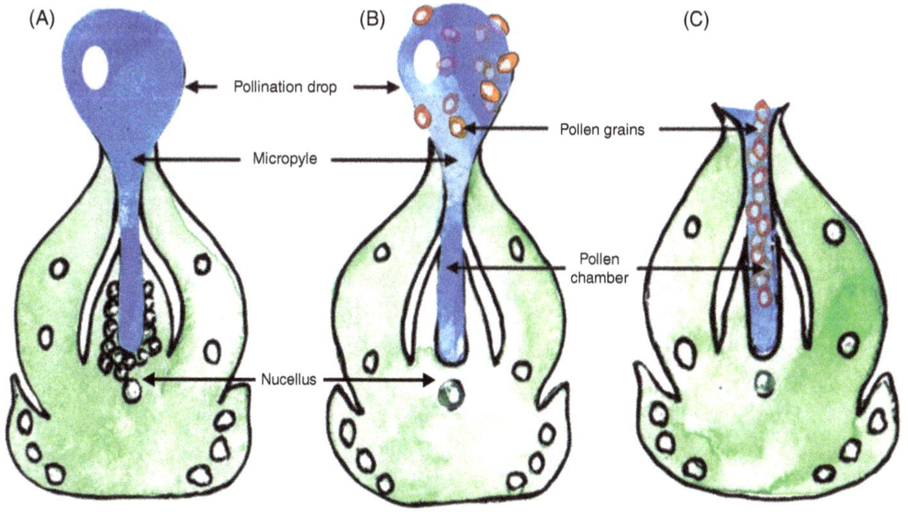

Fig. 1.2. Gymnosperm pollination and fertilization. The pollination drop in gymnosperms is a precursor to angiosperm nectar. The sugary solution extends beyond the micropylar opening of an ovule (A). After airborne pollen lands on it (B), the droplet recedes back into the pollen chamber (C), facilitating pollination. The pollination drop is secreted by cells in the *nucellus* – ovular tissues that contain the embryo sac. Redrawn from Jin *et al.*, 2012.

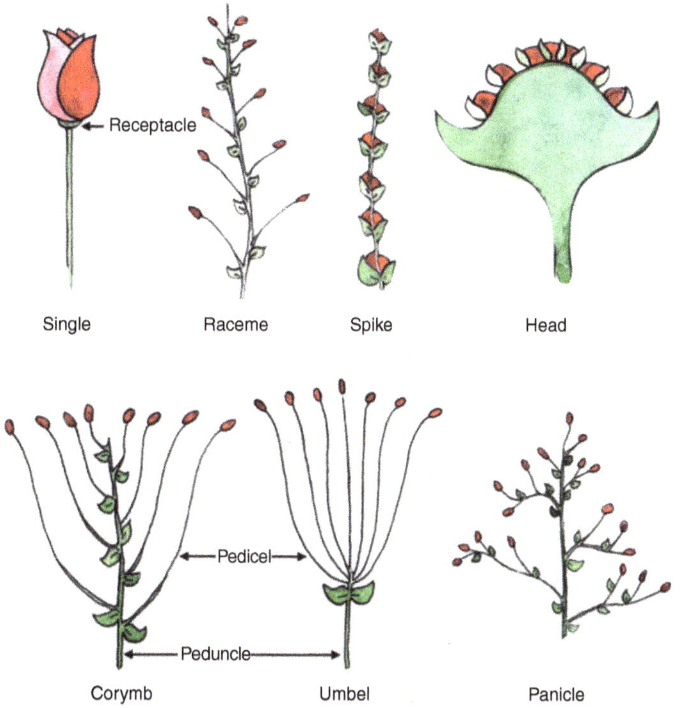

Fig. 1.3. Some examples of inflorescence designs. The floret in each example is indicated in red.

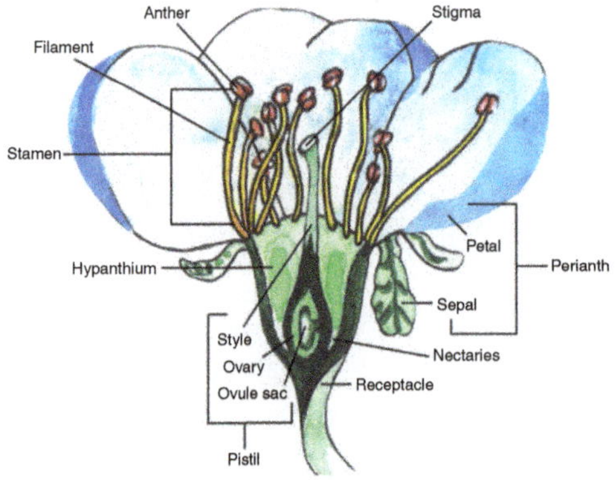

Anther Stigma
Filament
Stamen
Hypanthium
Petal
Perianth
Sepal
Style
Ovary
Ovule sac
Nectaries
Receptacle
Pistil

Fig. 1.4. General morphology of a hermaphrodite (also called perfect or bisexual) flower. The pistil (female part of the flower) in this case has only one carpel and is an example of a perigynous ovary (see Fig. 3.2, this volume).

Male parts of a flower are called the *stamens*, each made up of a slender *filament* holding an *anther* at the tip. When it is mature, the *anther* opens and exposes or releases *pollen grains* which contain the equivalent of animal sperm. In both gymnosperms and angiosperms each pollen grain consists of two cells – a vegetative cell and a reproductive cell – all encased in a tough covering of biopolymer that protects the pollen during its adventures in the environment.

The female part of a flower is called the *pistil*, made up of an *ovary* with *ovules* and a stalk-like *style* terminating with a sticky *stigma*. Ovaries are frequently enclosed or compartmentalized into units called *carpels*. An ovary with one carpel is called *simple* and an ovary with more than one is called *compound*. Recognizable examples of carpels are the five compartments of seeds found at the centre of an apple or the sections of an orange. Carpels may (e.g. orange) or may not (e.g. cucurbits) be separated by a thin partition called a *septum*.

A flower with both stamens and a pistil is called a *perfect*, *bisexual* or *hermaphrodite* flower. A flower expressing only one sexual function is called *unisexual* or *imperfect* and is further distinguished as *staminate* if male and *pistillate* if female. If staminate and pistillate flowers both occur on the same plant, the species is called *monoecious*; this term applies also to species with bisexual flowers (Beentje, 2016). If sexes segregate so that any one plant bears exclusively male or

exclusively female flowers, the species is called *dioecious*. Dioecy is by far the minority condition in the angiosperms, occurring in only about 6% of species (Renner and Ricklefs, 1995). Complicating things is the fact that some species defy these categories by producing partial exceptions. So-called *andromonoecious* species have both male flowers and bisexual flowers on the same plant. A *gyno-monoecious* plant has both female and bisexual flowers on the same plant. *Androdioecious* species have individual plants with exclusively male flowers and other individual plants with bisexual flowers, and *gynodioecious* species have individual plants with exclusively female flowers and other plants with bisexual flowers. It clearly helps to keep in mind the Greek *andro* for male and *gyno* for female!

Yet another form of sexual segregation happens with species whose bisexual flowers express *dichogamy* – a functional sex change over time. In *protandrous* flowers the stamens activate first, then senesce, leaving the pistil to activate next. In *protogynous* flowers the reverse happens – the female pistil activates first. This process is more generally called *sequential hermaphroditism*, a term that makes clear the fact that the flowers involved are still morphologically bisexual.

All these forms of sexual segregation, or lack thereof, have enormous implications in the breeding systems and pollination requirements of a crop plant (Table 3.1).

Once a pollen grain lands on a receptive stigma, the vegetative cell inside the grain elongates to rupture the biopolymer coating and grows a *pollen tube* that penetrates the length of the female style to reach an ovule. The pollen's reproductive cell divides to yield two sperm cells which are pushed along by the growing pollen tube and delivered to an ovule (Fig. 1.5).

Like pollen, the angiosperm ovule is not a single-celled haploid gamete. It is rather a sac-like structure containing an egg cell plus other cell-like components (Fig. 1.5). Inside the ovule a curious 'double fertilization' event ensues once a pollen tube delivers its sperm cells. One of the sperm fertilizes the egg cell in the ovule which becomes the *zygote* and *embryo*. The second sperm fuses with two ovule components called *polar nuclei*, resulting in a triploid *endosperm* – the starchy seed matrix that nourishes the embryo, being the functional equivalent to yolk in a chicken egg. All are encased in a tough coating derived from ovary tissue. After

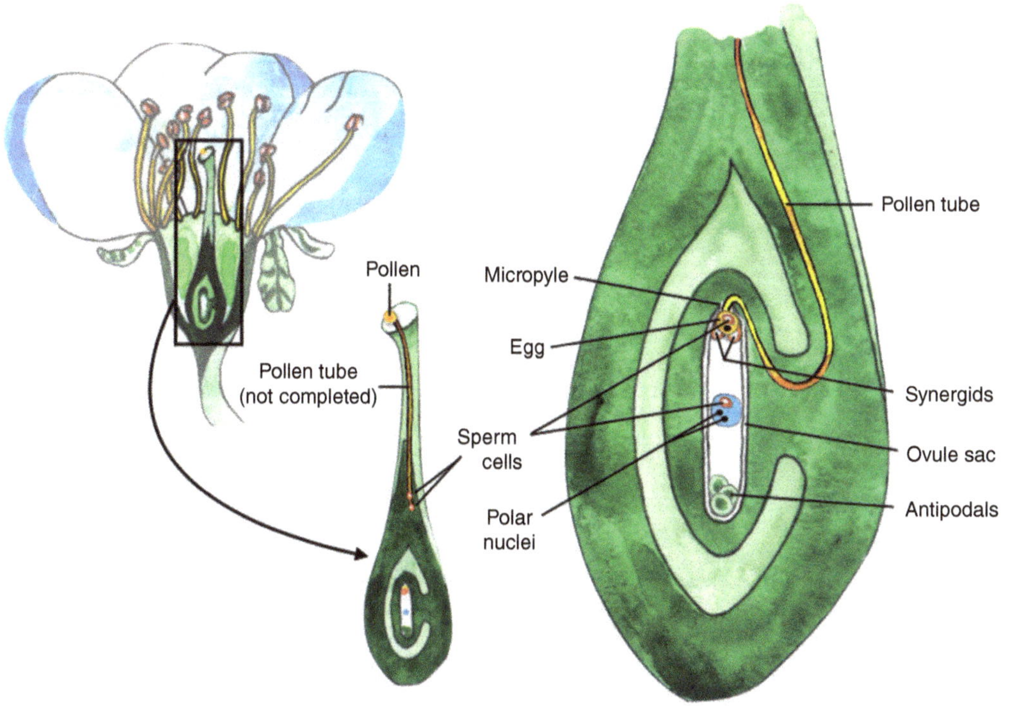

Fig. 1.5. Pollen deposition, pollen tube growth and fertilization in the angiosperm ovary. Synergids occur at the opening, or micropyle, of the ovary sac and help guide a sperm to the egg. Antipodals occur at the opposite pole of the ovule and are involved in nourishing the embryo and helping to grow the endosperm.

these developments, an ovule may properly be called a *seed*. Endosperm is the nutrient basis of such important grains as wheat and barley, giving cause for celebration to eaters of bread and drinkers of beer everywhere. In angiosperms, fertilization of one or more ovules further stimulates the ovary wall to develop into either a fleshy *fruit* or, in the case of nuts, a tough shell.

1.4. Evolution of the Flower

We have diverged into these details about plant sex to set the stage for an understanding about how and why flowers evolved in the first place. Make no mistake, flowering was a wildly successful innovation. Angiosperms today are among the most successful of all life forms. They comprise up to 400,000 named species, making up nearly 90% of all terrestrial plants (Jarvis and Linné, 2007). They embrace virtually every known plant body form and growth strategy and occupy every terrestrial habitat on Earth (Crepet and Niklas, 2009). The flower was certainly front and

centre of these developments, so let us think a moment about the innovations a flower does, and does not, represent.

To begin, flowers do not mark the beginning of pollen or ovules; these were already here millions of years earlier, entering the fossil record together (not uncoincidentally) in the late Devonian around 365 million years ago (mya) (Fairon-Demaret, 1996; Wang *et al.*, 2016), fully double the antiquity of the flowers (183 mya; Fig. 1.1). Pollen and ovules are rather the common legacy of all *seed plants* – the group that includes the angiosperms and the gymnosperms. Neither do flowers signal the beginning of insect pollination. Evidence for that comes from as early as 320–300 mya (Crepet, 1979), predictably after the arrival of ovules and pollen but long before the angiosperms.

Flowers do not mark the beginning of nectar. The pollination drop described in the previous section is living evidence for the kinds of pre-angiosperm ovule secretions that evolution would later coopt into sugar rewards for flower visitors. Other evidence for pre-angiosperm nectar comes from a comparative fossil analysis of 11 scorpionfly species of the Mesozoic that had tubular sucking mouthparts and were able to feed on nectar-like secretions of five extinct gymnosperm taxa, a case representing the earliest known example of plant and pollinator coevolution (Ollerton and Coulthard, 2009; Ren *et al.*, 2009).

Given that the gymnosperm pollination syndrome was the ground plan for what was to come, we can expect that between the two there are similarities – characters ancient and shared – as well as differences in innovations unique to the angiosperms. For example, let us go back to pollen and ovules, structures common to both groups. For both, pollen is released from male structures that are morphologically distinct from female structures. But when it comes to the location of ovules and mode of pollination, the angiosperms have something totally new. Seeds are no longer naked but enclosed in an ovary. The ovary affords protection and spatial concentration of ovules and probably represents efficiencies in their metabolic development. However, secreting ovules away inside a protective ovary necessitated a new mode for bringing pollen and ovules together, and natural selection's answer was the style and its terminal stigma. Lloyd and Wells (1992) argue that the angiosperm stigma evolved from precursors associated with the ancient gymnosperm pollination drop (Fig. 1.2).

What successful innovations did flowers thus contribute to the project of angiosperm evolution? A generality weaving through the literature is the idea that flowers represent improvements in plant mating and reproductive efficiency (Barrett, 2010).

One efficiency may be the elevated number of pollen grains per ovule made possible by the association of multiple ovules per stigma, in contrast to the single ovule/single pollination drop model in the gymnosperms. The surface of animal-borne pollen grains tends to be heavily reticulated and sticky, a feature referred to as *pollenkitt*, which causes insect-adapted pollen to clump (Faegri and Van der Pijl, 1979), doubling in the case of honeybees the amount of pollen the insect can carry (Amador *et al.*, 2017). This clumping not only encourages large depositions of pollen on a stigma (Pacini, 2000), but also promotes fitness gains through pollen tube competition (Erbar, 2003) and lowers the chance of individual pollen grains drying out (Dafni and Firmage, 2000). In general, a stigmatic pollination surface seems to optimize opportunities for ovules being fertilized.

Another efficiency may be the fruit's role in seed dispersal. Fruit design is variable, and one ovary/fruit may encase one seed or many. A common feature of fruits, however – not unnoticed by the frugivores that eat them – is their palatability. By ingesting sweet, fatty or proteinaceous fruits and excreting their seeds, frugivores across a range of taxa – mostly mammals, birds and ants (Eriksson, 2008) – have partnered with angiosperms to disperse their seeds, expand their population ranges and promote their species diversity. However, seed dispersal by animals is also common in the gymnosperms (Leslie *et al.*, 2017), so any benefits of fruit-assisted seed dispersal are contextual. Tiffney and Mazer (1995) show that species diversity in woody dicots is positively associated with seed dispersal by vertebrates, whereas the reverse is true in herbaceous monocots and dicots in which species diversity is associated with abiotic seed dispersal. The authors point out that these results are consistent with a traditional hypothesis that says selection will favour large seeds with large stores of starchy endosperm in woody species, in which seedlings must germinate and grow in conditions of low light. Only large vertebrates can successfully disperse such large seeds. Thus, at least for woody angiosperms and forest communities, fruit-assisted seed dispersal seems to be adaptive.

Another efficiency may derive from floral hermaphroditism, which is the majority condition among the angiosperms (Klinkhamer and de Jong, 2002). Compared to gymnosperms with their sexually segregated cones and profligate expenditure of pollen, floral hermaphroditism reduces energetic demand for pollen production and increases likelihood that a stigma will receive pollen. The fact that much of that pollen is self-pollen suggests an evolutionary trade-off in the form of inbreeding depression by selfing. However, as we will see in later chapters, many flowers have desynced maturation sequences in floral sexual tissues so that selfing is minimized and out-crossing maximized. Alternatively, a species may segregate the sexes on to separate flowers (unisexual or imperfect flowers) or different plants (dioecy), or it may evolve varying degrees of self-compatibility (Schemske and Lande, 1985), a not uncommon option. In fact, of all hermaphroditic plants, 62–84% of temperate species and 35–78% of tropical species are at least partially self-compatible (Arroyo and Uslar, 1993). The frequency of self-compatibility seen across the angiosperms is thought to indicate a selective response to pollinator scarcity (Lewis, 1973). It demonstrates how strongly pollinator availability has shaped the breeding systems of angiosperms (Lloyd, 1965; Schemske, 1978). In the end, the majority condition of hermaphroditism suggests that there were efficiencies to be gained by concentrating the sexual tissues in one space.

The sexual parts of the angiosperm flower – the carpels and stamens – evolved once in the angiosperms, meaning that the bewildering varieties of carpels and stamens we see today are homologous variations of that one innovation. The non-reproductive perianth organs, on the other hand, have arisen independently multiple times so that perianth tissues in one angiosperm branch may not be homologues to those in another branch (Eames, 1961; Takhtajan, 1991). However, this tidy story was upended by subsequent genomic analyses suggesting in avocado (*Persea americana*) that petals derive from staminal tissue (Chanderbali *et al.*, 2006). It is likely that other surprising homologies await discovery, especially in the basal angiosperms and monocots.

Additional reconstructions suggest that the ancestral state of the earliest flowers included an extended stigma and stamens with lateral pollen sacs; however, whether the ancestral flower state was bisexual or unisexual remains unresolved (Endress and Doyle, 2009).

1.5. Coevolution of Animal Pollinators and the Flower

We have reviewed some of the putative improvements in mating and reproduction made possible by the flower, but let's face it, these efficiencies seem marginal and inadequate for explaining the rapid expansion and dominance of angiosperms across ecosystems. This expansion happened over a span of 46 million years, from about 130 to 84 mya, and was directly responsible for large-scale replacements and extinctions of gymnosperms and ferns (Coiffard *et al.*, 2012). Charles Darwin, perplexed at the suddenness of their appearance in the fossil record, famously lamented in a letter to Joseph Dalton Hooker dated 22 July 1879, that 'The rapid development as far as we can judge of all the higher plants within recent geological times is an abominable mystery' (Darwin and Seward, 1903, p.539).

Down the decades, legions of commentators have weighed in on the matter, and the consensus is that flowers are indeed major actors in angiosperm diversification and niche penetration (Barrett, 2010) – but not owing to any one-sided selection responding to 'reproductive efficiency'. Instead, flowers are adaptations by plants in response to animal visitors –

sexual organs that encourage beneficial flower visitors and exclude ineffective ones. Most of those flower visitors are insects. The insects, foraging for food in the forms of pollen and nectar, pollinate their host plants as a matter of course, ensuring those plants' sexual reproduction and succession to another generation. In the case of bees in particular, the relationship has engendered adaptations for nectar and pollen feeding, such as the famous pollen basket or *corbiculum*. It is textbook *coevolution* – the reciprocal evolution of interacting species driven by natural selection (Thompson *et al.*, 2017). It is not an altruistic relationship: evolution operates unwaveringly under the principle that genes act selfishly to optimize their own reproduction (Dawkins, 1989), but it is equally true that the way to optimize one's reproduction often means partnering with other genomes to form alliances of shared interests (Bourke, 2011). One suspects one is on the trail of coevolution when the cause and effect thread becomes difficult to untangle: did the pollinator select for the flower or the flower select for the pollinator?

What energized the diversification and ecological success of angiosperms is the synergy between the morphological plasticity of the flower's *bauplan* and the ecological agility of its insect pollen collectors. The floral design of 'sexual parts + perianth' is a versatile template that allowed for near-infinite variations in form, colour and scent. For their part, the 5.5 million species of insects (Stork, 2018) occupy virtually every terrestrial niche and many aquatic environments as well, which means that the insects' tight association with angiosperm pollen vectoring afforded flowering plants directional expansion into most of Earth's terrestrial ecosystems.

Almost immediately, the dynamics of coevolution set the stage for *specialization* in pollination systems. Specialization can be considered from the perspective of the plant or the perspective of the pollinator. For the plant it means adaptations that enhance delivery of conspecific pollen to receptive stigmas. It can be things like the petals of orchids that mimic female bees, duping sexually aroused males into vectoring pollen when they were expecting to have sex (Ciotek *et al.*, 2006; Schiestl, 2010). Adaptations can be long tubular corollas (Rodríguez-Gironés and Santamaría, 2007) that limit their pollinators to long-tongued visitors such as hummingbirds (Fenster, 1991), bees (Inouye, 1980), moths (Alexandersson and Johnson, 2002) or butterflies (Bauder *et al.*, 2011) – a strategy for foiling nectar thieves

(Rodríguez-Gironés and Santamaría, 2005) or in Darwin's (1862) venerable opinion, a strategy by which the plant compels a pollinator to probe deeply to reach nectar, touching the anthers and stigma with its face, effecting pollination with subsequent floral visits, an hypothesis substantiated with 21st century field data (Alexandersson and Johnson, 2002). For the pollinator, specialization means obligatory commitment to visiting flowers within a narrow range of plant taxa or floral morphologies.

Specialization is only significant in context to *generalization* – a pollination syndrome that seems to get less attention than it deserves. For a plant this means a floral morphology that is openly available to a variety of flower visitors or even abiotic vectors. A good example is the flower of oilseed rape (*Brassica napus*) that has a simple, open morphology that admits pollen vectoring by a variety of insects as well as wind (Langridge and Goodman, 1982). For the pollinator this means a flower visitor who visits a range of plant taxa. Owing to their long colonial lifespans and season-long foraging activity, social bees often fit the bill for a generalist flower visitor (Heithaus, 1979; Westerkamp, 1991). Analysis of the fossil record (Crepet, 2008) and extant angiosperm divisions (Hu *et al.*, 2008) shows a trend from generalized to specialized pollination systems over geological time.

In pollination, the adjective *polylectic* is applied to a flower visitor that visits a wide taxonomic range of plants. The term *oligolectic* is used for a pollinator that visits a narrow taxonomic range of plants. In an effort to accommodate cases ascribed to 'broad oligolecty' or 'narrow polylecty', Cane and Sipes (2006) proposed the term *mesolectic* to apply to those bee taxa that collect pollen from numerous species and genera of plants drawn from the same few plant families or tribes.

Although the idea of specialization in pollination systems has merit, it can be overinterpreted. Nature, as always, is messier than our categories for it. For one thing, specialization is not symmetrical: there are more plants that specialize on one or a few pollinators than pollinators that specialize on one plant (Vázquez and Aizen, 2004). This would appear to be dissolutive to coevolution; but what actually happens is that individuals of even 'generalist' flower visitors can become flower specialists, visiting a narrow range of flowers in their lifetimes (Chittka *et al.*, 1999), the result being a 'complex geographical mosaic of coevolutionary interactions' (Johnson and Anderson, 2010, p.32) at the local level. Similarly, even 'specialized' plants receive flower visitors of many taxa at a local scale. The upshot being – it is more productive to consider specialization and generalization as dynamics that operate at different timescales. Specialization – the accurate delivery of conspecific pollen – is plainly adaptive in the short term; but in the long term, generalization is more favourable to ecosystem-wide species diversity and stability (Brosi, 2016).

The dynamism sparked between floral morphological adaptability and insect ecological dominance invited the proliferation of a near-infinite variety of reproductive and ecological strategies adaptive to both players. This is captured in the concept of *pollination syndromes* – that floral evolution acts in response to the most effective pollinator or group of pollinators (Rosas-Guerrero *et al.*, 2014; Ashworth *et al.*, 2015). Pollinator syndromes are seen as evidence of convergent evolution among flowering species jockeying for access to an available cohort of effective pollinators (Rosas-Guerrero *et al.*, 2014).

The result is a suite of pollinator and flower characteristics that cluster together (Table 1.1). For example, bird-pollinated flowers are often long-tubed and red. It is no accident that these long corollas exclude ineffective pollinators; neither is it accidental that birds can see red easily whereas this colour can be difficult for bees (Spaethe *et al.*, 2001). On the other hand, bee-pollinated flowers are overwhelmingly represented by flowers that are blue, white or yellow – colours comfortably within the visible spectrum for bees. Moreover, in the bees there are potential 'subsyndromes' such as the mechanical keel tripping mechanism in alfalfa (lucerne) that repels ineffective flower visitors, or the anthers of blueberries whose pollen is only released by bees capable of sonicating the flower (De Luca and Vallejo-Marín, 2013). Bumble bees famously partition themselves so that bee species with long tongues forage on flowers with deep corollas while short-tongued bees specialize on those with shallow corollas (Heinrich, 1976). In cropping systems, morphological flower constraints such as these often segregate effective from less effective bee pollinators.

However, if it is true that specialization in pollination systems must be handled cautiously, the same holds for pollination syndromes. Studies formally addressing the matter have concluded in favour of pollination syndromes (Rosas-Guerrero *et al.*, 2014; Ashworth *et al.*, 2015) while others

Table 1.1. Pollinator syndromes, compiled from Thien *et al.*, 2000; Rodríguez-Gironés and Santamaría, 2004; Wolfe and Sowell, 2006; Fleming *et al.*, 2009; Rosas-Guerrero *et al.*, 2014; and Varatharajan *et al.*, 2016.

Syndrome	Pollinator	Floral colour	Floral shape	Floral odour	Reward
generalized	many, including abiotic	variable	open, easily accessible	variable	nectar, pollen, none
chiropterophily	bats	white, green	suspended on long stalks, tubular, radially symmetrical	musky	nectar, pollen
melittophily	bees	yellow, blue, white	corolla enlarged for landing pad, bowl, tubular	sweet	pollen, fragrance, oil, resin, low-volume concentrated nectar
cantharophily	beetles	brown, green, red, white	bowl	fruity, musky	heat, nectar, pollen
ornithophily	birds	red	tubular	imperceptible	high-volume dilute nectar
psychophily	butterflies	blue, orange, pink, red, yellow	small, medium, large, bell, tubular	fresh	nectar
myophily	flies	brown, green, white, yellow	small	sweet, fruity, sour	nectar
saprophily	flies (carrion)	brown, green, purple	bell, dish, trap	putrid	none
rhinomyophily	flies (long-tongued)	pink, purple	tubular	imperceptible	nectar
phalaenophily	moths	white	tubular	sweet, scented in evening	nectar
therophily	non-flying mammals	brown, green, white	dish	fruity, musky, sour	nectar, pollen
thripophily	thrips	white, yellow	medium size	sweet	nectar, pollen
sphecophily	wasps	brown, green, purple	bell, dish	sour, sweet	nectar

have concluded that pollination syndromes are unreliable at predicting the most effective pollinators for a focal plant group (Li and Huang, 2009; Ollerton *et al.*, 2009). In spite of these cautions, the general robustness of a coevolutionary relationship between flowers and flower visitors has never been seriously challenged.

1.6. Insect Flower Visitors and the Significance of Bees

It is safe to say that the flower has spectacularly exploited insects – those widely available pollen vectors capable of directional motility that promoted angiosperm diversification, range expansion and niche penetration. That insects were involved at an early stage is supported by the fact that wind pollination prevails in ancient and modern non-angiosperm seed plants (Ollerton and Coulthard,

2009); however, by the emergence of basal angiosperms we see insects as the primary mode of pollination in 86% of families (Hu *et al.*, 2008). Moreover, molecular phylogenetics infers that animal pollination is the ancestral state of the angiosperms (Hu *et al.*, 2008); in other words, the coevolution of flower morphology and animal flower visitors was the defining innovation of the most recent common ancestor of the angiosperms. Flowers and animal flower visitation are not accidental to the angiosperms; they are essential to the angiosperms. Today, 78–94% of angiosperms rely on animal pollinators (Ollerton *et al.*, 2011), the vast majority of which are insects (Grimaldi, 1999).

The panoply of insect flower visitors is wide and deep, taxonomically speaking. Figures 1.6 and 1.7 show a variety of pollinator syndromes documented from several tropical and temperate sampling sites. A cursory glance shows that bees,

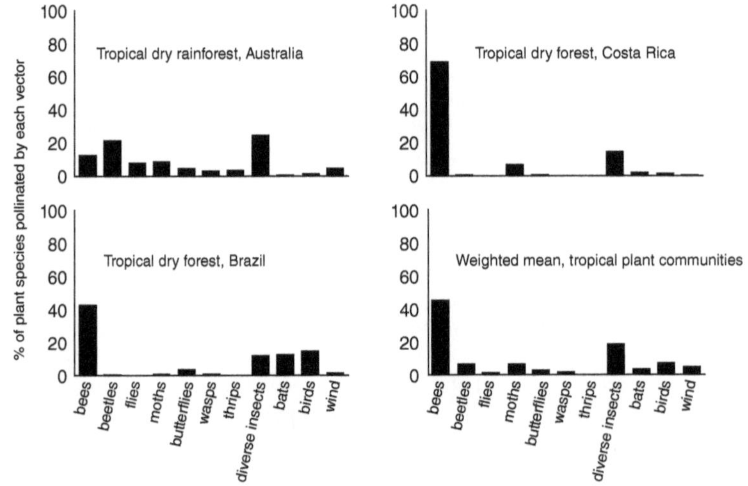

Fig. 1.6. Primary pollinator syndromes in tropical communities (Wardhaugh, 2015).

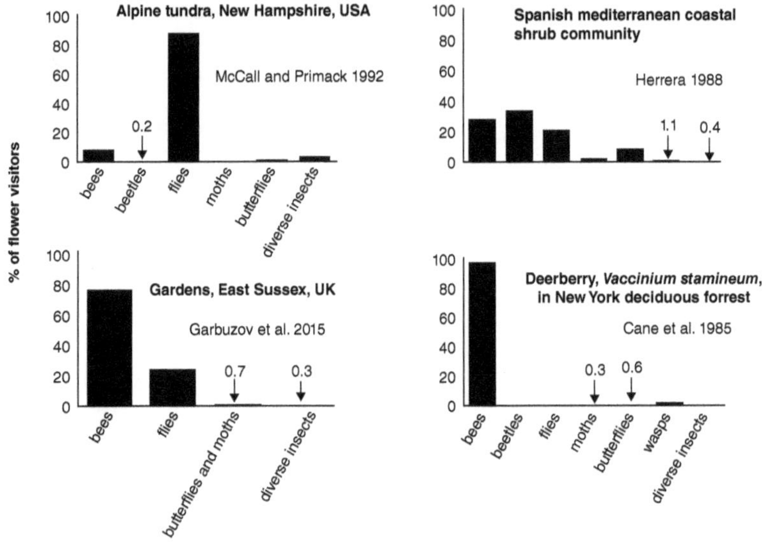

Fig. 1.7. Primary pollinator syndromes in temperate communities (Cane *et al.*, 1985; Herrera, 1988; McCall and Primack, 1992; Garbuzov *et al.*, 2015).

beetles and flies predominate. Of these, beetles are by far the most species-rich insect order, and their density and taxonomic diversity at flowers in the tropics can rival that of the bees (Wardhaugh *et al.*, 2013); in fact, it has been argued that beetles, with 400,000 described species and potentially millions more awaiting discovery, may constitute the most species-rich group of flower visitors on Earth (Wardhaugh, 2015). Flies, with a species count approaching that of beetles, are the most ecologically diverse insect order with blood feeders, vertebrate endo- and ectoparasites, gall makers, leaf miners, wood borers and pollinators among their ranks (Grimaldi *et al.*, 2005). Surely the contribution of beetles and flies to plant and crop pollination is substantial, underappreciated and understudied.

However, among all the animals that visit flowers, it is the bees who have attained pre-eminence as the paradigmatic animal partner with flowers in their shared coevolution. As we have already seen, insect flower visitation is ancient and predates the angiosperms; however, it is noteworthy that among the 'big three' – the beetles, the flies and the bees – the bees are by far the youngest taxon. Beetles and flies are higher taxa, each constituting its own order whereas bees are a monophyletic offshoot of a different order, the Hymenoptera. By *monophyletic*, we mean a taxon with all members sharing a derived set of characters inherited by genetic descent from the group's most recent common ancestor.[i] The mid-Triassic marks the emergence of the flies (240 mya) and beetles (230 mya) (Grimaldi *et al.*, 2005), while the bees diverged from their crabronid wasp sisters much later, 113–132 mya, in the mid-Cretaceous (Cardinal and Danforth, 2013).

What makes that last timepoint so astonishing is its literal synchrony with the emergence of the eudicots – the largest and most taxonomically rich branch of the angiosperms, including most of the crop plants covered in this volume (Fig. 1.1). Among the big three, the relative youth of bees means that only bees were there at the beginning of eudicot divergence, literally coevolving with and synergizing that divergence. To underscore this point, the derived shared characters that define the monophyly of bees are adaptations associated with flower foraging – plumose hairs (the better for picking up pollen), pollen-feeding larvae, and the hind basitarsus (first foot segment) broader than subsequent segments (a precursor to the later pollen basket) (Michener, 2000). Such foundational,

integral, obligatory and coinfluencing generalizations cannot be applied to any other taxon. Even though many, especially the bats, post-date the emergence of angiosperms and have members who evolved pollinating behaviours, for all of them flower feeding is derived, not ancestral. For example, the emergence of bats (64 mya; Teeling *et al.*, 2005) post-dates the angiosperms; however, the ancestral state of bats is insectivory, not flower feeding, and pollination in bats is a specialization of only some families (Simmons *et al.*, 2008).

It remains that the relationship between angiosperms and bees is exceptional for its mutuality, codependence and diversity of cospecializations. Among the big three, only the bees express universal adaptations and obligations to flower foraging. Bees are universally committed to flower feeding and morphologically adapted for it.[ii] For these compelling reasons, bees will be the focus of our attention for the rest of this volume.

Notes

[i] Monophyletic taxa are considered 'natural' in that they categorize groups based on natural descent and not classification errors. For example, a category based on 'winged animals' would not constitute a natural taxon of insects and birds because wingedness evolved independently in each and their modern states of wingedness have no relationship based on shared descent. It is the work of phylogenists to reconstruct relationships by identifying evolutionarily informative characters and determining whether they are inherited by descent or represent independent evolutionary convergences.

[ii] The only exceptions are three species of *Trigona*, stingless bees in Central and South America that have abandoned pollen feeding for carrion. This is a derived condition and essentially the exception that proves the rule of obligatory flower foraging in the bees.

2 Biology of Bees

Bees are a monophyletic branch off the wasp group. Their most recent ancestors are the wasp family Crabronidae, in particular two of its subfamilies the Philanthinae and Pemphredoninae (Fig. 2.1; Debevec *et al.*, 2012). As described in the previous chapter, the emergence of bees 123 million years ago (mya) (113–132 mya) (Cardinal and Danforth, 2013) coincides with the emergence of the eudicots, the largest and most diverse group of angiosperms which includes the majority of crops pollinated by bees today.

2.1. Bee Fundamentals

The seven recognized bee families are the Melittidae, Andrenidae, Halictidae, Stenotritidae, Colletidae, Megachilidae and Apidae. As proper names, each is capitalized, but the adjectival form of the word is not. Hence, a bee belonging to the Apidae is an *apid* bee. The literature is full of references to 'the six major families' of bees, owing to the fact that the numerically diminutive Stenotritidae is limited to Australia and includes only two genera and 21 species. For convenience, bees are often distinguished as the five short-tongued families and the two long-tongued families (Fig. 2.1). The two long-tongued families are also the most speciose, with Megachilidae containing over 20% of all bee species and Apidae over 30% (Michener, 2000, Table 16-1).

Bees and wasps belong to the insect order Hymenoptera which also includes the sawflies and ants. Unlike other hymenopterans, bees specialize on vegetarian diets. However, the label 'vegetarian' that has traditionally distinguished the bees from the carnivorous wasps has itself come under qualification in recent years. For one thing, not all bees are vegetarian. Three species of the *Trigona* stingless bees in Central and South America are obligate carnivores that feed on animal carrion (Camargo and Roubik, 1991) or even live prey such as larvae

of abandoned wasp nests (Mateus and Noll, 2004). These rare examples of bee carnivory must be thought of as late derivations from the ancestral state of pollen feeding. To keep things interesting, a recent analysis of trophic positions among the six major bee families suggests that bees in general qualify properly as omnivores: their dietary protein comes not only from primary producers, plant-derived pollen, but also from the microbial communities that colonize and feed on bee-collected pollen (Steffan *et al.*, 2019).

These exceptions and qualifications notwithstanding, the dietary connection between bees and flowering plants is ancient and tight – a firm foundation on which to build the principles of crop pollination developed in this volume. For protein, both immature bees and adults eat pollen and its associated microbes, and for energy they imbibe nectar. In solitary species these dietary building blocks are typically eaten as they are, with modifications developed little beyond dampening pollen balls with nectar. In the long-lived colonies of perennially *eusocial* species, however, the protein fraction of diet fed to the young is first metabolically converted into glandular secretions produced by nurse bees, conceptually like milk in lactating mammals. Perennial colonies also dehydrate nectar into *honey*, a process which preserves the nectar for long-term storage over periods of nectar dearth such as winter.

Bees have body adaptations for plant foraging. The body hairs on bees are finely branched so that pollen grains cling to them. Most bees have external body structures specialized for carrying pollen. The four apid tribes comprise the so-called *corbiculate* bees, of which a segment of each hind leg has a highly modified structure called the *corbiculum* or *pollen basket* for holding loads of pollen while the bee is foraging (Fig. 2.2). Groups outside the corbiculates carry pollen on structures called *scopae* (singular *scopa*) which are patches of long, parallel

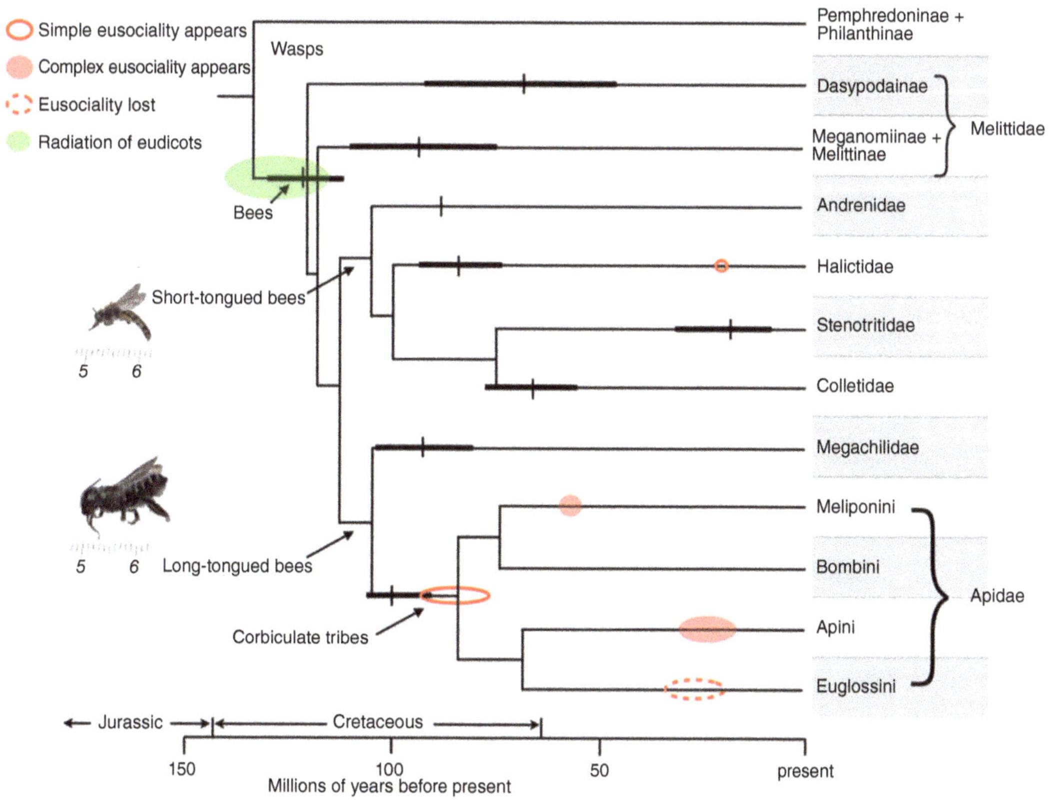

Fig. 2.1. Phylogeny of extant bee families modified from Cardinal and Danforth, 2013. Bold lines indicate 95% confidence intervals on divergence times and vertical tick marks indicate divergence chronology for the crown taxon. Dates of emergence of apid eusociality by Cardinal and Danforth (2011) and for halictid eusociality by Brady *et al.* (2006). Dates for emergence of eudicots by Bell *et al.* (2010). The short-tongued bee photograph is an Arizona melittid (*Hesperapis* sp.) and the long-tongued bee example is an Arizona megachilid (*Megachile* sp.). Identification: Conor Fair.

hairs attached to their hind legs or underside of the abdomen (Fig. 2.3).

Developing immature bees go through a *complete metamorphosis* (Fig. 2.4). Individual bees start life as a single *egg* laid by their mother. After a few days the egg hatches into a *larva* (plural *larvae*) which is a grub-like, rapidly growing feeding stage. As they grow, larvae shed their skin several times by a process called *moulting* to advance to the next larger stage, or *instar*. Eggs are laid and larvae develop in cells of varying complexity, ranging from excavated dead ends in earthen tunnels to constructed cells of wax or resin. The mother or siblings provision each larval cell with food. Some species add pollen and nectar to the cell regularly as the larva needs it (called *progressive*-provisioning), while others feed

it all at once as a large moist lump at the time the egg is laid (*mass*-provisioning). As bee larvae literally live *in* their food, defecation is a problem. Larvae resolve this by postponing defecation until their feeding career is over. When it finishes its feeding period, the larva defecates, stretches out (at which point it is called a *prepupa*), and transforms into a *pupa* (plural *pupae*) which is a quiet stage during which larval tissues are reorganized into those of an *adult*. Finally, the pupa moults into an adult complete with six legs and four wings, and breaks out of its cell. Species vary in the amount of time immature forms spend in each stage.

Female bees control the sex of their offspring. They store sperm from their matings in the *spermatheca*, an organ connected to the *oviduct* which is the passage

Fig. 2.2. The corbiculum, or pollen basket, is the unifying feature of the corbiculate tribes of Apidae, including the Apini, Bombini, Euglossini and Meliponini. The corbiculum is comprised of the concave outer side of the hind leg tibia with associated structures on the widened basitarsus (see Michener, 2000, p.48; Winston, 1987, p.24). For the bumble bee (*Bombus* sp., identification: Conor Fair) (A), the corbiculum shown in the box has the beginnings of pollen accumulation at the bottom of the tibia. The live honeybee (B) shows the corbiculum filled with pollen.

down which eggs pass during oviposition. Females have muscular control over the spermatheca. By opening it and releasing sperm on to a passing egg the female can fertilize the egg, causing it to develop into a female. If she withholds sperm the resulting unfertilized egg becomes a male. This ability to regulate sex of offspring is important in solitary tunnel-nesting bees because they tend to lay male eggs near the nest entrance so males can precede the females in spring emergence. For social species it is important to time male production according to seasonal food availability. Male bees are visibly distinguishable from workers and queens; their only known function is to mate with females.

The adult stage of bees is dedicated to dispersal, nesting, foraging and reproduction. Bees do this with a variety of life strategies and nesting habits, ranging from solitary to social, from simple burrows to elaborately constructed nests. Both solitary and social species are important crop pollinators.

2.2. Solitary Bees

For the majority of people, the word 'bee' conjures images of the stinging makers of honey who live in white wooden boxes and their mysterious handlers whose activities are as opaque as the rites of an extinct Sumerian cult. These casual observers are thinking of the highly social honeybee, *Apis mellifera*, which indeed figures prominently in these pages. The truth is, however, that of the 19,900 described bee species in the world (Asher and Pickering, 2013), at most 15% (Batra, 1984) match the description of a eusocial, or 'truly' social species. The rest, the vast majority, are *solitary* species which means that their females single-handedly make a nest and produce the next generation of fertile offspring. Most solitary species produce only one or two generations per year. Mounting evidence is showing that it is these unsung, cryptic and overlooked majorities that constitute the real workers of crop pollination.

(A)   (B)

Fig. 2.3. For bees outside the four corbiculate tribes, pollen is carried externally on more generalized structures called scopae which are long parallel hairs under the abdomen (A: *Megachile*, Megachilidae) or on the hind legs (B: *Xylocopa*, Apidae). Identification: Conor Fair.

Just as the majority of bee species are solitary, the majority of solitary bees are soil nesting. As a nesting medium, soil has many benefits, chiefly its ubiquity, its plasticity to design modification, its relative security against predators, and its temperature insulating properties. In fact, soil nesting is the ancestral state for all bees, and fossil bee tunnels are known from as early as the late Cretaceous (Genise *et al.*, 2002). The basic tunnel design admits variations, but a recurring feature is a single main shaft with lateral individualized brood cells, each large enough for one pollen clump and its attendant larva (Fig. 2.5A). Complexifications on this basic design include lateral tunnels leading to more brood cells and vertical dead-end tunnels constituting cesspits for accumulating waste and managing inundation (Fig. 2.6A). Nest apertures often have a tumulus, or mound of earth from excavation activity (Fig. 2.5B). Some bee species prefer digging into vertical earthen banks or cliffs, perhaps an adaptation for resisting inundation (Fig. 2.6B).

Soil-nesting bees often build nests near one another in large aggregations. To a human observer, the bees' sudden spring-time emergence in a concentrated space can give the false impression of a large social colony. Closer examination, however, shows the ground peppered with individual burrows (Fig. 2.7): it is not the coordinated activity of a colony but instead the simultaneous life frenzy of scores, hundreds or even thousands of individuals. There is everything to be done – and in short order. It is usually the males of the species who emerge first, a phenomenon called *protandry*, an adaptation for heightening mating success of the females who will emerge shortly thereafter (Fig. 2.8). Males may exercise territoriality, patrolling a patch of the aggregation and fighting off other males. In other species territoriality is subdued, and females mate on the ground in – there is no other word for it – orgies with clumps of males, each struggling for his turn to mate with her. After mating, each female busies herself with excavating a nest, constructing one or more brood cells and lining each with her glandular secretions to ward off moisture and predatory soil nematodes. Into each brood cell goes a pollen clump, each clump representing the foraging labours of roughly 1 day. On top of the pollen clump the female deposits one egg (Figs 2.4–2.6). She then seals the cell shut with soil and begins another brood cell. The larva will consume its pollen ball, moulting along the way into successive instars, before moulting into a pupa or adult.

Other solitary bee species nest above ground in hollow reeds or stems, in which case their nests

Fig. 2.4. Bees share with other holometabolous insects the feature of complete metamorphosis in which immature forms progress from an egg (A) to an active feeding stage, the larva (B), which grows through a series of episodic moults until elongating to a prepupal stage (C). The final immature step is the pupa (D), a quiescent stage in which the larval tissues are reorganized into those of an adult. Photos and taxa identifications courtesy of: (A) *Augochlora pura* eggs, Jason Gibbs; (B) *Nomia melanderi*, second instar larva, James Cane; (C) *Augochlora pura* prepupa, Jason Gibbs; (D) *Augochlora pura* pupa, Jason Gibbs.

have a fixed linearity with brood cells placed successively in the reed or stem, separated from one another by partitions of mud (Fig. 2.9), cut leaves (Fig. 2.10), or other materials. Others nest in pre-existing tunnels excavated into wood by previous occupants, often wood-boring beetles. The carpenter bees, *Xylocopa* spp., famously earn pest status from human homeowners for carving original tunnels in solid wood (Fig. 2.11). While maintaining the strong linearity of brood cells, carpenter bees deviate from reed nesters by separating their brood cells with the material closest at hand – masticated wood shavings. One tribe of solitary megachilids, the Anthidiini, takes the prize for the most diverse nests among the bees, in some species accepting pre-existing cavities as variable as hollow stems, snail shells or abandoned insect galleries, and in

other species constructing nests of resin and plant fibres (Fig. 2.12), supplemented with pebbles, pieces of leaf, sand, animal fur or other debris (Litman *et al.*, 2016).

Regardless of nesting mode and material, in *univoltine* species, that is, species who produce only one reproductive generation per year, mating and nesting activities are concentrated into a single brief interval coinciding with the flowering of an associated guild of suitable food plants. *Multivoltine* bee species undergo this process more than once, producing more than one reproductive generation per season. Regardless of life strategy, the long-lived, quiescent overwintering stage can be either as a new generation unemerged adult as in the megachilid orchard mason bees *Osmia* spp. (Sgolastra *et al.*, 2012); as a new generation female who

Fig. 2.5. Basic configuration of a solitary bee subterranean nest (A) consists of a vertical main shaft with one or more lateral tunnels and individualized brood cells. Inside each brood cell the female forms a ball of pollen and deposits an egg on it. The egg will hatch and the pollen ball provides the larva its entire food requirements during development. Many bee burrows have a mound or tumulus of excavated soil at the entrance. A 'cityscape' view (B) of a small aggregation of bee nests and their tumuli.

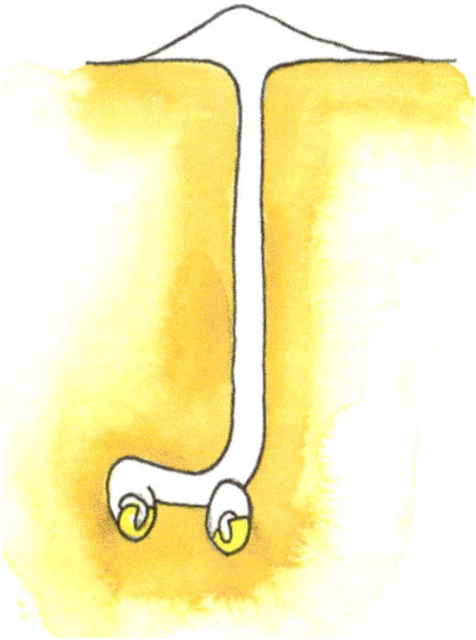

emerges, mates, enters diapause and emerges to nest the following spring as in the halictid *Halictus rubicundus* (Yanega, 1988); or a prepupa as in the andrenid *Perdita portalis* (Danforth, 1991).

Transcending these variations, the life strategy of a solitary bee is that of a reproductively autonomous individual. His or her intraspecific social interactions are ephemeral and consist of little more than brief contacts with coemerging siblings, territorial fighting (in the case of males) and mating. Females bear the burden of single-handedly nesting, foraging and offering what little parental care is given the young. Their aggregation areas, despite appearances, are mirages of sociality. The proximity of one nest to another is not architecture born out of cooperation but rather an emergent outcome of favourable nesting sites and the fact that their makers themselves emerged from not very far away.

In spite of, or perhaps because of, these stark simplicities of life history, solitary bees are the real champions of crop pollination, a subject covered in more detail in Chapter 3 (this volume).

2.3. Social Bees

To call solitary bees *reproductively autonomous* as mentioned in the previous section implies the possibility of an alternative life history, and *sociality* is that alternative. Some bees are not reproductively autonomous and can only contribute to the next reproductive generation by cooperating with groups of family members. This condition is called *eusociality*, or 'true' sociality, and a qualifying species must express all three of the following criteria simultaneously at least at some point in its life cycle (Wilson, 1971): (i) there must be cooperative care of the young; (ii) there must be reproductive division of labour such that some individuals are fertile and reproduce while others refrain from reproducing and perform work for the group; and (iii) offspring stay at the nest to help their mother produce more siblings.

It seems a law of the universe that nature resists categories and welcomes exceptions and qualifications. In the case of eusociality, we have eusociality that is *simple* and eusociality that is *complex*. Species

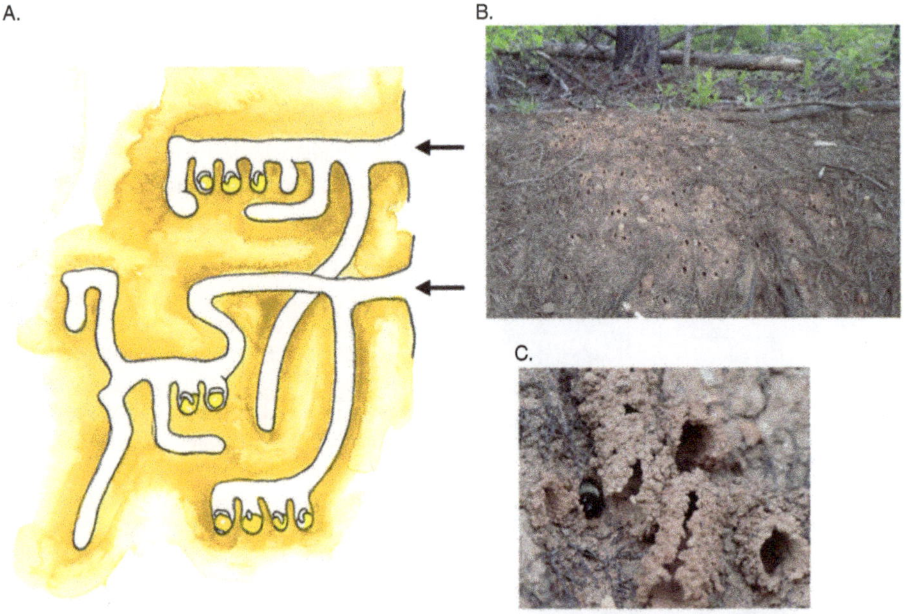

Fig. 2.6. This modification of the basic soil burrow is a horizontal shaft opening on to a vertical cliff face (A, openings indicated with arrows). Increasing numbers of secondary shafts and cells suggest increasing expression of social behaviour. Vertically terminating shafts trap nest refuse and act as a sump to resist inundation. An aggregation of bee burrows opening on to a soil cliff (B) – a roadside embankment in north Georgia (USA). An occupant is visible (C).

Fig. 2.7. An aggregation of solitary bee soil nests.

expressing simple eusociality cycle in and out of a eusocial state over the course of a year. The best example is the bumble bees, one of the most economically important crop pollinators who will be treated in some detail in this volume. Young queens emerge late in the season, mate, and enter into a solitary diapause in some sheltered nook. The following spring the overwintered queen emerges from her *hibernaculum*, or overwintering shelter, and begins a classically solitary life, finding a suitable nesting cavity, foraging for pollen, laying eggs and incubating the eggs with her own body heat. Often the first batch of daughters to emerge is small and stunted but still helpful at maintaining the nest while the mother continues foraging. Only after enough daughters are present to take over foraging and nest duties, is the queen freed to focus solely on egg laying, and the little family enters a eusocial state. By the end of the season, new queens and males are reared, mating occurs, and all nest members die except for the newly mated queens who diapause and re-enter the solitary phase (see Fig. 8.1, this volume).

In contrast to simple eusociality, species expressing complex eusociality sustain a eusocial state year round. There is no solitary phase which means in the case of temperate species the colony must overwinter as a homeothermic group. This is most famously accomplished by the European honeybee, *Apis mellifera*, whose entire life cycle can be thought of as an elaborate strategy for surviving cold winters (see Chapter 7, this volume). The European races of *A. mellifera* have been exported all over the world and are valued as producers of honey, pollinators of crops and gardens, and a source of pleasure and recreation to hundreds of thousands of beekeepers everywhere.

Fig. 2.8. A mating pair of *Andrena* sp. in north Georgia (USA). Identification: Conor Fair.

Fixity of the eusocial state allows for evolution of distinct *castes* – morphological or behavioural variants within sex. We have, for example, worker honeybees who are smaller and qualitatively different, morphologically, from their mothers. Workers are well adapted for work, whereas queens are specialized for mating and laying eggs (Fig. 2.13). In the simple eusocial species, caste is less fixed, more fluid. We see in the bumble bees, for example, that workers differ from queens primarily in size. Queens are little more than bigger, more fertile workers, and workers are little, subfertile queens. Complex eusociality is universal in the honeybees (Apini) as well as the stingless bees (Meliponini) – a large pantropical tribe that includes species cultured by humans for their honey and many that are important pollinators (see Chapter 11, this volume). In contrast, simple eusociality predominates in the bumble bees (Bombini) with the exception of 30 species who have abandoned simple eusociality in order to pursue life as social parasites (Williams, 1998).

2.4. A Word About Pollinator Efficacy and its Labels

By now the reader has seen enough hints to realize that not all bee pollinators are equal when it comes to vectoring pollen between compatible flowers. The difference between solitary and social life cycles certainly plays a part in this (see Chapter 3, this volume). If we may hazard a generalization, solitary bees as individuals are more *specialized* and therefore more *efficient* at pollinating a focal

Fig. 2.9. A hollow reed nest of probably *Osmia* spp. with cell partitions of mud. Photo courtesy of Josh Fuder.

Fig. 2.10. A hollow reed nest of *Megachile* sp. with cell partitions of cut leaf. Photo courtesy of James Cane.

crop than complex eusocial species; whereas complex eusocial species can field a large forager force of generalists available for many crops across a single growing season.

Alternatively, we could categorize bee pollinators by their manageability. Chief to the evolution of animal domestication has been a range in genetic tolerance to human management (Driscoll *et al.*, 2009). Such variable tolerance toward human interference explains why guinea pigs make good pets and opossums do not. In the case of bees, we can discern a similar range of tolerances to human intervention. A case in point is the relative ease by which pollen-storing bumble bees accept human management while their pocket-making cousins resist it (see Chapter 8, this volume).

Figure 2.14 categorizes some of the bee species covered in this volume according to their position along the two continua of pollinating efficiency and manageability. One can see that solitary bees predominate in the high efficiency pole, whereas social bees cluster at the high manageability pole. One also notices, however, that manageability is not the monopoly of sociality. Culture and management of some solitary species, chiefly the alfalfa leafcutting bee (*Megachile rotundata*), the alkali bee (*Nomia melanderi*), and to a lesser extent the

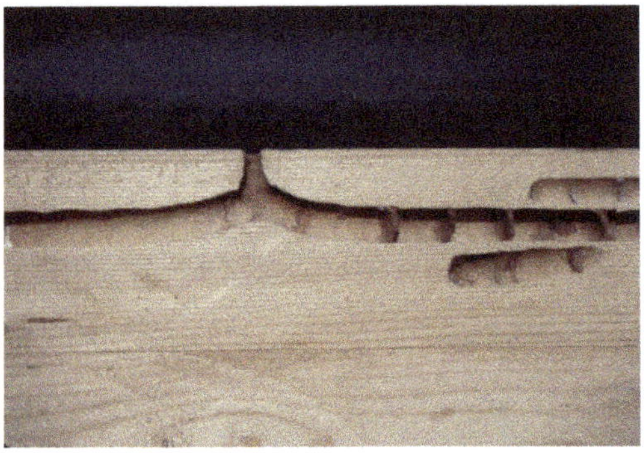

Fig. 2.11. A nest of the carpenter bee, *Xylocopa virginica*, showing cell partitions of masticated wood.

orchard mason bees (*Osmia* spp.) are sophisticated and rival the standards of honeybee keeping for numbers of pollinators fielded and infrastructure for servicing growers.

So, if both solitary and eusocial species occupy both poles of manageability and efficiency, where does this leave us?

Expanding on the association of honeybees with honey, Adams and Senft (1994) proposed the term 'pollen bees' to describe all pollinating bees that are not honeybees. Bees in this proposed grouping are unified by their concentration on pollen foraging, compared to honeybees who collect large quantities of nectar. The term 'pollen bees' is certainly more pleasing than other choices (i.e. 'non-honeybees'), but it is ambiguous because both honeybees and pollen bees collect pollen and pollinate crops. Similarly, the label 'native bees' loses relevance at a time when modern agricultural landscapes are themselves loaded with increasingly non-native crop plants and are barely recognizable as natural ecosystems.

Since the first edition of this volume when I rather chauvinistically embraced 'non-honeybees', use of the couplet 'wild' versus 'managed' bees has gained traction. 'Wild bees' has somewhat fewer restrictions than other choices. Wild bees can be both native or exotic, solitary or eusocial, and they have the advantage of instant context recognition for the purposes of management considerations. When we speak of wild bees we are in the realm of land management and conservation. When we talk about managed bees we are in the realm of husbandry. An increasingly integrated landscape approach to pollinator management seeks to maximize pollinator numbers, diversity and health. When this happens naturally as dividends from a healthy agroecosystem, the grower will be rewarded with a sustainable population of resident pollinators. When this situation is not the case, imported managed pollinators are the only recourse. Weighing heavy on these considerations is the reality of spatial scale in modern agriculture. Monocultures measured in tens of thousands of acres are rarely conducive to thriving resident pollinator populations. Although a crop

Fig. 2.12. The 'wool carder bees' partition their cells with plant fibres. This example is *Anthidium formosum*, family Megachilidae. Photo courtesy of James Cane.

Fig. 2.13. The paradigmatic example of complex eusociality in bees, the western honeybee (*Apis mellifera*). The queen (centre) and workers (surrounding) represent two female castes. In complex eusocial species, the castes within sex differ qualitatively in morphology. Worker honeybees are not only smaller than their queen, but they possess functional corbicula, barbed stingers, more complex brains and higher cognitive capacities, all of which are secondarily vestigial in the queen.

may offer an instant blast of nectar and pollen, the very nature of its monoculture means a nectar and pollen desert for the rest of the season. Nevertheless, the foundation of a sound pollinator strategy should involve conservation and investment in resident pollinators. Where resident wild bees are numerous enough to service a crop there is no need to bring in managed bees.

2.5. Effects of Non-Native Bee Species

Human beings have moved bees outside of their native ranges, some species intentionally for their perceived economic value. Russo (2016) estimates that 73% of exotic bee introductions have been accidental, 18% intentional and 5% the result of natural range expansions. The majority (69%) of introduced species are above-ground nesters in hollow twigs, stems or bored tunnels in wood – an artefact of the ease by which these kinds of materials can be accidently transported. In contrast, soil-nesting bees comprise only 26% of introduced species. Of the 80 known non-native bee introductions

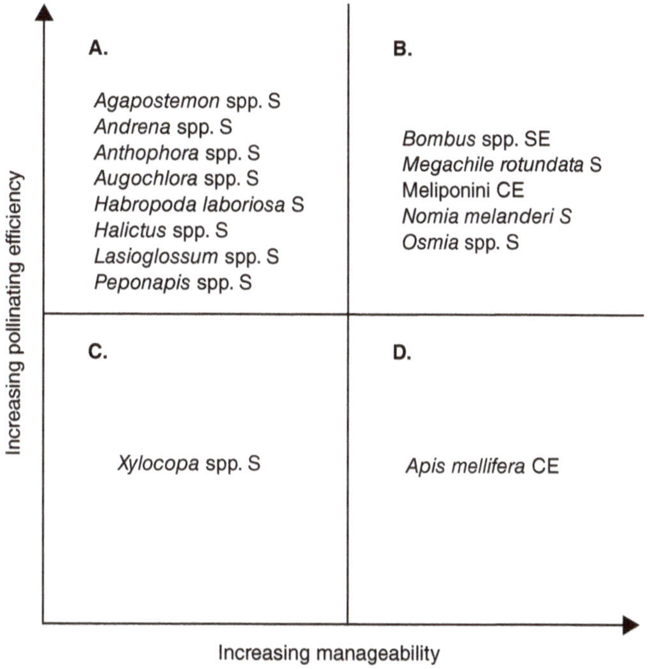

A.

Agapostemon spp. S
Andrena spp. S
Anthophora spp. S
Augochlora spp. S
Habropoda laboriosa S
Halictus spp. S
Lasioglossum spp. S
Peponapis spp. S

B.

Bombus spp. SE
Megachile rotundata S
Meliponini CE
Nomia melanderi S
Osmia spp. S

C.

Xylocopa spp. S

D.

Apis mellifera CE

Increasing pollinating efficiency

Increasing manageability

Fig. 2.14. Comparative positions of some important bee groups along the two continua of manageability and pollinating efficacy. S=solitary; SE=simple eusociality; CE=complex eusociality.

worldwide, their numbers are dominated by the families Megachilidae (33 introduced species) and Apidae (30 introduced species).

Most notable among the intentional introductions are three superlative Europeans – the honeybee *Apis mellifera* (see Chapter 7, this volume), the bumble bee *Bombus terrestris* (see Chapter 8, this volume), and the alfalfa leafcutting bee *Megachile rotundata* (see Chapter 9, this volume). A sizeable literature has emerged investigating the extent to which these and other exotic species displace native bees, disrupt pollination of native plants, and help propagate introduced weeds. Curiously, the overarching conclusions are equivocal, in part because flower visiting behaviours of bees are often generalized so that fitness changes to native bees or plants are affected only modestly or not at all. Investigators are hindered with low site replication or inability to find suitable control plots. And finally, complicating judgements is the fact that many pollinated crops of agricultural importance are themselves exotic species outside their native ranges. This thread runs through the following discussion – introduced bees are a welter of negative, neutral and even positive effects.

Honeybees have established feral populations everywhere they have been introduced. For much of the 20th century, it was these feral populations along with managed ones that were credited with the majority of crop pollination, especially in the temperate world. It was not unusual for specialists to treat exotic honeybees as full participants in natural ecosystems, pollinators of native plants, and sustainers of natural trophic webs that support native species (Barclay and Moffett, 1984). It has been suggested that, as a generalist, honeybee flower visiting behaviour does not differ qualitatively from that of other bee taxa (reviewed in Butz Huryn, 1997) and therefore it is to be expected that honeybees can pollinate many native plants effectively, as shown by Horskins and Turner (1999) with native *Eucalyptus costata* in Australia.

Probing specifically into the idea of non-difference among generalists, Giannini *et al.* (2015c) did a network analysis of two 'supergeneralists' in Brazil – the exotic *Apis mellifera* and native meliponine *Trigona spinipes*. The authors found no differences between the bees in their abundance, number of plant species with which each interacts, or the sum of their dependencies. However, at the network level, *A. mellifera* expressed higher *nestedness* – the property of participating in a tight core of generalist interactions to which other species in the community

can attach. *T. spinipes*, on the other hand, expressed higher niche overlap with other bee species. *A. mellifera* was more sensitive to high temperature extremes under which conditions native bees are more likely to displace this species. Although these results did not implicate negative or positive effects of *A. mellifera*, they did underscore that it is naïve to presume that the ecological impacts of an exotic generalist are indistinguishable from those of native pollinators.

Paini (2004) reviewed 28 studies that examined effects of introduced honeybees on native bees, with indirect ecological measures such as floral resource overlap, visitation rates and resource harvesting. He also reviewed nine studies that addressed the matter with direct measures on native bee fecundity, survival or population density. Of the combined 37 studies, the authors of 35% of these concluded that the impacts of introduced honeybees on native bees are neutral, while 65% of studies concluded the effects are negative, and 5.4% of studies concluded the effects are positive. Of the nine studies making direct measures of honeybee impacts, the authors of four (44%) concluded that impacts of honeybees are neutral, while three (33%) reported effects that are negative, and two (22%) concluded that effects are neutral or positive. The reviewer noted problems in these studies, many times unavoidable, with low site replication or absence of no-invader controls and urged future investigators to concentrate on direct measures of native bee population demographics.

Dohzono and Yokoyama (2010) reviewed the literature on the impacts of non-native bees on plant–pollinator relationships. These authors predict that impacts of non-native bees will include: (i) reduced pollen transfer because of morphological mismatch between native flowers and exotic bees; (ii) competitive exclusion of native bee species who share similar niche requirements with the exotic species; and (iii) changes in native bee pollination efficiency owing to disruptive behaviours by the exotic species such as nectar or pollen theft. These authors concluded that bird-pollinated plants are relatively unaffected by non-native honeybees. For bee-pollinated native plants 'honeybee impacts on reproduction may be pervasive' (Dohzono and Yokoyama, 2010, p.37), but the literature is again impeded with a shortage of control sites for making comparisons of plant effects with and without exotic honeybees. In contrast, the reviewers judged that impacts of exotic bumble bees are strong. This effect derives from the fact that all bumble bees (*Bombus* spp.) are congeners, and thus natives and exotics alike share similar flower visiting behaviours and niche requirements, thus competition between them will be keen. It must be stressed, however, that the same principle holds for *Apis mellifera* where it has been introduced into the natural range of its Asian congeners. Yang (2005) reports that since *A. mellifera* was introduced to China from Italy in 1896 the natural range of native *A. cerana* has decreased by over 75%.

Hanley and Goulson (2003) analysed the effects of non-native bees on the pollination and spread of non-native weeds. These investigators reviewed the literature and performed original surveys in New Zealand where four non-native species of European *Bombus* were introduced from 1885. The authors' New Zealand surveys showed that introduced *Bombus* overwhelmingly prefer visiting introduced plants. Only three of 36 plant species visited by the exotic *Bombus* were native species. Of all flower visits recorded, only 1.2% were to native flowers. Similarly, introduced honeybees in New Zealand rely almost exclusively on non-native introduced plants (Pearson and Braiden, 1990). The best explanation for these biases is the principle of *coadaptation* which in this context presents two sides of one coin: just as non-native pollinators have greater probability of possessing coevolved adaptations favouring non-native plants (with which they may be sympatric in their native ranges), the same exotic pollinators will have no or few adaptations favouring native plants. As the non-native plants are also freed from the constraints of their natural herbivores, the presence of non-native pollinators creates a disproportionate reproductive advantage to the exotic weeds. In short, there is clear evidence of an association between the spread of exotic weeds and presence of introduced bee pollinators.

When it comes to bee species other than *Apis* or *Bombus*, the alfalfa leafcutting bee *Megachile rotundata* has received the most scrutiny for its impacts as an introduced species. Empirical evidence exists for its negative impacts as a spreader of pathogens to native bees, and hypothetical concerns are plausible for its impacts as a competitor for nesting sites and spreader of exotic weeds. True to pattern, however, studies exist that refute some of these concerns (reviewed in Russo, 2016).

Among developments in more recent years is an increasing awareness of the risk of pathogen movement from managed to wild bee populations. Agricultural intensification has caused a surge in the use of managed bees to meet the pollination demands of high-acreage plantations. Although this does not necessarily involve the admixture of native and non-native species, in practice it virtually always does because 'high manageability' (Fig. 2.14) is a feature of those species that are deliberately relocated for their economic value. The high bee densities achieved in managed systems is a recipe for the evolution of parasite virulence (Brosi *et al.*, 2017), emergent infectious diseases, and pathogen spillover (Graystock *et al.*, 2016a). One team reviewed case histories of parasite transmission between managed and wild bees in Japan, North America, the UK and Ireland and concluded that managed bees have negatively impacted wild bees through parasite *spillover*, *spillback* and *facilitation* (Graystock *et al.*, 2016a). Pathogen or parasite spillover occurs when a reservoir population (in this case, the managed bees) transmits disease to a sink population (the wild population) (Daszak *et al.*, 2000). In pathogen or parasite spillback, a parasite present in the wild population infects the managed population where high densities and optimum conditions amplify its prevalence. The pathogen then spills back into the wild population at higher than natural levels (Hatcher and Dunn, 2011). In facilitation, a densely managed population increases nutrient and competitive stressors in the resident wild population, increasing its susceptibility to infections (Graystock *et al.*, 2014). Whereas populations of managed bees can be rapidly restored after parasite outbreaks, the same is not true for wild populations which in some cases suffer irreparable harm (Graystock *et al.*, 2016a). Indeed, declines in bumble bee relative abundance and ranges in North America (Cameron *et al.*, 2011; Meeus *et al.*, 2011) are frequently associated with an abnormally high prevalence of pathogens. Hence, we have another example where odds are biased against wild populations.

Appraising the assets and liabilities of introduced pollinators is a difficult and controversial issue. It inevitably pits the economic assets of the bees as crop pollinators against their ecological liabilities as exotic species. It is naïve to think that introduced bees have had no effect on native plants and animals, but the bulk of experimental evidence suggests that the effects have been, in the main, inconsistent and subtle.

3 What Makes a Good Pollinator?

Chapter 1 of this volume covered flower fertilization. The rest of this volume focuses on one subset of that process – pollination, the vectoring of pollen from one plant to a fertile receptive stigma on the same or different plant. I tell my students that fertilization is the realm of botany, whereas pollination the realm of entomology (or at least when it involves insects). That is a bit simple, but not far off.

To begin understanding what it means to be a *good* pollinator (or a bad one), we must recognize that the impact of any pollinator is an interaction between two dynamics – the pollen vectoring capacity of the flower visitor and the genetic obligation, or responsiveness, of the plant to pollen deposition on its stigmas. Successful pollination is a product of many components: first of all, does the plant need pollen to set fruit? Where must the pollen come from – self or other? How much pollen does the stigma need? How many stigmas are there per inflorescence? How evenly must the pollen be spread? How much pollen does the pollinator need to carry and where on its body? Is the pollinator carrying the right kind of pollen? How should the pollinator behave when it is gathering pollen or depositing it? And finally, what does *success* look like?

3.1. Pollinator Efficiency

Let us talk about pollination success. For a reproductive biologist, success means pollen deposition sufficient to fertilize most or all of the ovary's ovules. For a grower, success means pollen deposition sufficient to produce a marketable seed or fruit. The pollen deposition threshold for a marketable fruit can be as variable as 60 pollen grains, the work of 3–4 bee visits in the case of almond (Henselek *et al.*, 2018) or as difficult as a dozen successive bee visits to service all 200 pistils of strawberry. It is through this connection between number of visits, pollen deposition and resulting fruit that we reach the concept of pollinator efficiency. A system in which

one bee visit results in a marketable fruit is more efficient than a system that takes 20 bee visits. As the results depend on the number of visits, a fair comparison of any two flower visiting species must be done on a per-visit basis. This is carried out with exclusion bags and all-weather flower labels, as described in the following paragraphs (Delaplane *et al.*, 2013).

An investigator wants to compare the pollinating efficiency of two bee species. If the investigator is interested literally in pollen vectoring, then she bags an unopened virgin flower and once the flower opens, removes the bag and observes the flower until one bee completes one flower visit without interruption. The bee species is noted, then the flower is removed and the stigma microscopically examined to count pollen grains deposited per single visit. Replicating this enough times between the two bee species will give a robust comparison of innate differences in pollen depositing efficiency.

However, one could argue that a better test of a pollinator would consider the viability or compatibility of the pollen it vectors. Pollen deposited on a stigma is no good if it is the wrong kind or species; in fact, such a situation may clutter the stigma and render it unavailable to 'good' pollen. So, this time the investigator repeats the process above, but after observing the single bee visit, she re-bags the flower until the flower is no longer fresh, removes the bag, and lets the fruit develop. She then harvests and counts the seeds.

An elaboration of this method is the direct pollinator effectiveness metric of Spears (1983). It requires the addition of two types of experimental control: a negative control in which the investigator bags a number of virgin flowers for the entirety of anthesis, removes the bags, lets the fruit develop, and counts the seeds (if any); and a positive control in which a number of unbagged and labelled flowers are allowed to open and freely receive any or all visitors. The pollinating effectiveness (E) of pollinator species z is determined from:

$$E_z = \frac{(S_z - N)}{(U - N)} \qquad (3.1)$$

where S_z = number of seeds set per flower resulting from a single visit from pollinator z, N = number of seeds set per flower from negative control, and U = number of seeds set per flower from positive control.

An even more robust form of this test is to germinate the seeds to confirm genetic compatibility of the pollen vectored by species z.

For fruit growers, however, it is not always seed numbers nor their viability that matter. In fact, for many growers, seeds are at best the physiological impetus for fruit formation or at worst, detractions from their crop's palatability. In these situations, the better way to compare pollinators is to measure *fruit-set*. The bagging and single visit observations are the same as above except that the investigator harvests and counts marketable fruit, not its seeds. Fruit-set (*f*) is derived from:

$$f = 1 - \left(\frac{M}{F}\right) \qquad (3.2)$$

where M = the number of marketable fruit and F = the number of initial virgin flowers. This is a relative value that must apply to some identifying unit of plant: for example, fruit-set per *raceme*, per *branch*, per *plant* or per *plot*. The number is rarely 100% owing to fruit abortion from numerous causes including causes unrelated to pollination. As with Eqn 3.1, increasing fruit-set per single visit of species z indicates increasing pollinator efficiency.

More crude and less informative measures of pollination efficiency are common in the literature (Delaplane *et al.*, 2013). These include flower visitation rate (bee visits per minute); size of pollen load per bee and per cent composition of target pollen; proportion of bees collecting pollen versus nectar (pollen foragers are presumed better pollinators); and per cent composition of target pollen from pollen collected from bees' corbicula with hive entrance pollen traps.

3.2. Pollination Performance from the Perspective of the Bee

Not all bees are equally effective pollinators on a per-bee basis. Compared to generalist foragers, wild bees pollinate many crops more efficiently because of distinctive behaviours, morphology or life habits

(Kuhn and Ambrose, 1984; Westerkamp, 1991; Stanghellini *et al.*, 2002). Wild bees and other pollinators often match honeybee abundance on crop flowers (Garibaldi *et al.*, 2013; Kleijn *et al.*, 2015), and crop yields more strongly correlate to flower visitation by wild bees than by honeybees (Winfree *et al.*, 2007; Blitzer *et al.*, 2016). However, managed bees also deliver advantages by virtue of their sheer numbers and manageability. Let us consider some of these nuances.

Even though honeybees are regarded a long-tongued taxon (Fig. 2.1, this volume), many wild bee species have longer tongues, and this enables them to effectively pollinate crops with long tubular flowers. This is conspicuously the case when comparing honeybees to *Bombus* spp. in lavender (Balfour *et al.*, 2013) or in *Vaccinium* spp. (Cane and Payne, 1993).

There is a strong positive association between bee size and stigmatic pollen deposition, with larger bees depositing significantly more pollen per single flower visit, as shown by Willmer and Finlayson (2014) with bumble bees. There is also a strong positive association between bee hairiness and stigmatic pollen deposition, as shown by Stavert *et al.* (2016);

Fig. 3.1. This corbiculate bee (*Melipona colimana*) will comb much of this pollen off its body during a foraging trip and pack it into corbiculae on its hind legs, after which the pollen is considered unavailable for pollination. However, ample pollen remains on 'safe sites' on the bee's body that will contact subsequent stigmas. Photo courtesy of José Octavio Macías-Macías.

there were even crop-specific differences with higher pollen deposition in pak choi (*Brassica rapa chinensis*) associated with hairiness on the bee's face and deposition in kiwi fruit (*Actinidia deliciosa*) associated with hairiness on the bee's thorax.

Bumble bees and many solitary species are able to *sonicate* or *buzz-pollinate* blossoms – an adaptation by which the bee dislodges pollen from anthers with high-frequency vibrations. The bee grasps the anther with its mandibles and vibrates it with its thoracic muscles or alternatively, as noted in the Australian apid *Amegilla murrayensis*, taps the anther with its head (Switzer *et al.*, 2016). Buzz-pollination is a derived state for both plant and bee and represents a kind of coevolution in which the plant limits its visitors to the most effective pollinators. In crop plants, it figures prominently in blueberry, cranberry, kiwi fruit, chilli peppers, aubergine and tomato, and for these crops a buzz-pollinator is far better than a more generalist visitor, like the honeybee, who lacks this adaptation. Buzzing thoracic muscles is a preadaptation performed by bees in a number of unrelated contexts – to intimidate enemies, compact soil, warm a nest (Buchmann, 1983), or signal alarm about predators (Larsen *et al.*, 1986). Plants exploited this pre-existing character in their flower visitors by evolving anthers that are *poricidal* – housing dry, powdery pollen accessible only through a small aperture at the tip. The pollen is released only when the anther is vibrated. It is believed that buzz-pollination evolved in plants to limit pollen loss from opportunistic pollen thieves – visitors able to remove the large quantities of pollen available from more open designs of non-adapted anthers – and to improve pollination efficiency by aiming pollen at so-called *safe sites* on the bee – areas of the bee's body more likely to contact a stigma on a successive flower visit (De Luca and Vallejo-Marín, 2013). In any case, floral sonication has independently evolved 32–58 times in bees, with numerous reversals back to non-sonicating states, which suggests both that the trait is easy to evolve and that there is frequent selection for this complex pollen collecting behaviour (Cardinal *et al.*, 2018). Today, poricidal buzz-pollinated anthers are a feature of more than 22,000 species of angiosperms across at least 72 families and 544 genera (Buchmann, 1983). Among the bees, over 3400 species representing all six of the major families express floral sonication (Cardinal *et al.*, 2018).

Corbiculate bees incorporate vigorous auto-grooming during their foraging bouts, using specialized hairs on their bodies called *combs* (single rows) or *brushes* (multiple rows) to concentrate pollen on to their corbiculae. The pollen is further moistened with nectar or plant oils to consolidate the powdery substance for transport (Thorp, 1979). Once deposited in corbiculae or scopae, the pollen is considered unavailable for plant pollination (Westerkamp, 1996). Consequently, it is thought that pollination by corbiculate bees is largely an effect of pollen carried on safe sites that are difficult for the bees to reach and groom clean. Koch *et al.* (2017) showed with corbiculate *Apis mellifera* and *Bombus terrestris* that these safe sites are concentrated at the bee's 'waist' (the constriction between thorax and abdomen) and dorsal areas of the thorax and abdomen. The stigmas of *Salvia* spp. and *Borago officinalis* contact these safe sites during a flower visit, supporting the hypothesis that pollination is effected by safe site pollen. A variation of this theme was found by Tong and Huang (2018) who, working with *Bombus friseanus* and a sympatric host plant *Pedicularis* spp., showed that no more than 30% of this corbiculate bee's pollen load is contained in safe sites, demonstrating the grooming and packing efficiency of these bees. This could set the stage for conflict between plant and pollinator, but what happens instead is an evolved cooperation between sympatric species: even though the majority of aggregate pollen is selfishly packed away by the bee for its own use, the site of the single largest pollen deposition on its body corresponds to a site easily contacted by the plant's stigmas. Consequently, pollen distribution on the bee's body arrives at an optimum for both species. It is likely that similar accommodations happen generally across the corbiculate bees and their main host plants. In the end, even corbiculate bees get thoroughly covered in pollen (Fig. 3.1), and no case has ever been made that the corbiculum is a serious impediment to pollination.

These pollen-grooming and pollen-concentrating behaviours are less pronounced in the non-corbiculate species who, as a consequence, are 'dirtier' in appearance, with pollen looser and more liberally scattered across their ventral surfaces (Thomson and Plowright, 1980). Moreover, the corbiculate bees' habit of dampening pollen for transport damages the pollen's viability for setting seed. Parker *et al.* (2015) swiped virgin stigmas of (*Brassica rapa*) with pollen recovered from either the bodies or the corbiculae or scopae of representative corbiculate and non-corbiculate bees. For the corbiculate bees, there was a significant deterioration in viability of pollen from the corbiculae compared to pollen from the body. No such difference was found for the non-corbiculate bees. This supports the premise that the in-field practice of packing and dampening corbicular pollen for transit deteriorates its viability. Moistening may cause physiological injury to the pollen, increase clumping, or reduce its adherence to stigmas, and tight packing may reduce the transfer of pollen to stigmas.

The strongly seasonal life history of solitary bees serves in many cases to enhance their effectiveness as pollinators. These bees have a simple life cycle in which adults emerge, fly, mate and provision brood cells, during the few weeks of peak bloom in their area. Thus, solitary bees have evolved specializations for those sympatric plant taxa blooming in synchrony with the bees' brief flight season, and these specializations work in favour of the grower if those plants happen to be economically important crops.[i] Social honeybees and bumble bees, on the other hand, visit many flowering plants over the course of a season. They are generalists, not specialists, and they are more easily lured away from the crop of interest. The urgency with which solitary bees work during their short active season also works in favour of the grower; compared to honeybees, wild bees work longer hours, work faster, visit more blossoms per day and fly more readily during inclement weather.

Solitary bees are less likely to sting than the social honeybees and bumble bees, which is especially important in situations where labourers or self-pick customers are on the farm. The reason for their docility is traced to simple evolutionary trade-offs between benefits and costs. A stationary colony of a social species represents the costliest investment the species can make. A social colony can afford to lose a few sterile workers in defence of the group. For a solitary mother facing a dangerous predator, however, it is far safer to simply abandon her simple burrow and start another one. Bee handlers who propagate solitary pollinators such as alkali bees, alfalfa leafcutting bees and orchard mason bees routinely do so with little or no sting-protective clothing.

Not only are some bees inefficient pollinators, some of their flower visits are outright antagonistic to pollination. One of the most egregious bee behaviours in the context of crop pollination is nectar thievery. Evolution, ever opportunistic and never standing still, has produced bees that can manipulate a flower to harvest its nectar without 'paying'

the plant by pollinating it. This exchange is frequently a complete loss for the plant (Maloof and Inouye, 2000). The pay-off for the bee is a net savings in the calories spent foraging compared to the calories gained (Dedej and Delaplane, 2005). In my part of the world the best example of this is the native carpenter bee (*Xylocopa virginica*) which steals nectar from bush-type blueberries (*Vaccinium* spp.) by piercing the corolla with its sharp mouthparts and sucking nectar from the nectaries, without making contact with the sexual parts. This behaviour is bad enough, but the floral perforations made by *Xylocopa* attract honeybees who, unable to pierce the corolla themselves, become secondary nectar thieves. In this manner, the pollination efficacy of honeybees is significantly diminished (Dedej and Delaplane, 2004).

There are some uncertainties with wild bees. First, it is well known that wild bees pollinate many crops more efficiently than honeybees on a per-bee basis. However, it has also been shown that colonial social bees can effect satisfactory pollination by virtue of their numbers; in other words, enough successive visits by an inefficient pollinator can add up to satisfactory fruit-set (Dedej and Delaplane, 2003). Populations of wild bees are sometimes too small to support commercial pollination needs (Morrissette *et al.*, 1985; Parker *et al.*, 1987; Scott-Dupree and Winston, 1987) or vary considerably between years and geographic regions (Cane and Payne, 1993). Just like honeybees (Aizen and Harder, 2009; Pettis and Delaplane, 2010), wild bees have diseases, predators and parasites that limit their natural populations (Cameron *et al.*, 2011; Goulson *et al.*, 2015; Meeus *et al.*, 2018), or for their managed counterparts must be controlled by the beekeeper (Pitts-Singer and Cane, 2011). Nevertheless, where they occur in sufficient numbers, wild bees can replace honeybees entirely (Garibaldi *et al.*, 2013), or synergistically complement them (Greenleaf and Kremen, 2006b) for commercial pollination. A recurring thread in this volume is that the essential foundation of a farm scale pollination scheme is to conserve and enhance wild pollinators. Only to the extent that those efforts come up short should the grower import managed pollinators.

This volume covers five groups of pollinating bees that are prominent in the crop pollination literature: honeybees (Chapter 7); bumble bees (Chapter 8); managed solitary bees including the alfalfa leafcutting, alkali and orchard mason bees (Chapter 9); wild bees (here, Chapter 2 and Chapter 10); and the tropical stingless bees (Chapter 11).

3.3. Pollinator Dependency from the Perspective of the Plant

3.3.1. Breeding systems

Owing to their diverse breeding systems, the angiosperms vary tremendously in their pollination requirements. Some new terms are introduced below, first reviewing a few that were introduced in Chapter 1 (this volume). Flowers can be *hermaphrodite* (bisexual or perfect) or *unisexual* (imperfect). If flowers are unisexual, they are called *staminate* if male or *pistillate* if female. If unisexual flowers of each sex occur on the same plant the species is *monoecious* (for example, the cucurbits); if sexes segregate so that any one plant bears flowers of only one sex, the species is *dioecious* (kiwi fruit). A plant with bisexual flowers is considered monoecious.

Apomictic plants can develop seeds or fruit asexually without requiring fertilization of the ovule; the seed, if any, derives solely from maternal tissues. Apomixis can happen when an embryo forms from the unfertilized egg within the diploid embryo sac (as in blackberry and dandelion) or from the diploid nucleus tissue surrounding the embryo sac (some *Citrus* spp., some mango varieties) (Delaplane *et al.*, 2013). When fruit forms without ovule fertilization it is called *parthenocarpy*, and the seeds are either non-existent or atrophied. This is the basis behind such high-value speciality crops such as seedless cucumber, orange and watermelon. There are two types of parthenocarpy. The first is *vegetative* parthenocarpy and occurs without pollination under the action of plant growth hormones on ovarian tissue. There are synthetic plant growth hormones that can be applied to induce crops, even crops that are normally cross-pollinated, to develop fruit parthenocarpically. The growth regulator gibberellic acid is used to induce parthenocarpic fruit under conditions of poor pollination or chill damage in apple, blueberry, citrus and pear. In the second type of parthenocarpy, pollination or another stimulant is required to induce formation of fruit; pollen acts as a stimulant to the process, but seeds are not fertilized and subsequently atrophy. An example of such *stimulative* parthenocarpy is seedless watermelon whose small white seeds pose little impediment to palatability. Finally, in some apomictic plants apomixis is inconsistent or partial, so that sexual reproduction can also happen (Delaplane *et al.*, 2013). In general, positive responsiveness of apomictic plants to pollen vectors is limited to those crops expressing stimulative parthenocarpy or incomplete apomixis.

It is to be stressed, however, that sexual reproduction is the norm in the angiosperms. However, given the diverse means by which sexual tissues are parsed among members of a species, sexually reproducing angiosperms have essentially two mating systems – *selfing* (also called *autogamy*) or *out-crossing* (also called *xenogamy*). Selfing happens from the transfer of pollen within the same flower (called *self-pollination*) or between flowers of the same plant (called *geitonogamy*). Out-crossing is the business of *cross-pollination* – defined as the transfer of pollen from flowers of one plant to flowers of a different plant or different variety. Layering on the possibilities, plant species can be strictly autogamous (*self-fertile*), strictly xenogamous (*self-infertile*), or *mixed*, employing both mating systems and even apomixis as well.

In self-fertile plants it is important to distinguish between self-fertility and self-pollination. Bees and other insects may still be necessary or helpful in moving pollen to stigmas. Viable self-pollination is necessarily limited to self-fertile species and is a hallmark of the beans (soy and groundnuts) and peach. As plants can express mixed mating systems and a range of self-fertility, facilitated out-crossing may provide corresponding increases in yield. Swede rape (canola) and highbush blueberry typically respond well to cross-pollination even though each is considered self-fertile. In such systems, interplanting of varieties may be helpful to facilitate crossing with non-self pollen.

Non-self pollen is obligatory for self-infertile or self-*sterile* plants which require pollen from a different plant or even a different variety of the species. If the plant requires different varieties, the grower must interplant *pollenizer* varieties with the *main* variety. *Cross-compatible* varieties are receptive to each other's pollen, whereas *cross-incompatible* varieties are not. Extension services and seed/nursery catalogues provide tables that cross-list compatible varieties. Many crops benefit from cross-pollination in what appears to be an expression of *heterozygosity* or *hybrid vigour* – a common feature across biology in which genomic allele diversity improves a range of fitness characters in the individuals who possess it.

To summarize, pollen deposition on stigmas is helpful or necessary for inducing fruit in stimulatively parthenocarpic plants. In sexually reproducing plants, the majority condition among angiosperms, pollen deposition on stigmas is obligatory. Pollen vectors, especially bees, are necessary for moving pollen between dioecious plants, between xenogamous plants, and between staminate and pistillate flowers in monoecious plants. In monoecious plants, the degree to which that pollen must come from a different plant varies in exact proportion to the plant's expression of self-fertility. Even in self-fertile or stimulatively parthenocarpic plants, a bee pollen vector may be necessary to move pollen to stigmas.

When it comes to the value of animal flower visitation, these contingencies are captured by the concept of a crop's *pollinator dependence ratio* (*D*), the fraction of the crop's fruit-set attributable to animal pollination (O'Grady, 1987; Melathopoulos *et al.*, 2015) which is derived from:

$$D = 1 - \left(\frac{f_{pe}}{f_{op}} \right) \tag{3.3}$$

where f_{pe} = fruit-set achieved under conditions of pollinator exclusion, such as insect-excluding field cages, and f_{op} = fruit-set achieved under open pollination. One sometimes sees the value D_{max} which is derived by substituting f_{op} for f_{pmax} – a measure of fruit-set achieved under saturative bee numbers, usually by caging plants with large numbers of pollinators. D_{max} is useful for understanding the plant's physiological reproductive limits, whereas *D* is more representative of reproduction one can expect from ambient densities of flower visitors. Even though the equation is usually posed in terms of fruit-set *f*, in practice the literature may offer other measures of plant reproduction such as number of pods or number of seeds. For the purposes of understanding the plant's obligation to animal flower visitors these values can be used to derive *D* as long as the data are collected for both insect-accessed and insect-excluded experimental plots.

The value *D* has gained notice in recent years as agricultural economists have produced national and global scale estimates of the economic value of insect pollination (Tables 3.1 and 4.1, this volume). Most of these estimates use the so-called *bioeconomic* approach by which cultivated acres of a crop are multiplied by *D* and market prices to derive a fraction of the crop's value attributable to insects.

Finally, purely from the perspective of human economy not plant reproduction, we can consider the concept of a crop's *direct* versus *indirect dependence* on pollinators. We regard a crop as *directly pollinator-dependent* if the product of pollination is the fruit or seed that is consumed. A crop is considered *indirectly pollinator-dependent* if the product of

pollination is the seed used to produce the crop whose consumable parts are vegetative, not reproductive. Apples, almonds, pomegranate and cantaloupes are examples of directly pollinator-dependent crops, whereas alfalfa (lucerne) hay, onion and lettuce are indirectly pollinator-dependent. The inclusion or exclusion of indirectly pollinated crops has huge influence on estimates of pollinator economic value, especially for leguminous forages fed to livestock. When animal agriculture is included in pollinator valuations, it invariably constitutes the largest share of value (Melathopoulos *et al.*, 2015). Animal agriculture, and a similar case with oilseed-based biodiesel production, encourage debate over how extensively such multiplier benefits should be included in pollination valuation studies. The most recent estimates of pollinator valuation (see Table 4.1, this volume) tend to categorically exclude such multipliers.

3.3.2. Flower and fruit morphology

Within the fundamental *bauplan* of a flower (Fig. 1.4, this volume) can be built an infinitude of variations, the complexity of which has direct consequences on fruit morphology. By extension, the more complex the flower/fruit morphology, the more difficult the job for the pollinator at delivering 'good' pollination. By way of explaining this, we will talk about some important categories of flower/fruit morphology.

To the flower morphology presented in Fig. 1.4 (this volume), we can now add considerations of the ovary's relative position on the flower (Fig. 3.2). If the ovary is above the point of its attachment with the hypanthium, the ovary is called *superior*, and if the ovary is below the point of its attachment with the hypanthium it is called *inferior*. Additionally, a superior ovary is considered *hypogynous* if the hypanthium attaches at the base of the ovary and *perigynous* if the hypanthium assumes a more cup-like configuration with corolla, sepals and stamens attaching along its rim.[ii]

Fruits are postpollination elaborations of the ovary wall or surrounding non-reproductive parts of the flower. Fruit with tissues originating from non-ovarian tissue – the receptacle or hypanthium – is called *accessory* fruit. When the ovary wall is involved in forming fruit, it is called the *pericarp* which in turn can be divided into three layers, the outer *ectocarp*, the middle *mesocarp* and the innermost *endocarp*. The structural origin and arrangement of these tissues is the basis for the botanical categories of fruit summarized here (Fig. 3.3).

Table 3.1. Taxonomic categories, breeding systems[a] and pollinator dependence of bee-pollinated crops covered in Volume II of this work and the plant phylogeny of Fig. 1.1 (this volume). Pollinator dependence ratio (D, 0–1, 1=highest positive response) is a measure of the responsiveness of a crop's fruit-set to animal flower visits. Except where otherwise referenced, D derives from Klein et al., 2007; Kasina et al., 2009; Majewski, 2014; Giannini et al., 2015b. Where reported values differ, the literature means are presented. This table includes crops Klein et al. (2007) identify as having a response to animal pollination 'modest, great or essential', or if for seed, a response to animal pollination that 'increases' yield.

Plant order in Fig. 1.1	Major division in Fig. 1.1	Crop species	Crop	Breeding/flowering system	Type of pollinator dependence	Pollinator dependence ratio (D)
Laurales	magnoliids	Persea americana	avocado	dichogamous, bisexual, self-incompatible	direct	0.65
Arecales	monocots	Cocos nucifera	coconut	monoecious, partially self-compatible	direct	0.25
Zingiberales	monocots	Elettaria cardamomum	cardamom	bisexual	direct	0.65
Asparagales	monocots	Allium cepa	onion, shallot	bisexual, self-incompatible	indirect	0.73 (Kumar et al., 1985; Yucel and Duman, 2005)
Asparagales	monocots	Asparagus officinalis	asparagus	dioecious	indirect	0.87 (Eckert, 1956; Ito et al., 2011)
Asparagales	monocots	Vanilla planifolia, V. pompona	vanilla	bisexual, mostly self-incompatible	direct	0.95
Proteales	eudicots	Macadamia ternifolia	macadamia	bisexual, mostly self-incompatible	direct	0.95
Fabales	eudicots	Dolichos biflorus	bean, hyacinth	bisexual	direct	0.25
Fabales	eudicots	Vicia faba	bean, broad, faba	bisexual, self-compatible	direct	0.25
Fabales	eudicots	Glycine max	bean, soy	bisexual, self-compatible	direct	0.25
Fabales	eudicots	Canavalia ensiformis	bean, jack	bisexual, self-compatible	direct	0.25
Fabales	eudicots	Medicago sativa	alfalfa (lucerne)	bisexual, mostly self-incompatible	indirect	0.82 (Free, 1993)
Rosales	eudicots	Malus domestica	apple	bisexual, self-incompatible	indirect	0.75
Rosales	eudicots	Prunus dulcis, syn. Amygdalus communis	almond	bisexual, self-incompatible	direct	0.77 (Godini et al., 1992; Socias et al., 2004)
Rosales	eudicots	Rosa spp.	rose hips	bisexual, mostly self-compatible	direct	0.65
Rosales	eudicots	Rubus idaeus	raspberry	bisexual, self-compatible	direct	0.68
Rosales	eudicots	Rubus fruticosus	blackberry	bisexual, self-compatible	direct	0.65
Rosales	eudicots	Sorbus aucuparia	rowanberry	bisexual, self-compatible	direct	0.95
Rosales	eudicots	Sorbus domestica	service apple	bisexual, self-incompatible	indirect	0.25
Rosales	eudicots	Zizyphus jujuba	jujube	bisexual, self-compatible	direct	0.25
Rosales	eudicots	Pyrus communis	pear	bisexual, self-incompatible	direct	0.78
Rosales	eudicots	Prunus persica	peach, nectarine	bisexual, self-compatible	direct	0.65
Rosales	eudicots	Eriobotrya japonica	loquat	bisexual, self-incompatible	direct	0.65
Rosales	eudicots	Prunus domestica	plum	bisexual, self-compatible and self-incompatible	direct	0.53

Table 3.1. Continued

Plant order in Fig. 1.1	Major division in Fig. 1.1	Crop species	Crop	Breeding/flowering system	Type of pollinator dependence	Pollinator dependence ratio (D)
Rosales	eudicots	Prunus armeniaca	apricot	bisexual, self-compatible and self-incompatible	direct	0.65
Rosales	eudicots	Prunus avium	sweet cherry	bisexual, mostly self-incompatible	direct	0.80
Rosales	eudicots	Prunus cerasa	sour cherry	bisexual, mostly self-compatible	direct	0.63
Rosales	eudicots	Fragaria spp.	strawberry	bisexual, self-compatible	direct	0.55
Fagales	eudicots	Castanea sativa	chestnut	monoecious, largely self-incompatible	direct	0.25
Cucurbitales	eudicots	Citrullus lanatus	watermelon	monoecious, self-compatible	direct	0.95
Cucurbitales	eudicots	Cucumis sativus	cucumber, gherkin	monoecious, andromonoecious, self-compatible	direct	0.65
Cucurbitales	eudicots	Cucumis melo	cantaloupe, melon	monoecious, andromonoecious, self-compatible	direct	0.95
Cucurbitales	eudicots	Cucurbita spp.	pumpkin, squash	monoecious, self-compatible	direct	0.95
Oxalidales	eudicots	Averrhoa carambola	starfruit	bisexual, self-incompatible	direct	0.65
Malpighiales	eudicots	Passiflora edulis	passion fruit	bisexual, mostly self-incompatible	direct	0.73
Malpighiales	eudicots	Mammea americana	mammee	androdioecious	indirect	0.25
Myrtales	eudicots	Feijoa sellowiana	feijoa	bisexual, mostly self-incompatible	direct	0.65
Myrtales	eudicots	Punica granatum	pomegranate	bisexual, andromonoecious, partly self-incompatible	direct	0.25
Myrtales	eudicots	Psidium guajava	guava	bisexual, self-compatible	direct	0.45
Myrtales	eudicots	Pimenta dioica	allspice, pimento	dioecious	direct	0.65
Malvales	eudicots	Gossypium hirsutum, G. barbadense, G. arboreum, G. herbaceum	cotton	bisexual, self-compatible	indirect	0.25
Malvales	eudicots	Durio zibethinus	durian	bisexual, monoecious, mostly self-incompatible	direct	0.65
Malvales	eudicots	Abelmoschus esculentus	okra	bisexual, self-compatible	direct	0.25
Malvales	eudicots	Theobroma cacao	cocoa	bisexual, mostly self-incompatible	direct	0.95
Brassicales	eudicots	Brassica oleracea	cabbage, cauliflower	bisexual, mostly self-incompatible	indirect	0.11 (Stanley et al., 2017)
Brassicales	eudicots	Brassica napus	oilseed rape	bisexual, self-compatible	direct	0.28
Brassicales	eudicots	Brassica rapa (formerly B. campestris)	turnip rape	bisexual, mostly self-incompatible	direct	0.60 (Atmowidi et al., 2007)
Brassicales	eudicots	Brassica alba, B. hirta, B. nigra, Sinapis alba, S. nigra	mustard	bisexual, self-compatible	direct	0.25
Sapindales	eudicots	Anacardium occidentale	cashew	andromonoecious	direct	0.45

Order	Clade	Species	Common name	Sexual/breeding system[a]	Pollination	Value
Sapindales	eudicots	Mangifera indica	mango	andromonoecious, mostly self-compatible	direct	0.33
Saxifragales	eudicots	Ribes nigrum, R. rubrum	blackcurrant, redcurrant	bisexual, mostly self-incompatible	direct	0.55
Caryophyllales	eudicots	Fagopyrum esculentum	buckwheat	bisexual, self-incompatible	indirect	0.65
Caryophyllales	eudicots	Opuntia ficus-indica	prickly pear	bisexual, mostly self-incompatible	indirect	0.25
Ericales	eudicots	Vaccinium corymbosum, V. angustifolium, V. ashei, V. myrtillus	blueberry	bisexual, mostly self-compatible	direct	0.65
Ericales	eudicots	Vaccinium macrocarpon	cranberry	bisexual, self-compatible	direct	0.65
Ericales	eudicots	Bertholletia excelsa	brazil nut	bisexual, self-incompatible	direct	0.95
Ericales	eudicots	Arbutus unedo	tree strawberry	bisexual, self-compatible	indirect	0.25
Ericales	eudicots	Actinidia deliciosa	kiwi fruit	dioecious	direct	0.95
Ericales	eudicots	Vitellaria paradoxa	shea nut	bisexual	direct	0.63 (Stout et al., 2018)
Asterales	eudicots	Helianthus annuus	sunflower	dichogamous, variable self-compatibility	direct	0.49
Apiales	eudicots	Daucus carota, D. sativus	carrot	bisexual, mostly self-incompatible	indirect	0.86 (Davidson et al., 2010; Howlett, 2012)
Apiales	eudicots	Carum carvi	caraway	andromonoecious, dichogamous, self-compatible	direct	0.25
Apiales	eudicots	Coriandrum sativum	coriander	bisexual, self-compatible	direct	0.65
Apiales	eudicots	Foeniculum vulgare	fennel	bisexual, andromonoecious, dichogamous, self-incompatible	direct	0.65
Dipsacales	eudicots	Sambucus nigra	elderberry	bisexual, self-compatible	direct	0.25
Solanales	eudicots	Lycopersicon esculentum	tomato	bisexual, self-compatible	direct	0.32
Solanales	eudicots	Solanum quitoense	naranjilo	bisexual, mostly self-compatible	direct	0.65
Solanales	eudicots	Solanum melongena	aubergine	bisexual, self-compatible	direct	0.25
Lamiales	eudicots	Sesamum indicum	sesame	bisexual, self-compatible	direct, breeding	0.25
Gentianales	eudicots	Coffea arabica, C. canephora	coffee	bisexual, mostly self-compatible	direct	0.25

[a] dichogamy = functional sex change over time; monoecious = single plant has both staminate and pistillate flowers or exclusively bisexual flowers; andromonoecious = single plant has both male flowers and bisexual flowers; dioecious = single plant bears flowers of exclusively one sex, androdioecious = single plant bears exclusively male flowers while other plants bear bisexual flowers

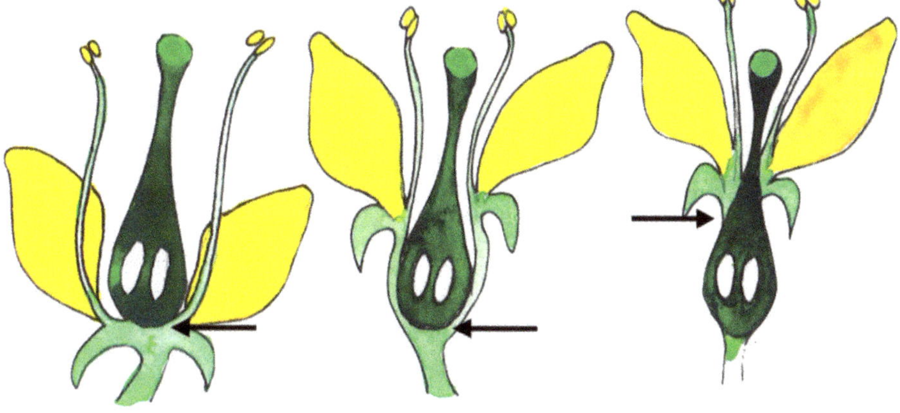

Fig. 3.2. Flower classification according to relative position of ovary to the hypanthium. Arrows indicate point of attachment of the hypanthium to the ovary.

A *berry* derives from one flower with one ovary of one or more carpels. The number of stigmas serving the single ovary can range from one to over 40. The 'true' berries derive solely from superior ovaries, whereas 'false' berries derive from inferior ovaries and include non-ovarian tissue in the fruit. In true berries the ovarian mesocarp assumes the fleshy layer and surrounds one or more small seeds. The ectocarp makes up the skin. A berry with a leathery separable skin and interior divided by septa into segments, like *Citrus* spp., is called a *hesperidium*, and a berry with a tough inseparable rind and an interior lacking septa divisions, like the cucurbits, is called a *pepo*. True botanical berries include banana, coffee berry, aubergine, grape, kiwi fruit and tomato. Examples of false berries include blueberry, cranberry, cucurbits and pomegranate.

In multi-seeded berry fruits, there is sometimes a positive relationship between the number of ovules fertilized and the size of the resulting fruit. Therefore, with berries the primary demand on a pollinator is to deposit 'enough' viable pollen on the flower's stigmas. Morphological variations on stigmatic shape and number make the job for a pollinator easier or harder. For example, the stigma is a relatively complex structure in pepos, with three pronounced lobes in watermelon and two in squash, pumpkin and gourd. In these stigmas, the pollinator, or subsequent pollinators, must not only deposit enough pollen to fertilize the many ovules but also distribute the

pollen evenly among the stigmatic lobes to achieve a well-shaped fruit. A similar challenge happens when the ovary's style branches into multiple stigmas. It may require a succession of flower visitors to deliver pollen to all of them.

A *pome*, such as apple and pear, is an accessory fruit in which much of the tissue derives from the hypanthium, although some authorities state that receptacle tissue is also involved. During fruit maturation, the hypanthium joins and subsumes the ovary wall, pushing the ovarian tissues inward to become the tough inner core. It is the hypanthium layer that comprises the edible fleshy part. Flowers of some important pomes, namely apple and pear, have five stigmas. A good pollinator must deliver pollen to all five of them to produce a well-shaped fruit.

A *drupe* or *stone* fruit, such as peach, has a fleshy outer wall surrounding one stony pit which contains one seed. In those species with more than one seed, only one fully develops postpollination. The fruit is entirely of ovarian origin with the ectocarp comprising the skin, the mesocarp forming the fleshy middle, and the endocarp forming the hardened stony cover of the seed. Representative drupes include coffee, olive, and all members of the distinctive genus *Prunus* which includes almond, apricot, cherry, nectarine, peach and plum.

Each fruit type we have talked about so far is a *simple* fruit, by which we mean a fruit that derives from only one ovary. With *aggregate* fruits, we have

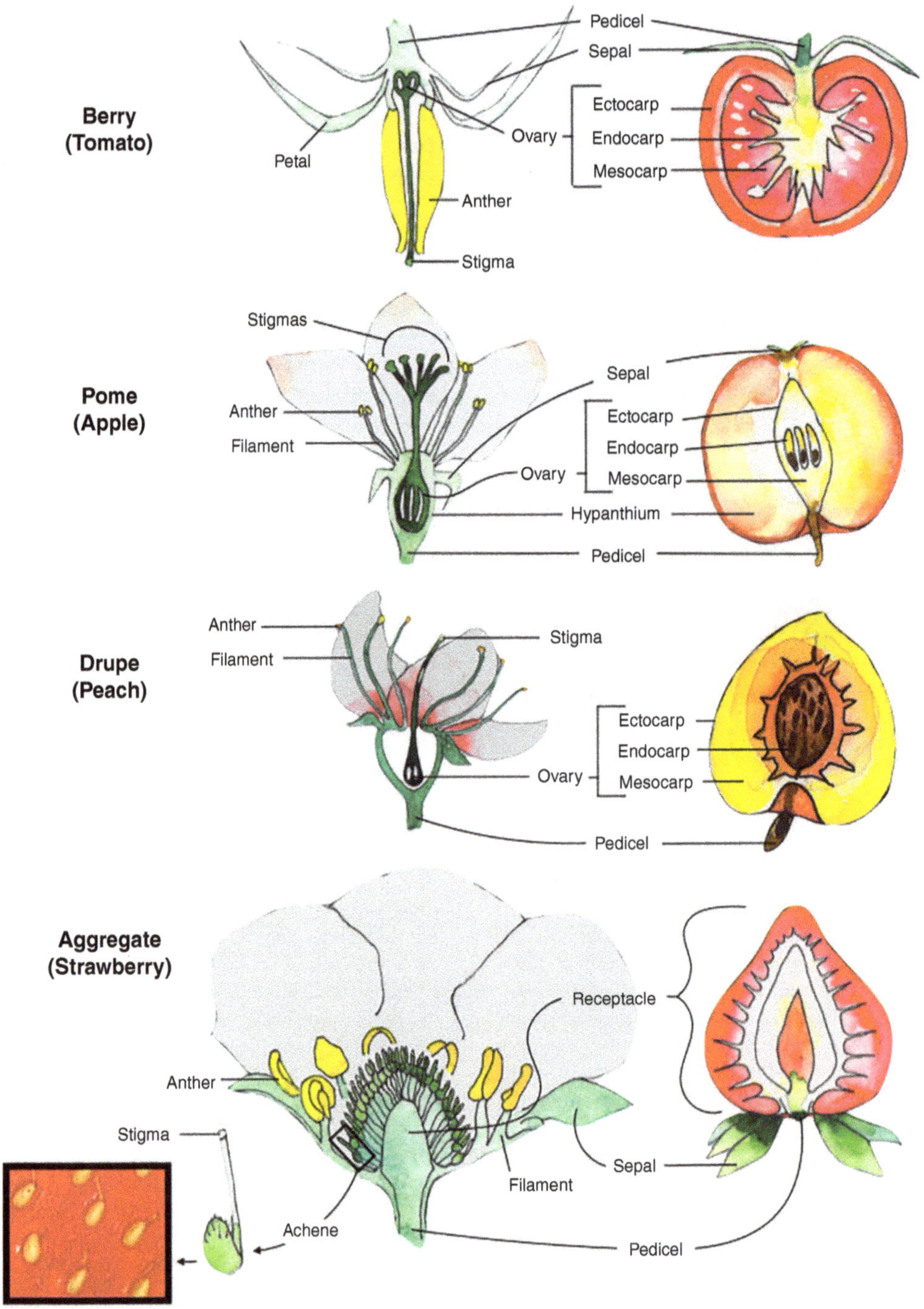

Fig. 3.3. Four common fruit types and corresponding structures of their flowers. Fruits are oriented in the posture relative to the flower appropriate to their development. Photo insert shows appearance of achenes on surface of ripe strawberry. Photo insert courtesy of Pilar Delaplane.

one flower with numerous pistils joined to one receptacle; each pistil is functionally independent. The tiny fruits that merge to form a ripe raspberry or blackberry is each a morphological drupe; hence they are called *drupelets*. In blackberry and raspberry, the receptacle is elongated and becomes part of the whole fruit; these are thus examples of *aggregate accessory* fruits.

Another well-known aggregate accessory fruit is strawberry. As an aggregate fruit, its flower is comprised of many massed pistils, and as an accessory fruit it is the enlarged receptacle that comprises the mature strawberry. The only part remaining of the ovaries are the numerous 'seeds' on the strawberry's surface. Each of these 'seeds' is one whole ovary. They are in fact *achenes* – defined as a single ovary with a single carpel and seed, the dry shell of which is the hardened ovary wall.

Functionally, aggregate fruits make the greatest demands on commercial pollinators. Every pistil must receive at least one viable pollen grain for the receptacle to enlarge and form a full-sized symmetrical fruit. To accomplish this, either one highly efficient flower visitor must physically contact the entire receptacle, or a succession of flower visitors must eventually carpet the receptacle with pollen. Small, misshapen, asymmetrical fruits are a sign of inadequate pollinator visitation and surface coverage.

3.4. Pollinator Performance from the Perspective of Foraging Ecology

3.4.1. Theoretical foundations

Although bees and bee-pollinated angiosperms depend on each other, each operates selfishly. For each, there is a cost/benefit equation that must balance in its favour. Nectar and pollen production are costly to a plant and must be balanced for maximum return (reproduction) for the energy spent to produce them. For the bee, foraging is energetically costly, and the bee will not forage unless calories gained exceed calories spent. Ecology is the empirical study of these kinds of interspecific relationships. A large subset of ecological scholarship, called *optimal foraging theory*, predicts that foraging animals will forage efficiently, moving between food patches and lingering in patches in such a way as to get the most calorific return for their effort. When such hypotheses are tested in the field, they very often prove true.

In the case of bees, optimal foraging theory is informed by the fact that bees are *central place foragers* (CPF), a condition common enough across the Animal Kingdom to engender *central place foraging theory* – that branch of foraging economics that accounts for travelling costs to and from a permanent point in space, such as a nest (Schoener, 1979). The presence of bees at a patch or crop, and by extension their pollinating efficacy there, will be a product not only of the nutrient richness of the patch but its distance from the nest (Cresswell *et al.*, 2000; Olsson *et al.*, 2015), and these dynamics operate independently according to complexity of the landscape and physiological state of the forager. Implicit to a CPF model is the idea that for a patch of any quality, there is a limit to the distance a forager will travel to get there. Nearer the nest, a forager will visit sites of varying quality, but at greater distances it will visit only the richest ones.

The independence of site quality and distance from nest was illustrated in work by Olsson *et al.* (2015) who compared two models of bee foraging – one a simple model that assumes that all patches within a bee's foraging range are equal and that bees disperse from the nest in a diffusion pattern, and the second a CPF model that assumes patches vary in quality and that bees increase their fitness by optimally choosing among patches according to their distance and quality. The two models give nearly identical predictions for forager numbers in patches of similar quality across distances; bee numbers decrease with distance from the nest – a simple effect of diffusion. Likewise, both models predict more foragers at high-quality patches near the nest than poor ones near the nest. However, when the models consider high-quality patches further from, and poor sites nearer to, the nest, the two give very different results. Whereas the simple model slavishly keeps bees diffusing from their nests, the CPF model finds fewer foragers at poor patches near the nest and more foragers at rich patches far from the nest. When the quality difference of patches is amplified even more, the CPF-modelled bees skip the near, poor patches altogether in order to concentrate their efforts at the nearest parts of the richest patches. These models prime us to expect what may, anthropocentrically, seem obvious: that foragers will do what is necessary to get the calories they need – but if available, the nearest calories are always best. Flying a little further may be worthwhile if the patch is sufficiently rich; however, if a really good patch is too far away, it may not be worth the effort.

From the perspective of a crop grower, it is worthwhile to make one's crop a nutritionally attractive choice for a selective forager.

3.4.2. Taxon-based differences in bee flight distance

CPF models apply generally to bees who are universally nesters. However, differences in cognitive ability, behavioural repertoires and life history commitments among bees affect how closely any bee taxon will comply with CPF predictions.

First, some data on flight distances. Social honeybees are capable of flying several km to forage if necessary, but they prefer to forage near their nest if resource richness permits (Gary and Witherell, 1977). In the case of solitary bees, forced flight maxima under experimental conditions have been recorded up to 1100 m in the small bee *Hylaeus punctulatissimus*, 1275 m in the medium-sized bee *Chelostoma rapunculi*, and up to 1400 m in the large species *Hoplitis adunca*. However, these are unrealistic distances for foraging, as only 50% of females of *Hylaeus punctulatissimus* and *Hoplitis adunca* forage at distances longer than 100–225 m and 300 m, respectively (Zurbuchen *et al.*, 2010). In general, increasing body size is a good predictor of increasing foraging distance, and the relationship is nonlinear such that large bees forage disproportionately longer distances than small bees (Greenleaf *et al.*, 2007; but see Hagen *et al.*, 2011).

When it comes to patch quality, bees are able to adjust flight distance according to quality of reward, as has been shown in honeybees (Beekman and Ratnieks, 2000) and alfalfa leafcutting bees (Bacon *et al.*, 1965) – a phenomenon consistent with CPF theory. Similarly, conditions of scarcity can force bees to extend their foraging distances, as shown for honeybees (Steffan-Dewenter and Kuhn, 2003) and likely true for others as well (Osborne *et al.*, 2008).

In general, social bees, by virtue of their colony size and recruitment behaviours, are both buffered against forage deficiencies at a near scale and capable of exploiting rich resources at a distant scale (Steffan-Dewenter *et al.*, 2002). Meliponines (Sánchez and Vandame, 2013) and honeybees (von Frisch and Chadwick, 1967; Dyer, 2002) excel among the social bees for their ability to direct nestmates to profitable foraging sites. Honeybees do this by performing symbolic recruitment dances encoded with information on patch quality, distance and direction from the nest (von Frisch and Chadwick, 1967), the

discovery of which earned the Austrian Karl von Frisch a Nobel Prize in 1973.

Meliponines perform dances that superficially resemble those of *Apis*, but investigators have been unable to show that they communicate distance or direction (Nieh, 2004). Instead, recruitment in meliponines is driven by pheromones. Successful meliponine foragers stop every 1–8 m on their flight home to deposit a pheromone mark on vegetation, thus laying a connecting trail from the nest to the patch (Nieh, 2004). Alternatively, some species deposit larger quantities of pheromone as they near the patch, which by extension creates a trail of increasing strength that recruits nestmates with high precision (Nieh, 2004). Finally, some meliponines deposit no trail at all but instead reserve pheromone deposition for the patch alone (Aguilar *et al.*, 2005). If all this seems crude compared to the elegant symbolic dance language of *Apis*, we should not forget that meliponine colonies meet or exceed the size and social sophistication of any *Apis*, and the group is pantropical, diverse and vastly understudied. Many marvels of animal communication could lay hidden in its cryptic ranks.

In contrast to the communicative meliponines and *Apis*, the primitively eusocial bumble bees (*Bombus* spp.) seem incapable of communicating resource location to nestmates. That is not to say that information exchange about foraging does not happen. Rather, *Bombus* foragers are sensitive to socially facilitated information (Dornhaus and Chittka, 2004a). Foragers are alerted to the smell of food on incoming foragers, knowledge that shortens their subsequent search times. Increasing rates of incoming nectar have a stimulatory effect on the colony that elicits more bees to commence foraging. Finally, a successful forager distributes a tergal gland pheromone around the nest which elicits an increased foraging response in her sisters.

By such means, even the primitively social bumble bees can reproduce the kinds of foraging decisions predicted by CPF hypotheses. This dexterity was shown in a study with *Bombus terrestris* in which investigators placed hives of *B. terrestris* at 250-m intervals along a 1.5-km transect terminating at a highly profitable foraging resource – a 2 ha field of borage (Osborne *et al.*, 2008). Although all sites had food plants, the transect captured a range of site quality such that some were richer or poorer relative to their distance from the borage; in other words, forage quality was similar at a global scale (>1 km) but different at a micro-scale (≤500 m).

Bees from all sites visited the borage even though all sites had nearer resources, and at the end of the study colonies at all sites had equivalent bee populations, nest weights and nest volumes. It is worth bearing in mind that *B. terrestris* is on the high end of the flight distance range for the genus (Hagen *et al.*, 2011), and such an advantage may help explain the resilience of this species in agroecosystems.

The conclusion is that social bees possess the behavioural repertoires to support CPF hypotheses. A bee that can choose among patches of varying quality with some degree of independence from distance to the nest is a bee that can easily reject the farmer's uninteresting field 10 m away in preference to a very interesting weed 1000 m away. Solitary bees, lacking such a socially augmented information web on habitat quality, are more likely to satisfy predictions of a simple diffusion model based on patch distance from the nest. For bees of all types, we cannot overstate the importance of a positive balance on the books, i.e. an optimum ratio of calories gained over calories spent (Kacelnik *et al.*, 1986). Bees should be nesting as close as possible to the crop of interest. This crop should be as interesting as possible to the bees. A diversity of bee species increases the likelihood of the crop being visited.

3.4.3. Morphological considerations

Bee morphology, physiology and calorific energy budgets directly predict a bee's behaviour on a focal crop flower or whether it will even visit it. Corbet *et al.* (1995) modelled pollinator performance by inputting three predictors of bee nectar foraging patterns: (i) maximum depth at which nectar is accessible in a floral tube (dependent on bee tongue length); (ii) minimum necessary energetic rewards per flower (dependent on bee's body mass and foraging costs); and (iii) minimum temperature for flight. Maximum nectar depth and temperature are dynamic, changing over the course of a day. Nectar levels are higher in the morning and therefore available to bees across a range of tongue lengths. High visitation during early hours may mean the rate of nectar depletion exceeds its rate of excretion. As nectar levels drop, bees may have to visit more flowers per unit time to recover costs of foraging, until the point at which they have to abandon the patch altogether to find more suitable flowers. The timing and sequence of this species shift is a direct outcome of the interactions between tongue length and nectar levels. Eventually, usually by around midday, the only bees left are those with the longest tongues. However, during those preceding hours the flower has profited from a variety of floral visitors.

If tongue length is the constraint for deep flowers, then bee energy cost thresholds are the constraint for shallow flowers which are open, freely visited by many bees, and consequently contain smaller nectar rewards. Small-bodied bees with short tongues can profitably visit such low-reward flowers by increasing their visitation rate. Large-bodied, long-tongued bees, on the other hand, cannot recover enough calories from shallow flowers to sustain high rates of foraging, and over time large-bodied bees are competitively excluded from shallow flowers. The dynamism of the relationships shown here underscores the importance of a grower having a diversity of bee pollinators on their farm so that the pollinator workforce is comprised of morphologically diverse flower visitors.

From the scenarios above it is worth noting that from the plant's perspective it is adaptive to manipulate its visitors into making numerous successive flower visits. A limitless nectary, openly available, would only invite nectarivores who land, fill up and fly away with little incentive to visit another flower and effect cross-pollination. Some plants avoid such a scenario by putting heavy nectar loads in only a few (5–8%) of their flowers. These constitute the so-called 'lucky hits' for a bee (Southwick *et al.*, 1981) that motivate it to keep foraging on successive flowers.

3.4.4. Forager behaviour in rich and poor habitats

From a pollination perspective, bee foraging activity is more efficient in flower patches that are rich in nectar and pollen. At the most fundamental level, there is a strong – intuitively so – positive relationship between bee densities at a patch and the patch's richness in nectar and pollen (Westphal *et al.*, 2003; Carvell *et al.*, 2007; Heard *et al.*, 2007; Woodcock *et al.*, 2014). Beyond regulating the density of pollinators, however, food richness at a patch affects pollinators' behaviours as well. Animals forced to forage in resource-poor habitats forage more slowly, spending more time at each food site than do animals in rich habitats (Pyke *et al.*, 1977). It is advantageous for bees to be moving rapidly between flowers, accomplishing a high rate of pollination, rather than lingering on the few flowers in a patch that are yielding nectar. Southwick *et al.* (1981)

demonstrated that bee visitation rates increase in flower patches with higher numbers of nectar-bearing flowers, nectar volume and sugar concentration. Silva and Dean (2000) showed that the rate of honeybee visitation is positively correlated to nectar production in hybrid onion.

Not only do resource-rich plantings encourage rapid bee visitation between flowers, but they encourage pollinators to stay in that patch. It was shown that bumble bees and honeybees that have just visited highly rewarding flowers fly shorter distances before visiting another flower than do bees that have just visited less rewarding flowers (Pyke, 1978; Waddington, 1980). This behaviour increases the likelihood of the bee encountering another rewarding flower in a site which is shown to be profitable.

The earliest models of pollinator foraging behaviour tended to assume that a forager is naïve, operating with no prior knowledge or experience of the patch it is in. However, later research has shown idiosyncratic differences among foragers so that, for example, some individuals establish small foraging areas to which they return daily (Makino and Sakai, 2004). When experienced foragers make circuits through a patch in a predictably non-random order this is called *trapline foraging* (Manning, 1956; Thomson, 1996). This mode of foraging is cognitively sophisticated and requires memory of motor patterns (Collett *et al.*, 1993) and of the sequential order of flowers visited along a route (Chameron *et al.*, 1998). Bumble bee foragers are more likely to repeat a circuit when successive patches are not only nearest to one another but also most direct, that is,

requiring the fewest turns (Ohashi *et al.*, 2006). This is another way of saying that bees, having encountered a patch of profitable flowers, like to forage in straight lines, an adaptation that limits the chance of a bee revisiting a flower recently emptied of its nectar (Pyke, 1978; Cresswell *et al.*, 1995). From a crop pollination perspective, this theory strongly supports the practice of interplanting main and pollinizer varieties in the same row. The strong directionality afforded by a row maximizes the chance that a pollinator will successively visit a main variety and compatible pollinizer.

Collectively, these studies make strong arguments for improving the nectar and pollen production output of our important bee-pollinated crops. Optimal foraging theory predicts that if nectar output of a crop is high, bees pollinate more efficiently because they visit more flowers in a given period of time. Conversely, if the crop is nectar-poor, bees forage more slowly and visit fewer flowers. Bees are programmed to follow visually strong cues, such as a linear row, that optimize calorie acquisition rate from available flowers. These discoveries from foraging ecology are easily translatable to practices that can enhance the pollination efficacy of a farm's resident pollinators.

Notes

[i] A good example of such an economically important sympatry is the relationship between cultured blueberry (*Vaccinium* spp.) and a native solitary soil-nesting apid (*Habropoda laboriosa*) in the south-eastern USA.

[ii] Some authors refer to perigynous ovaries as 'half inferior' ovaries.

4 Economic and Ecosystem Benefits of Bee Pollination

We learned in Chapter 1 (this volume) that 78–94% of all angiosperms rely on animal pollinators (Ollerton *et al.*, 2011), so we can approach this chapter anticipating that where angiosperms have been pressed into agricultural service that their insect pollinators, especially bees, will be economically important. The various estimates of pollinator valuation focus on changes in crop yields, marketability or nutrient quality and occur at different geographic scales, global, national and local. A pattern in the studies cited below is that 'animal' pollination is repeatedly shorthand for 'bee' pollination. Economically important animal pollinators occur among the wasps, flies and small vertebrates, some of whom will be treated in Volume II of this work, but the general pre-eminence of bees as agricultural pollinators is otherwise uncontested.

Institutional record keeping in the agricultural sector makes it comparatively simple to calculate economic benefits of bee pollination when it comes to crop yield, value, nutrient quality and land use assignments among different crop choices. However, the benefits of bee pollination spill beyond agriculture and promote human and environmental well-being at many levels. Bee pollination is a classic *ecosystem service* – a gift of nature on par with rain, energy cycling, purification cycles for water and air, and ecological processes that restrict pathogens and parasites that would otherwise cause destruction.

Being so fundamental to our existence, ecosystem services ironically resist our attempts to evaluate them. What value rainfall? What investment return on protecting natural pest predators? What profit conserving carbon dioxide (CO_2)-absorbing ocean bacteria? These fundamental services are the basis of life as we know it and transcend any values mediated by human currency.

Let us not stop there. How to put a price on the beauty of flowers and the emotional recharge of working in one's garden? What about the pleasures of installing a pollinator habitat in one's suburban front yard? How about the solace a beekeeper takes donning a veil, lighting a smoker and hunting for the queen? It is strange that however much economic arguments drive the ways we order our societies, it is in these other domains that most of us find personal comfort and meaning. We find pollinators there too.

4.1. Worldwide Production Trends for Bee-Pollinated Crops

In a widely cited study of data from 200 countries, Klein *et al.* (2007) quantified the contribution of animal pollinators to the world's most important crops used for human food. These investigators showed that 67 commodity[i] species and 57 single crop species account for 99% of total global food production. Varying degrees of self-fruitfulness and self-pollination abound in the angiosperms, making necessary the use of the pollinator dependence ratio (D) for a focal crop as discussed in Chapter 3 (this volume, Eqn 3.3, Table 3.1). Adjusting for D, the authors showed that of the 57 leading single crops, production of 39 of them increases with the action of animal pollinators. Of the 67 commodity species, 48 respond positively to animal pollinators. Collectively on a volume basis, crops that are wind-pollinated or passively self-pollinated account for 60% of global food supply, whereas pollinator-responsive crops account for 35%.

Aizen *et al.* (2008) took a long view of pollination trends, analysing worldwide agricultural data from 1961 to 2006 across the developed and developing worlds. When they applied D to yields of pollinator-responsive crops, they estimated the fraction of world agricultural production attributable to pollinating animals to be between 14.7–22.6% (based on 2006 data).

Curiously, the contribution of pollinated crops to world agricultural output does not align with their contribution to gross human nutrients. Again, adjusting for *D*, Eilers *et al.* (2011) showed that among pollinator-responsive crops, the fraction of gigajoules (in metric tons) of the human diet directly attributable to animal pollinators is only 2.6% for gross energy, 3.0% for protein and 7.0% for fat.

This means that advocates of bees and other pollinators must understand that it is the crops that are not responsive to animal pollination that provide humanity with most of its calories, protein and fats – a fact that should moderate some of the hyperbole that makes its way into the public conversation about pollinators and their ecological challenges. One of the most common of these memes is that bees are responsible for every third bite of food we eat, which as far as I can tell makes its propitious entry into global consciousness in Samuel E. McGregor's USDA Handbook, *Insect Pollination of Cultivated Crop Plants* (1976, p.1), arguably the most influential and widely cited government agricultural publication of all time. With evidence unpacked below and in the next section, however, I show that McGregor's axiom is overgeneralized and poorly understood.

In their analysis of data from 1961 to 2006, Aizen *et al.* (2008) found divergent trends in cultivated area and production of pollinator-dependent and non-dependent crops. In both developed and developing countries, production of pollinator-dependent crops increased steadily over the 45-year period, exceeding the rate of increase for non-pollinator-dependent crops. In 1961, pollinator-dependent crops were responsible for 8.4% of total agricultural production in the developed world, and by 2006 this value had increased to 14.7%. In the developing world the trend was more pronounced, with fractional production increasing from 13.7% in 1961 to 22.6% in 2006. The overarching trend is that global investment in pollinator-dependent crops is increasing faster than investment in non-dependent crops, or as the authors put it, 'The net effect of these trends is that global agriculture has become increasingly pollinator-dependent over the last five decades.'

This increasing global dependence, however, is still weighted heavily toward high-value crops grown in developed regions such as China, Europe, Japan and the USA (Lautenbach *et al.*, 2012). The stark differences cast by pollination between the developed and developing worlds are covered in the next section.

4.2. Quality Properties Distinctive to Bee-Pollinated Crops

If a general increase in pollinated crops is a good thing, this raises the question, 'What is so valuable about pollinator-dependent crops?' The answer in three words: vitamins, fibre and minerals. Pollinator-dependent crops may take second place in calories, protein and fats, but when it comes to micronutrients and minerals, they make an outsized contribution to human nutrition. Calories obtained from nutrient-rich, fibrous and mineral-rich pollinated crops contribute more to health than calories from typical sustenance diets, heavy in grain which often leads to obesity and obesity-related health problems (Ciati and Ruini, 2011). Pollinated crops are the antidote to 'empty' calories.

Eilers *et al.* (2011) analysed data from the Food and Agriculture Organization (FAO) on worldwide production of more than 150 leading global crops and showed that among the vitamins, the fraction directly attributable to animal pollination is 41% for vitamin A, 9% for vitamin K and 20% for vitamin C. Of the carotenoids, the fraction directly attributable to animal pollinators is 38% for β carotene, 38% for α carotene, 42% for β cryptoxanthin and 43% for lycopene. Among vitamin E-active compounds, animal pollination is responsible for 7% of α tocopherol, 27% of β tocopherol, 14% of γ tocopherol, and 23% of δ tocopherol. Of the minerals, animal pollination contributes to 9% of dietary calcium, 6% of iron, 2% of magnesium and 20% of fluoride.

Ellis *et al.* (2015) published a direct test of the hypothesis that animal pollinators are crucial for human nutritional health. These investigators focused their attention on women and children in four developing countries – Zambia, Uganda, Mozambique and Bangladesh – where high rates of malnutrition make populations more vulnerable to pollinator declines. The authors narrowed their investigation to five of the most important global nutrients: vitamin A, zinc, iron, folate and calcium. Using individualized food consumption surveys and available data on crop pollination requirements and food nutrient content, they determined the proportion of populations at risk of nutrient deficiencies under 'full pollination' or 'no pollination' scenarios. The percentages of children newly at risk of vitamin A deficiency from a loss of pollinators would be relatively low for Bangladesh (2%) and Zambia (5%) but significantly higher for

Uganda (15%) and Mozambique (56%). Anything that threatens the availability of vitamin A is worrisome given that vitamin A deficiency is already the leading cause of preventable childhood blindness in the world (WHO, 2009) and responsible for nearly 800,000 yearly deaths in women and children (Rice *et al.*, 2004). Dietary availability of calcium, iron and zinc are relatively unaffected by a projected loss of pollinators, but folate makes a concerning, yet statistically non-significant, drop (23%) in Mozambique.

In an insightful analysis of the distinction between 'empty' versus 'rich' calorie diets, Smith *et al.* (2015) determined nutrient composition and pollinator dependence of 224 food types in 156 countries and, while keeping calorific intake constant with staple foods, modelled changes in public health measures under a scenario of complete pollinator loss. An hypothetical complete loss of pollinators was predicted to induce 41 to 262 million new cases of vitamin A deficiency, 134 million to 1.33 billion new cases of folate deficiency, and significant crop losses for fruit, vegetables, nuts and seeds – the sum of such changes accruing to 1.38–1.48 million additional human deaths yearly from non-communicable, malnutrition-related diseases.

An explanation for the disproportionate nutrient richness of pollinator-dependent crops traces back to evolutionarily driven resource allocation among reproductive versus non-reproductive tissues in angiosperms. It has been known for a long time that photosynthates are not evenly distributed among tissues and that flowers and fruits draw heavily on metabolites from adjacent leaves (Wardlaw, 1968). Fruits often act as sinks, removing and sequestering photosynthate and other materials from the translocation system, as shown for apple (Hansen, 1967) and grape (Meynhardt and Malan, 1963). As the consumed parts of pollinated crops are most often the embellished ovary and seeds, it follows as a matter of course that these tissues will have proportionally more nutrients.

The nutrient richness of pollinated crops; the prevalence of the ripe ovary as the part consumed, often in a succulent state shaped by natural selection to be palatable to seed dispersers; the shelf life challenges that that succulence demands of handlers of fresh fruits and vegetables; and the reticence of governments to extend price supports to fruit and vegetable producers, all mean that the calories from pollinator-responsive crops are more expensive than calories from grain staples. As such,

these are bellwethers of the 'good life' – blueberry muffins, cranberry scones, almond milk, watermelon wedges for wedding receptions, kiwi fruit smoothies, and those special gifts of the gods, coffee and chocolate. They are the mustard, ketchup and cucumbers with dill that garnish our grilled hamburgers, and if we are generous we can multiply their benefits to the meat and dairy products fuelled by pollinated legume forages. Pollinated crops represent the difference between eating for sustenance and eating for pleasure. They are inseparable from the quality of life enjoyed in wealthy countries.

Pollinated crops are also caught up in a web of drivers operating at an epochal scale, however, altering our societies and imperilling entire ecosystems across this planet. Long-range trends toward rising incomes and greater urbanization are making unprecedented demands on industrial agriculture, with generally negative consequences for ecosystems and mixed results for human populations. Today over one-third of the ice-free arable land on Earth has been pressed into agricultural service (FAO, 2003), with the greatest increase in crop acreage since 1961 going to pollinator-responsive crops (Aizen *et al.*, 2008). Demand for these 'icons of the good life' has contributed not only to the general increase in farmed acres but to wide-scale tropical deforestation, nowhere more conspicuous than the plantations of thrips-pollinated (Syed, 1979) oil palm in South-east Asia (Carlson *et al.*, 2012) and cocoa smallholdings in Latin America, Africa and Asia (Asase *et al.*, 2008).

Environmental problems associated with pollinated crop culture are not restricted to the tropics. Tree fruit poses concern when long-lived orchards occur in area-wide monocultures. The resulting habitat homogeneity can be vast in time and space. Such conditions prevail in almond plantations in Australia (Luck *et al.*, 2014) and California (Klein *et al.*, 2012), turning such plantations into 'pollinator deserts' incapable of sustaining resident wild pollinators except during the brief flowering window of the one prevailing crop (Traynor, 2017). Such extreme examples of habitat simplification compel such drastic measures as the annual mechanized importation of over 1.5 million honeybee hives into California every winter from all quarters of the USA to meet the almond's insatiable demand for pollination (Pettis and Delaplane, 2010).

If production of pollinator-dependent crops is universally increasing between the developed and developing worlds, this is not to say that the benefits of pollinated crops are universally spreading between these worlds. Indeed, the dual dynamics of delocalized food production (Pelto and Pelto, 1983) and globalized markets (Hawkes, 2006) interact to advantage wealthier nations and funnel nutrient-rich pollinated food away from the developing countries where it is grown. Although calories have steadily increased in human diets of the developing world since the early 1960s, a significant fraction of those calories are empty, derived from cheap vegetable oils and sugars (non-dependent on pollinators), and conducive to health problems such as obesity and type II diabetes (Drewnowski and Popkin, 1997; Tilman and Clark, 2014). Whether pollination contributes one bite in every three (McGregor, 1976) or four (Aizen *et al.*, 2008) or seven (Aizen *et al.*, 2008) depends very much on where one lives.

The situation was summarized by Chaplin-Kramer *et al.* (2014) who point out that global economic valuations of pollination do not move in tandem with nutrient dependence on pollination. Economic valuations are driven by high-value crops produced in rich countries, whereas nutrient dependence on pollination is driven by such hotspots of pollinator vulnerability as India, Southeast Asia, and central and southern Africa.

Understanding the special advantages of pollinator-dependent systems can help remove barriers to more liberal delivery of their benefits at a global scale. This prospect speaks directly to the optimization that already exists between nutrient benefits of pollinated crops and their environmental costs. Take, for instance, the impressive per-calorie punch in nutrient quality contributed by fruits and vegetables compared to cereals. The fact that it is these nutrient-rich powerhouses prevailing in the global increase in cultivated acreage is a good thing compared to such empty-calorie alternatives as sugars and oils. Similarly, of the 25% of all human-caused greenhouse gas emissions attributed to agriculture (Smith *et al.*, 2014), pollinated crops make up only a modest fraction. The grams of carbon emissions per kilocalorie of vegetables (0.68 ± 0.25), temperate fruits (0.1 ± 0.02) and tropical fruits (0.14 ± 0.04) are higher than for maize (0.03 ± 0.004) and other cereals (0.05 ± 0.005), comparable to butter (0.33 ± 0.03), eggs (0.59 ± 0.03) and dairy (0.52 ± 0.04), but orders of magnitude below those for poultry (1.3 ± 0.05), pork (1.6 ± 0.1) and ruminant meats (5.6 ± 0.41) (Tilman and Clark, 2014). Pollinator-responsive crops, therefore, occupy an optimization niche – with better nutrient quality than cereals and lower environmental costs than meats which, along with refined sugars, refined fats and oils, are implicated with raising the world obesity rate to 2.1 billion (Popkin *et al.*, 2012; Ng *et al.*, 2014). Evaluations such as these should help environmental policy makers and consumers prioritize pollinator-responsive crops as a good compromise between nutrient demands of a surging human population and finite ecosystem resources.

Moreover, the fact that pollination is a dynamic of two actors, the pollinator and the plant, means that pollination carries the seeds (pun unavoidable) of its own optimization – the complementary pillars of bee management and conservation. In short, pollination can always be improved by inputting bees and, unlike inputs in other cropping systems such as pesticides and fertilizers, these inputs are not alien to the system but liberal with collateral environmental benefits. Consider the fact that whereas pesticides are conservative inputs that limit losses, bees are positive inputs that increase gains. This means that optimizing pollinators has the corollary benefit of increasing per-acre yields, putting a check on the demand for unbridled requisition of wild lands for agricultural production (Aizen *et al.*, 2008). Chapters 6 and 10 (this volume) will develop these ideas further.

Great challenges remain to spreading the benefits of crop pollination more equitably. For one thing, except for some intensively managed plantations organized for export markets, pollination-for-hire is virtually unknown in many developing countries. In my experiences in Albania (Dedej *et al.*, 2000), Azerbaijan, Belize, Cuba, Honduras, Mexico and Nepal, I have found the honeybee-keeping industries in these places overwhelmingly focused on production of honey, propolis or other hive products for human consumption. Neither market demand nor infrastructure exists for renting beehives for pollination, although gains in this enterprise are happening with stingless bee culture (Chapter 11, this volume). This trend may reflect the disconnect, discussed above, that exists between the production and consumption of pollinator-responsive crops in poor countries. It also points out that education and capacity building for inputting bees at cottage scales must be part of any overarching strategy for improving human life in poor communities.

4.3. Value of Optimizing Pollination in Bee-Pollinated Crops

In Chapter 3 (this volume), the idea that pollination occurs along a continuum was introduced, with 'good' pollination and 'bad' pollination depending on the number of flowers pollinated, the availability of suitable pollenizers, the number of pollen grains deposited per stigma, or the distribution of pollen grains on a stigma. Yield is the criterion most often at stake in questions about pollination quality, and indeed much of the current work, especially in Volume II, addresses pollination from that vantage. Suffice it to say that examples are legion for the yield-enhancing properties of strong pollinator visitation, much of which accrues to increased farmgate revenue for the grower. Here, I touch upon only a few examples.

Good pollination and yield increase are often inseparable from other positive measures of consumer acceptance and grower revenues. Good pollination improves firmness, shape, size, colour, flavour and shelf life of strawberries, increasing their farmgate value by 92% (Klatt *et al.*, 2014; Wietzke *et al.*, 2018). Good pollination is associated with increased oil content in oilseed rape (*Brassica napus*) (Bommarco *et al.*, 2012); increased sweetness in mandarin oranges (Wallace and Lee, 1999); and increased fruit diameter in pear, translating to a US$427/acre increase in farmgate value (Currie *et al.*, 1992b; Naumann *et al.*, 1994). Good bee pollination has been linked to increased yield, size and sweetness in cantaloupe (Eischen and Underwood, 1991) in the Rio Grande valley of Texas; larger and heavier sweet peppers in Spain (Serrano and Guerra-Sanz, 2006); and increased size and weight of apples in the UK (Garratt *et al.*, 2014a) with a corresponding increase in farmgate value of more than £14,000/ha. It has been traced to a 41% increase in cranberry yield (Currie *et al.*, 1992a) with a corresponding increase in farmgate revenue of US$8804/ha. In Burkina Fasso (Stein *et al.*, 2017), bee-mediated out-crossing in cotton increases yield (fibre + seed) by 27–31% with a corresponding increase in farmer earnings of US$98–113/ha, and for sesame, out-crossing increases yield by 37–42% with an increase in farmer earnings of US$32–37/ha. In Ecuador, a fourfold increase in bee density is associated with a 78% increase in coffee yield, translating to an 816% increase in farmgate revenue (Veddeler *et al.*, 2008). For one farm in Costa Rica, forest-based wild bees increase coffee yields by 20% within 1 km of forest margins, reducing incidence of 'peaberries', small misshapen beans, by 27%, the combined effects of which translated to an annual value of US$60,000 (Ricketts *et al.*, 2004).

In the previous section we touched upon distinctive contributions made by pollinator-responsive crops toward the pool of vitamins and minerals in the human diet. Research has shown that even within pollinator-responsive crops nutrient quality of the fruits increases with the quality of pollination the crops receive. This was shown for almonds by Brittain *et al.* (2014) who controlled whole-tree inputs including combinations of pollinator exclusion, hand cross-pollination, open pollination, and normal or reduced levels of fertilizer and water. Effects of self-pollination on nutrient quality of nuts was greater than the effects of reduced fertilizer and water. Nuts from experimentally cross-pollinated trees had higher oleic to linoleic acid ratios than trees that were self-pollinated. A high oleic acid ratio is desirable in almonds for human consumption because it is associated with reduced risk of coronary disease (Jalali-Khanabadi *et al.*, 2010) and improves stability of fats against rancidity, thus improving almond shelf life (Kodad and Socias I Company, 2008).

Mechanisms by which pollination affects fruit nutrient quality remain unknown, but Brittain *et al.* (2014) suggest that a slowed development rate in self-pollinated fruit may be one explanation. Self-pollen in almond exhibits slower pollen tube growth compared to cross-pollen (Certal *et al.*, 2002), and the resulting delays in nut development may hinder accumulation of desirable constituents such as oleic acid.

4.4. Efforts at Valuing Bee Pollination Across Geographic Scales

We have discussed that global estimates of pollinator economic valuation are overinfluenced by markets for high-value crops concentrated in mostly wealthy countries. These market-driven estimates tend to be blind to other considerations of equal or exceeding importance such as the contributions of pollinators to human nutrition and ecosystem stability. For better or for worse, however, economic valuations drive legislatures. The worldwide hardship of pollinators, widely covered in the popular and scientific press since the late 2000s, has sparked interest in quantifying the human stakes in healthy bee populations. It is to be hoped that demonstrating the value of pollinators in dollars will translate to policies amenable to broadening the availability of pollinators' wider benefits.

Examples of each of the following estimation methods for valuing pollination services are given in Table 4.1. All but one are market driven with data from traditional supply, demand or price drivers.

4.4.1. Economic value of insect pollination

Table 4.1 gives a summary of economic valuation of animal pollinators across a range of geographic scales. The vast majority of these studies have used the *economic value of insect pollination* (EVIP) method which, as its name implies, seeks to attach dollar values to the agricultural production, or marketable output, traceable to insect pollinators. This approach assumes that yield rises or falls according to the quality of pollination received.

The earliest efforts at pollinator valuation simply summed the products of each crop's production by its value, resulting in a crude measure of *economic value* (EV). The only criterion was that each crop was known or suspected to be responsive to pollinators. Summing these terms by crop i and region x yields the equation:

$$EV = \sum_{i=1}^{I}\sum_{x=1}^{X}\left(Q_{ix} \times P_{ix}\right) \qquad (4.1)$$

where Q_{ix} = quantity produced (in metric tons) of crop i in region x, P_{ix} = price received per unit crop i in region x, I = all crops considered, and X = all regions considered.

By ignoring differences in crop dependence on flower visitors, however, the EV approach grossly inflated the contribution by pollinators to agricultural economies. The obvious improvement was to include a measure of crop i's relative pollinator dependence, D_i (Eqn 3.3, this volume), resulting in EVIP (O'Grady, 1987):

$$EVIP = \sum_{i=1}^{I}\sum_{x=1}^{X}\left(Q_{ix} \times P_{ix} \times D_i\right) \qquad (4.2)$$

where D_i is the fraction of the crop's fruit-set or yield (Q_i) attributable to animal pollination. Melathopoulos *et al.* (2015) critiqued current methods of pollinator valuation and noted problems in particular with the parameter D. Published estimates of EVIP have drawn heavily on estimates for D compiled in one paper, the work of Klein

et al. (2007). These authors reviewed the literature on 178 cropping systems and classified each into one of five categories of pollinator dependence: (i) essential $D \geq 0.9$; (ii) high $D = 0.89$–0.4; (iii) modest $D = 0.39$–0.1; (iv) little $D < 0.1$; or (v) no increase $D = 0$. Subsequent authors, myself included in Table 3.1 (this volume), have tended to preserve the ranges of Klein *et al.* (2007) as midpoints; thus, essential=0.95, high=0.65, modest=0.25, little=0.1 and no increase=0. Such clunky categories of D miss the nuance and range of natural D which varies continuously from 0–1. This variation occurs not only by crop but by region, cultivar within a crop, spatial arrangements in a field, environmental conditions, and management practices (Melathopoulos *et al.*, 2015). The solution to this problem, therefore, is more research to determine localized variants of D and to better resolve regional estimates of EVIP. In terms of Eqn 4.2, this means replacing D_i as much as possible with D_{ix}.

If the fraction of total pollination performed by a subset of insects is known (ρ), then the accompanying EVIP equation can be used to estimate economic value attributable to that group. A value for ρ can be estimated by recording the relative distribution of taxa visiting a crop's flowers (Vázquez *et al.*, 2005). The value of all pollinators is $\rho = 1$. If the relative fraction (0–1) of flower visits performed by taxon z is known (ρ_z) across the range of focal crops and regions, then the economic value of that taxon ($EVIP_z$) is:

$$EVIP_z = \sum_{i=1}^{I}\sum_{x=1}^{X}\left(Q_{ix} \times P_{ix} \times D_{ix} \times \rho_{ixz}\right) \qquad (4.3)$$

4.4.2. Attributable net income

It does not take long to realize that even the most elaborate forms of the EVIP equation are still simplistic. EVIP takes no consideration of *variable costs* (VC) incurred by growers, and it makes no distinction between the pollen deposited by a pollinator taxon and the amount of pollen actually needed by the plant to set fruit. The *attributable net income* (ANI) method was offered by Winfree *et al.* (2011) as a means to address these shortfalls.

The ANI method begins similarly to EVIP, that is, with gross value $Q \times P$. It then subtracts all known variable costs to production (VC) to arrive at net income. Good estimates of variable costs can be found in farming budgets maintained by state extension services.

Next, the ANI method narrows the pollinator's value explicitly to the minimum necessary stigmatic pollen deposition required to set a marketable fruit. There is no value for deposited pollen in excess of the plant's requirement. The ANI method is also sensitive to the relative contribution of any pollinator taxon z, as long as two other variables are known: α, the maximum proportion of set flowers that the crop can sustain to maturity; and ρ_z, the fraction of flowers that are fully pollinated by pollinator z. Winfree *et al.* (2011) use watermelon as an example. The crop can sustain only 50% of its set flowers; hence α=0.5. The watermelon flower needs $\geq$1400 pollen grains to produce a marketable fruit; hence, ρ_z = fraction of flowers that receive $\geq$1400 pollen grains from pollinator z. The value of pollinator Vz is thus:

$$V_z = \left((P \times Q) - VC\right) \times D \times \min\left(\frac{1}{\alpha}\rho_z, 1\right) \quad (4.4)$$

If one runs this equation in a spreadsheet with fixed D (for watermelon 0.9 or 1.0) and values of 0–1 for ρ_z, one sees that the value of taxon z asymptotes when its pollinating performance ρ is equal to α. There is no additional value gained for setting more fruit than the plant can rear to maturity. Fully exploiting this equation would therefore require local knowledge of the following terms: field sampling of local pollinators, the amount of fruit-set achieved by those natural populations, and knowledge of α for the focal crop. To the extent that local populations are meeting the crop's needs, importing additional pollinators would be superfluous.

One elegant outcome of the ANI method, Winfree *et al.* (2011) point out, is that it draws a direct connection from the ecosystem service of bees (deposition of pollen) to an outcome of economic value to people – marketable fruit.

4.4.3. Replacement value

The *replacement value* (RV) method begins with an hypothetical scenario of total loss of pollinator z and asks what it would cost farmers to replace that service, whether by substituting existing pollinators with new ones (RV$_{subst}$) or by replacing insect pollinators with mechanized or manual labour alternatives (RV$_{alt}$).

Winfree *et al.* (2011) used an RV method (Eqn 4.5) to show what it would cost farmers to replace a total loss of native bees (RV$_{nb}$) with rented honeybees.

$$RV_{nb} = A \times SR \times RC \times FA_{nb} \quad (4.5)$$

where A = area (hectares) of crop in question, SR = recommended stocking rate of honeybees (hives per hectare), RC = annual rental cost per hive, and FA_{nb} = fraction of farms at which native bees alone are pollinating the crop. The term FA_{nb} is locally derived through research and may not always be available; without it, RV valuations of pollinator z will be inflated.

4.4.4. Consumer surplus

Consumer surplus (CS) is defined as the price increase for a good or service that consumers are willing to pay rather than go without it (Willig, 1976). CS valuations take into account the diverse parties affected by the significant loss of a resource such as pollinators (Mburu *et al.*, 2006). For example, if crop production in region a is sufficiently large to affect wider markets, then a pollinator loss in region a can potentially affect two other entities: the growers outside the region ($\neq a$) unaffected by loss of the pollinator, and the consumers. Lowering supply will raise price (P), thus lowering consumer welfare. However, raising price will increase welfare of growers in region $\neq a$. The extent to which growers in region a are affected will depend on the magnitude of the price effect. As pointed out by Winfree *et al.* (2011), the economic effect on region a is equal to the revenue gain from increased price, minus revenue lost from reduced yield, plus production cost savings from reduced yield, minus costs to replace lost pollinators.

When Southwick and Southwick (1992) made their highly influential CS valuation of honeybee pollinators in the USA, however, they essentially ignored producer welfare, noting that even small changes in price drive large changes in supply. Farmers constantly recruit new acres to one crop and away from others based on yearly price appraisals, rendering the producer stake in the equation zero in the long term. The CS valuations of Southwick and Southwick (1992) in Table 4.1 are therefore focused on economic benefits that accrue to US consumers from increases in crop yields and quality attributable to pollinators.

Table 4.1. Some estimates of economic value of pollinators on crop production. M, millions; B, billions; all dollars US. Directly pollinator-dependent crops are those for which pollination is necessary to produce the edible fruit, whereas indirectly pollinator-dependent crops are those for which pollination is necessary to produce the seed. Valuation methods employed are: economic value of insect pollination (EVIP); attributable net income (ANI); change in consumer surplus (CS) of pollinator-dependent crops; replacement value (RV) based on substituting existing insect pollinators with new ones (RV_{subst}) or replacing insect pollinators with non-insect alternatives (RV_{alt})[a]; computable general equilibrium (CGE) models; higher-order dependence (HOD) network analysis of multiplier effects on non-agricultural sectors; and tax burdens that a citizenry is willing to pay (WTP) to fund pollinator conservation schemes.

Geographic range or region	Valuation method	Pollinators considered	Crops considered	Annual valuation of pollination services	Dates	Reference
Australia	RV_{subst}	honeybees	25 crops pollinator-dependent	$16.4–38.8 M	2007	Cook et al. (2007)
Brazil	EVIP	all biotic	44 crops pollinator-dependent	$12 B	2006, 2011–2012	Giannini et al. (2015b)
China	EVIP	insects	44 crops directly pollinator-dependent	$52.2 B	2008	An and Chen (2011)
Egypt	EVIP	all biotic	directly and indirectly pollinator-dependent	$2.4 B	2009	Brading et al. (2009)
European Union[b]	EVIP	insects	directly and indirectly pollinator-dependent	€14.6±3.35 B	1991–2009	Leonhardt et al. (2013)
India	EVIP	insects	directly pollinator-dependent	€14.2 B	2005	Gallai et al. (2009)
India	EVIP	all biotic	plants of economic importance	$22.52 B	2000–2001, 2013–2014	Chaudhary and Chand (2017)
Iran	EVIP	bees	31 crops with pollinator dependence ratio ranging from 0.1–1.0	$6.59 B	2005–2006	Sanjerehei (2014)
Kenya (Kakamega)	EVIP	bees	bean (*Phaseolus vulgaris*), cowpea, green gram, bambara nut, pepper, tomato, sunflower, passion fruit	$3.2 M	2005	Kasina et al. (2009)
Korea	EVIP	honeybees	major fruits and vegetables	$5.8 B	2008	Jung (2008)
Pakistan	EVIP	insects	fruits, vegetables, nuts, oilseed, spices	$1.59 B	2006, 2011	Irshad and Stephen (2013)
Poland	EVIP	bees	apple, pear, plum, cherry (sour, sweet), oilseed rape and agrimony, currant, gooseberry	€825.1 M	2012	Majewski (2014)
	EVIP	insects	directly and indirectly pollinator-dependent	€470±170 M	1991–2009	Leonhardt et al. (2013)
South Africa	EVIP	insects	directly pollinator-dependent	€1.1 B	2005	Gallai et al. (2009)
Western Cape (South Africa)	RV_{alt}	insects	deciduous fruit	$338.3 M	2005	Allsopp et al. (2008)
	RV_{alt}	managed honeybees	deciduous fruit	$119.8 M	2005	Allsopp et al. (2008)
UK	EVIP	bees	crops with history of intentional importation of bee pollinators	£202 M	1998	Carreck and Williams (1998)

Region	Method	Taxa	Category	Value	Year	Reference
	EVIP	insects	directly and indirectly pollinator-dependent	€510±120 M	1991–2009	Leonhardt et al. (2013)
	WTP	bees	N/A	£842 M	2008	Mwebaze et al. (2010)
	WTP	bees	N/A	£379 M	2010	Breeze et al. (2015)
	CS	honeybees	pollinator-dependent	$1.6–5.7 B	1986	Southwick and Southwick (1992)
USA	EVIP	all insects	directly pollinator-dependent	$10.69–15.12 B[c]	1996–2009[c]	Calderone (2012)
	EVIP	all insects	indirectly pollinator-dependent	$11.80–15.45 B	1996–2009	Calderone (2012)
	EVIP	honeybees	directly pollinator-dependent	$8.33–11.68 B	1996–2009	Calderone (2012)
	EVIP	honeybees	indirectly pollinator-dependent	$5.39–7.33 B	1996–2009	Calderone (2012)
	EVIP	alfalfa leafcutting bee	alfalfa (lucerne) hay	$4.99–7.04 B	2003–2009	Calderone (2012)
	HOD	all biotic	pollinator-dependent	for ag sectors $14.2–23.8 B; for all other sectors $10.3–21.1 B	2007	Chopra et al. (2015)
New Jersey, Pennsylvania (USA)	EVIP	native insects	major fruits and vegetables, row crops	$3.07 B	2001–2003	Losey and Vaughan (2006)
	EVIP	bees	highbush blueberry, apple, cherry (sweet, sour), almond, watermelon, pumpkin	wild bees $1.5 B honeybees $6.4 B	2013–2015	Reilly et al. (2020)
	RV_{subst}	native bees	watermelon	$0.2–0.21 M	2005	Winfree et al. (2011)
	RV_{subst}	honeybees		$0.17–0.18 M	2005	Winfree et al. (2011)
	EVIP	native bees		$2.25±0.18 M	2005	Winfree et al. (2011)
	EVIP	honeybees		$1.38±0.18 M	2005	Winfree et al. (2011)
	ANI	native bees		$3.40±0.16 M	2005	Winfree et al. (2011)
	ANI	honeybees		$0.24±0.16 M	2005	Winfree et al. (2011)
Georgia (USA)	EVIP	all biotic	directly pollinator-dependent	$367 M	2015	Barfield et al. (2015)
Global	EVIP	insects	directly pollinator-dependent	€153 B	2005	Gallai et al. (2009)[c]
	CGE	insects	directly pollinator-dependent	$334.1 B	2004	Bauer and Wing (2010)
	WTP	all biotic	directly pollinator-dependent	$127–152 B	2004	Bauer and Wing (2016)

[a]Winfree et al. (2011) have shown that RV is a special case of EVIP and that the two models collapse into the same equation

[b]See Leonhardt et al. (2013, Table 3) for valuation for individual countries which ranges from €10±1 M (Malta) to €2.02±0.29 B (Italy)

[c]See Gallai et al. (2009), Table 3, for valuation for individual world regions which ranges from €700 M (central Africa) to €51.5 B (eastern Asia)

4.4.5. Computable general equilibrium

Computable general equilibrium (CGE) models are usually reserved for economy-wide analyses of the consequences of agricultural policies or external perturbations such as climate change. Their application to ecosystem services such as pollination is relatively new. Compared to other market-based valuation methods discussed so far, CGE methods are nimble enough to track price changes consistently across multiple related markets, to describe macroeconomic effects of market shocks with theoretically derived measures of welfare change, and to test multiple scenarios for substituting lost pollinators (Bauer and Wing, 2010). When Bauer and Wing (2010) applied a CGE analysis to a global valuation of pollinators, modelling the impact of total pollinator loss for each of 18 world regions. Their general equilibrium model predicted direct crop sector losses at US$10.5 billion and indirect non-crop sector losses at US$323.6 billion, for a total global cost of US$334.1 billion. Their regionalized approach allowed nuanced regional interpretations. For example, western Africa seems particularly vulnerable to pollinator losses because pollinator-responsive crops make up a large fraction of the region's agricultural output. However, for southern Africa the model predicted that pollinator declines would positively impact the value of crop production because increases in price more than compensate for losses in yield.

4.4.6. Higher-order dependence

Higher-order dependence (HOD) pollination analyses attempt to account for cascading dependencies of non-agricultural industry sectors on pollinator services. HOD estimates do not necessarily predict economic losses but rather focus on intersections of agricultural and non-agricultural sectors, ranking sectors according to their vulnerabilities to disruptions in pollinator availability. Chopra *et al.* (2015) performed Monte Carlo simulations and network analyses to identify and rank the top 15 industry sectors for their vulnerabilities to pollinator disruption. The order of support sectors, from highest dependency to lowest are: agriculture, agrochemicals, fertilizer manufacturing, farm machinery manufacturing, storage battery manufacturing, wood products, oil and gas extraction, cutlery and hand tool manufacturing, textiles, inorganic chemical manufacturing, paper bag and coated/treated paper manufacturing, non-metallic mineral mining, stone mining, ground or treated mineral manufacturing, and water treatment and sewage systems.

4.4.7. Stated preference or willingness to pay

The *stated preference* approach uses surveys to derive an estimate of consumers' *willingness to pay* (WTP) for a specified ecosystem service. It is considered a 'non-market' valuation because the data do not come from traditional supply, demand or price drivers, but rather from respondents' non-market values – things such as appreciation for pollinator conservation, a secure food supply, and the aesthetic and intrinsic value of the bees. Mwebaze *et al.* (2010) used this approach to determine how much UK respondents would be willing to pay in taxes for an hypothetical bee protection policy. Respondents were given basic information on the value of pollinators and shown pictures of bees visiting flowers to be sure respondents understood what they were being asked. They were then told the details of the proposed policy and, with the help of bias limiting survey methods, were asked how much they would be willing to pay in income taxes to support such a policy. It was found that the mean WTP value was £43 per household per year. Multiplying this by the number of UK taxpayers (30.6 million, 2009 values) and the fraction of working adults indicating their readiness to pay the tax increase (0.64) yielded an estimate of total national support for protecting bees of £842 million.

4.5. Other Ecosystem Services Provided by Bees

The value of bees as pollinators is proportional to that of the angiosperms whose bee-dependent reproduction adds value to the world. As this volume is about *crop* pollination, we cannot deviate long from food and fibre production. The value of angiosperms, however, is so multiplicative that it is worth acknowledging a few of the other ways that their pollinators enable Earth's version of terrestrial life.

Foremost, angiosperms are among the most successful of all life forms, occupying and dominating every terrestrial habitat in the world excluding polar regions (Crepet and Niklas, 2009). Hundreds of studies have converged on the general conclusion that species-diverse communities are at least twice as productive as monocultures, and this productivity increases over time (Tilman *et al.*, 2014).

The result is ecosystem stability, driven by interspecific complementarity, coevolved adaptations, mitigated disease and herbivory impacts, and nutrient-cycling feedbacks that secure nutrient stores over the long term. Diversity loss is as traumatic to a system, or more so, than herbivory, fire, drought or climate-changing elevated CO_2. The picture that reveals itself is that diversity is not just an interesting artefact of a stable ecosystem, but rather the emergent existential enabler of the ecosystem. It is hard to imagine a more important participant in this self-generative dynamic than the angiosperms – and bees as their reproductive partners. The existence of angiosperms yields dividends beyond reckoning – food and substrate for primary consumers; cover and materials for their predators; carbon sequestration and oxygen (O_2) generation for the atmosphere; and root networks to keep soil from washing into the oceans.

Among the webs of interactions that multiply systems stability, there are some collateral benefits of bee foraging that, strictly speaking, do not accrue to angiosperm reproduction. Tautz and Rostás (2008) were able to show that flight activity of honeybees around bell pepper (*Capsicum annuum*) and soybean (*Glycine max*) helps protect the plants from herbivorous caterpillars of the beet armyworm (*Spodoptera exigua*). The caterpillars, who feed almost continuously, respond to the airborne wingbeat vibration frequencies of their wasp predator by ceasing movement, regurgitating gut contents and dropping off the plant, all of which behaviours cause a cessation of feeding. It so happens that the airborne wingbeat frequencies of pollinating honeybees elicit the same defensive responses in these caterpillars. Thus, pollinating honeybees visiting these plants confer the collateral benefit of protecting the host plant from herbivory.

Another insight into collateral benefits of bee flower visitation began emerging when a postdoctoral scientist at my lab, Dr. Ohad Afik of Israel, got the unlucky assignment of sitting for hours inside a field tent with a beehive, counting bee visits to watermelon flowers. Ohad noticed the copious amounts of bee faeces (frass) that accumulated everywhere inside the 2×2×2 m tent, including on the clothing and person of the observer. Wondering whether this generous gift of nitrogen (N) occurred at plant-relevant quantities, he recruited soil scientists at our university who partnered with us to measure and characterize the N deposited on plants by bee frass. The results were conclusive: the quantity of inorganic N released from a colony of 20,000 bees foraging in a patch of 3.24 m^2 was estimated at 0.62–0.74 g/m^2/month, an amount considered significant at a community scale (Mishra *et al.*, 2013). The deposition of plant-available nitrogen by bees as they visit flowers is a collateral benefit of their primary activity as pollinators. As a co-author of this study, I was tempted to assert the title, 'And you thought it was pollination', but cooler heads prevailed.

Note

[i] 'Commodity' formally refers to goods with units that are interchangeable and indistinguishable from one another. In this study, the term commodity was applied to crops where data were pooled with similar crop species, for example, 'fresh vegetables'.

5 State of the World's Bee Pollinators and the Consequences for Crop Pollination

When I was a young boy growing up in Indiana, my father had frequent encounters with bumble bees. As we grew maize, soybeans and reared swine, we had grain in our outbuildings. And because we had grain, we had mice; and because we had mice, we had bumble bees who found access to old mouse nests – one of the bees' favourite nesting sites – through gaps in our shed walls. My father's encounters never ended well for the bees. He would grab a tin can – scores of which littered our sheds (I think he kept them for no other purpose) – pour in a dollop of gasoline, and throw it on to the buzzing heap of burlap. From age 13 when I began keeping honeybees in hives around the farm, I acquired experience of bee stings and began wondering if my father's reactions lacked a due proportionality. Are bumble bees not near cousins to honeybees? Are their cryptic nests not – interesting? And isn't something lost when that buzzy outpost of life is reduced to a silent, greasy scene of violence? You might say this represented the first stirrings in my consciousness about the intrinsic value of bees.

Poignantly, it is also testimony to our changing bee demographics that a living memory can be so full of bumble bees. They were everywhere, it seemed, something to be cautious of as one poked around barns or walked barefoot in soft grass. Today that memory is less accessible in Indiana where four bumble bee species are now in decline since my childhood and at least one is extirpated (Jean, 2010).

5.1. Bee Decline: Evidence Over Hyperbole

The notion that bees are in trouble is almost a cliché, on a par with the plight of amphibians (Houlahan et al., 2000), bats (Ingersoll et al., 2013), birds (Rosenberg et al., 2019) and monarch butterflies (Thogmartin et al., 2017). This is not a bad thing, as awareness of a problem is better than ignorance. Over my 30 years on faculty at an agricultural college, I have seen a general shift in the attitude of

homeowners who telephone our department about 'a bee problem'. The complaint is nearly always nestled in the language of apology, and callers express earnest desire for a resolution that does not involve killing the bees. I think people are very sincere about this.

This groundswell of good will has been translated into infusions of public funding for university- and government-level research on bees, bee conservation and pollination management (Buchanan, 2016), yielding rich returns in our knowledge of general principles of epidemiology, ecology, evolution, as well as food production. Research such as this ranks among the best investments in social benefit possible with public funds. No fewer than 35 economic analyses performed between 1965–2005 have shown that the average social return on the dollar for agricultural research is 53%. This means that a one-time investment of US$1 in agricultural research yields US$0.53 every year over a typical 35-year window, during which most scientific discoveries contribute to productivity improvement (Buchanan, 2016, p.142).

Maintaining these kinds of returns depends on governments having mechanisms in place to channel research funding toward evidence-based priorities. Grant administrators, stakeholder focus groups and scientific review panels are society's front-line defence against public funds being squandered on causes concocted by assumption, anecdote or narrative, however charismatic. In my experience this safety net works, not just well but extremely well.

Let me give an example – our topic, bee decline. It should be an encouragement to taxpayers everywhere to know that the greatest sceptics of bee decline come from within the scientific community itself, a demographic that has much to gain by receiving grants to study bee decline. In an influential paper, Jaboury Ghazoul (2005) cautioned scientists against overstating the matter. Localized and regionalized declines of certain bee taxa are verifiable and consistent, and Table 5.1 (below) summarizes a portion of this database. However, Ghazoul was pushing against what he saw as an untoward tendency to extrapolate regional conditions into global generalities. As Ghazoul pointed out, and as a perusal of Table 5.1 makes clear, either bee declines are inequitably reported or are inequitable in reality. Reports of declines are overwhelmingly concentrated in North America and Europe, and are profoundly represented by two groups: managed honeybees and local species of bumble bees. As borne out by subsequent studies, wild bees vary in the degree to which they respond to human-induced stressors (Cariveau and Winfree, 2015), and there is a robustness to pollination systems so that pollinators either adjust to changes in land use (Winfree *et al.*, 2009; Mandelik *et al.*, 2012), or different pollinators compensate for shortage or inefficiencies in others (Winfree *et al.*, 2007, 2008; Garibaldi *et al.*, 2013). In short, the state of a system with as many moving parts as a functioning agroecosystem cannot be described with global generalities. As all bee decline is local bee decline, all pollination crises are local crises.

Ghazoul's article was predictably met with pushback in the form of objections to methodology and interpretation (Steffan-Dewenter *et al.*, 2005), and today the weight of evidence sides with widespread occurrence of local and regional extinctions, population contractions and impoverishment of species richness. Our point for now, however, is that a conversation of the kind Ghazoul incited is an example of one of science's most important properties – a culture of self-correction. Since the first edition of this volume, the subdiscipline of crop pollination has experienced a maturing in its scientific rigour and resistance to sweeping simplifications and self-serving non-scientific narratives.

One of these narratives is a sentence attributed to the 20th century physicist Albert Einstein, quoting him as saying, 'If the bee disappears from the face of the Earth, man will have no more than 4 years left to live.' Any quick Internet search shows that this quote surfaced for the first time in the early 1990s, long after Einstein's death, and in contexts far removed from any possibility of verification. Another overworked trope is the one traceable to Samuel E. McGregor (1976, p.1) that bee pollination is responsible for every third bite of food we eat, which I went to some lengths in Chapter 4 (this volume) to explain is simplistic and applies at best to only the most wealthy citizens of the planet.

So, fortified with a bias for evidence over hyperbole and charisma, we can better approach the state of the world's bee pollinators.

5.2. Bee Decline Examined

To begin, let us consider some properties of bees as a taxon that elevate their environmental vulnerabilities over those of other insects. Except for the parasitic species, bees are central place foragers (see Chapter 3, this volume), a descriptor that takes in the totality of life with a nest – a point in space. Energy budgets take into account the energy expended in foraging to and from this central place. The bee must forage profitably, balancing energy spent against calories and protein gained per unit effort. The bee must make good foraging decisions, balancing opposing dynamics of distance and reward richness. This suite of tasks and decision points represents a huge draw on powers of cognition – learning, memory, information processing and navigation. Even low levels of biotic and abiotic stressors can impair the bee brain, with dire consequences on energy budgets, brood production, and ultimately on the colony's or population's ability to produce another generation (Klein *et al.*, 2017). These risks are exacerbated for a solitary nesting species whose females lack the environmental buffers of the social bumble bees and honeybees (Rundlöf *et al.*, 2015; Straub *et al.*, 2015). For a solitary female, the path between stressor exposure and peril to her brood is brief and direct. At a fundamental level, therefore, commitment to a nest consigns bees to the vulnerabilities of sessile organisms such as trees; escape from habitat perturbations is not readily available.

Second, the haplodiploid sex determination system that bees share with all Hymenoptera has the effect of rendering males sterile when close-kin matings occur, as increasingly happens when effective breeding populations are small. This means that a bee population, once reduced from other stressors, is vulnerable to a spiralling loss of genetic diversity that hastens its decline (Zayed and Packer, 2005).

Third, to the extent a pollinator is specialized to a narrow range of environmental conditions, it is expected to be more vulnerable to habitat perturbations and population declines. Consistent evidence for this as a general principle has been elusive, however. A comparative study (Biesmeijer *et al.*, 2006) of bee species exhibiting decline in the

Table 5.1. Overview of studies documenting changes, mostly negative, in local or regional bee populations.

Region	Bee taxa	Population change measures	Dates	Presumptive contributing cause(s)	Reference
Evidence of Decline					
São José dos Pinhais, Paraná, Brazil	all bees	• 22% decline in species richness • 50% reduction in large species • Previously abundant species *Bombus bellicosus*, *Gaesischia fulgurans* and *Thectochlora basiastra* now extirpated	1962–1963; 1981–1982; 2004–2005	• urbanization • habitat fragmentation • competition from honeybees	Martins *et al.* (2013)
UK, The Netherlands	all bees, native species	• Decline in species richness in 52% (UK) and 67% (The Netherlands) of sampled 10x10 km cells • Greatest loss in species with narrow habitat requirements • In solitary species in UK, greatest loss in oligolectic species • In The Netherlands, greatest loss in long-tongued species	pre- and post-1980		Biesmeijer *et al.* (2006)
Europe	all bees, native or naturalized	• Declining populations in 7.7% (150) species • 9.2% species threatened • 5.2% species near threatened	2014	• agricultural intensification • urbanization • increasing fires • climate change	Nieto *et al.* (2017)
Belgium, The Netherlands, Luxembourg, Denmark, Germany, Switzerland, Austria, Czechia, Slovakia, Hungary, Poland	Bombini (*Bombus* spp. and *Psithyrus* spp.)	• 80% of taxa threatened in at least one country • 30% taxa threatened throughout range • More extinctions 1951–2000 than 1900–1950 • 13 local extinctions between 1951–2000	pre- and post-1950	• habitat fragmentation • habitat homogenization from farming • agrochemicals • food shortage from mowing • competition from honeybees	Kosior *et al.* (2007)
Ontario, Canada	*Bombus* spp.	• *B. affinis*, *B. pensylvanicus* and *B. ashtoni* present in first interval, missing in second • Decreasing relative abundance for *B. fervidus*, *B. terricola*, *B. vagans* and *B. citrinus*	1971–1973 vs 2004–2006		Colla and Packer (2008)

(Continued)

Table 5.1. Continued.

Region	Bee taxa	Population change measures	Dates	Presumptive contributing cause(s)	Reference
China	*Bombus* spp.	• General decline in species diversity and population sizes	1976–1980 vs 1996–1980	• human-induced reductions in food plants	Yang (1999)
Hungary	*Bombus* spp.	• *B. elegans* and *B. serriquama* present in first interval, missing in second • 47.6% of all species declining	pre- and post-1953		Sárospataki *et al.* (2005)
Japan	*Bombus* spp.	• Decreasing relative abundance for native *B. hypocrite sapporoensis* and *B. diversus tersatus*	2003–2005	• ecological displacement by exotic *B. terrestris*	Inoue *et al.* (2008)
Illinois, USA	*Bombus* spp.	• *B. borealis, B. ternarius, B. terricola* and *B. variabilis* locally extirpated • Ranges decreased for *B. affinis, B. fraternus, B. pensylvanicus* and *B. vagans*	1940–1960	• agricultural intensification	Grixti *et al.* (2009)
New Hampshire, USA	*Bombus* spp.	• *B. affinis, B. fervidus, B. terricola* in drastic decline • *B. vagans* in significant decline • *B. terricola* being restricted to high elevations	since 1867	• habitat loss • climate change	Jacobson *et al.* (2018)
New York, USA	*Bombus* spp.	• *B. fervidus, B. pensylvanicus, B. terricola* and *B. affinis*, formerly abundant, not found	2003	• parasite spread from greenhouse trade in *Bombus* spp.	Giles and Ascher (2006)
Oregon, northern California	*Bombus* spp.	• *B. occidentalis* extirpated from San Francisco Bay area • *B. franklini* declining precipitously in southern OR and northern CA	1998–2004	• parasite spread from greenhouse trade in *Bombus* spp. • habitat loss • pesticides • tracheal mites • parasitoids (Diptera, Hymenoptera) • protozoans (*Crithidia* spp.) • microsporidia (*Vairimorpha* spp., formerly *Nosema*[a] spp.)	Kissinger *et al.* (2011); Thorp (2005)

Table 5.1. Continued.

Region	Bee taxa	Population change measures	Dates	Presumptive contributing cause(s)	Reference
USA, 382 sites in 40 states	*Bombus* spp.	• Extensive range reduction and relative abundance decline for *B. occidentalis, B. affinis, B. pensylvanicus* and *B. terricola*	extensive historic records vs 2007–2009	• gut pathogen *Vairimorpha bombi*	Cameron *et al.* (2011)
Austria	managed honeybees	• 9.3% average winter loss	2008–2009		Van der Zee *et al.* (2012)
Belgium	managed honeybees	• 18% average winter loss	2008–2009		Van der Zee *et al.* (2012)
Denmark	managed honeybees	• 7.5% average winter loss	2008–2009		Van der Zee *et al.* (2012)
Germany	managed honeybees	• 10.4% average winter loss	2008–2009		Van der Zee *et al.* (2012)
Germany	managed honeybees	• 3.8% average winter loss (2004–2005) • 15.2% average winter loss (2005–2006)	2004–2008	• *Varroa* mites • deformed wing virus • acute bee paralysis virus • aged queens • low autumn colony population	Genersch *et al.* (2010)
Ireland	managed honeybees	• 21.7% average winter loss	2008–2009		Van der Zee *et al.* (2012)
Israel	managed honeybees	• 11.2% average winter loss	2009–2010		Van der Zee *et al.* (2012)
Italy	managed honeybees	• 6.3% average winter loss	2008–2009		Van der Zee *et al.* (2012)
Macedonia	managed honeybees	• 12.9% average winter loss	2009–2010		Van der Zee *et al.* (2012)
The Netherlands	managed honeybees	• 21.7% average winter loss	2008–2009		Van der Zee *et al.* (2012)
Norway	managed honeybees	• 7.1% average winter loss	2008–2009		Van der Zee *et al.* (2012)
Poland	managed honeybees	• 11.5% average winter loss	2008–2009		Van der Zee *et al.* (2012)
Slovakia	managed honeybees	• 8% average winter loss	2009–2010		Van der Zee *et al.* (2012)
Spain	managed honeybees	• 18.9% average winter loss	2009–2010		Van der Zee *et al.* (2012)
Sweden	managed honeybees	• 14.6% average winter loss	2008–2009		Van der Zee *et al.* (2012)
Switzerland	managed honeybees	• 9.1% average winter loss	2008–2009		Van der Zee *et al.* (2012)
Turkey	managed honeybees	• 25.9% average winter loss	2006–2007	• weather	Giray *et al.* (2010)
Turkey	managed honeybees	• 25.8% average winter loss	2009–2010		Van der Zee *et al.* (2012)
UK	managed honeybees	• 16% average winter loss	2008–2009		Van der Zee *et al.* (2012)

(Continued)

Table 5.1. Continued.

Region	Bee taxa	Population change measures	Dates	Presumptive contributing cause(s)	Reference
USA	managed honeybees	• 28.6% average winter loss (2006–2020) • 21.6% average summer loss (2010–2020) • 39% average loss (2010–2020) • 32.8% loss by backyard beekeepers (2019–2020) • 31.8% loss by sideline beekeepers (2019–2020) • 20.7% loss by commercial beekeepers (2019–2020)	2006–2020; 2010–2020		Bruckner *et al.* (2020)
Little or No Evidence of Decline					
Ontario, Canada	*Bombus* spp.	• *B. citrinus, B. griseocollis, B. impatiens, B. ternarius, B. bimaculatus* persistent	1864–2009		Colla *et al.* (2012)
Europe	all bees, native or naturalized	• Stable populations in 12.6% (244) species	2014		Nieto *et al.* (2017)
Japan	*Bombus* spp.	• Relative abundance unchanged for *B. pseudo-baicalensis*	2003–2005		Inoue *et al.* (2008)
Nebraska, USA	*Bombus* spp.	• 19 of 20 species collected in 1962 still present in 2000	1962–2000		Golick and Ellis (2006)
Evidence of Increase					
Europe	all bees, native or naturalized	• Increasing populations in 0.7% (13) species	2014		Nieto *et al.* (2017)
Ontario, Canada	*Bombus* spp.	• *B. bimaculatus, B. impatiens* and *B. rufocinctus* show increasing relative abundance	1971–1973 vs 2004–2006		Colla and Packer (2008)
Hungary	*Bombus* spp.	• 14% of all species increasing	pre- and post-1953		Sárospataki *et al.* (2005)
São José dos Pinhais, Paraná, Brazil	general	• Relative abundance increase for *Augochlora iphigenia, Augochlora amphitrite, Augochloropsis iris* and *Bombus morio*	1962–1963; 1981–1982; 2004–2005		Martins *et al.* (2013)
UK, The Netherlands	general, native species	• Increase in species richness in 10% (UK) and 4% (The Netherlands) of sampled 10×10 km cells	pre- and post-1980		Biesmeijer *et al.* (2006)

[a]The microsporidian genus *Nosema* was recently revised to *Vairimorpha* (Tokarev *et al.*, 2020)

UK and The Netherlands showed that decline is concentrated in species that are dietary specialists, univoltine (one generation per year) and non-migratory, compared to species that are food generalists, multivoltine and mobile. Narrow niche breadth in food plants has been used to explain vulnerabilities of bumble bees (Goulson and Darvill, 2004; Goulson *et al.*, 2005), but this view has been contested in preference to describing declines in terms of climatic and habitat specializations (Williams, 2005). A team analysing museum specimens of 75 bee species from The Netherlands and their associated pollen loads concluded that species diet niche breadth and climate change sensitivity did not explain downward population trends in the second half of the 20th century; rather, it was loss of preferred plant species interacting with bee body size, with larger bees more vulnerable to food limitation and population declines (Scheper *et al.*, 2014). The conclusions of these authors may be taken to apply generally: 'These results highlight the species-specific nature of wild bee decline and indicate that mitigation strategies will only be effective if they target the specific host plants of declining species' (Scheper *et al.*, 2014, p.17,552).

In my opinion, these qualifiers do not detract from a general downward trajectory for bees in many parts of the world – and due to a fairly limited range of categorical drivers, albeit with local variants (Table 5.1). A pattern that is clear in Table 5.1 is that records of bee decline are concentrated on bumble bees and managed honeybees. No doubt there is sampling bias at work here as both groups rank high on the public's 'charisma' scale; each is conspicuous, and each is highly valued as an important crop pollinator. However, where drivers of decline have been elucidated for these groups there are patterns to those drivers that apply generally.

The stress drivers are numerous and known. When analysed, they always amount to some combination of: parasites and pathogens, natural or exotic; deteriorating habitats, nesting sites and natural food plants with associated dietary deficiencies; agricultural intensification; competition with non-native species; climate change; pesticides, both acutely acting and chronic; and for highly managed species like the honeybee, stress associated with intensified management.

For over a decade it has been axiomatic to say that bee stressors are multifactorial and interacting. While this is certainly true, it also highlights the fact

that no one has done the hard work of finding a solution to these stressors. Meeus *et al.* (2018) enumerate three requirements for disentangling webs of causation. Investigators must: (i) experimentally demonstrate the impact of single drivers; (ii) demonstrate the interaction of each, whether antagonistic, synergistic or neutral, with one or multiple other drivers; and (iii) demonstrate the impact of drivers, singly and in interaction, on each target bee species. Clearly, such an enterprise is daunting if not impossible, as the combination of potential stressor interactions borders on the infinite.

The truth is, such an infinite web of interactions represents the frontier where experimental science ends and modelling takes over. For the most mind-bogglingly complex systems – global climate change or even the daily weather forecast – running such models taxes the limits of modern computing power. The best models are only as good as the data that populate them, however, and for this we still rely on experimental science while acknowledging the paltry levels of interaction that even the best experiments can control for.

So we turn to some experimental studies that have identified stressors, singly or interacting, that directly contribute to bee morbidity. A sampling of this research is presented here, drawing from the review of González-Varo *et al.* (2013) and other studies. The majority of these studies have limited their interactions to no higher than two factors; the most I am aware of is three. Yet studies like these are the essential fodder for modelling simulations of multi-stressor effects on bee populations.

5.2.1. Interactions between landscape alteration and agricultural intensification

A typical experimental set-up is a 2×2 factorial with two levels of landscape alteration such as simple (low natural cover) versus complex (variable natural cover), and two levels of agricultural intensification such as organic versus conventional farming, or pesticide use versus no use. In general, simple landscapes exacerbate the negative effects of intensified farming on pollinator species richness and abundance. An extension of this result is that the net benefit of introducing organic farming practices may be greater in simple landscapes than in complex ones which already support a comparatively rich assembly of bees. A better return on investment may be found with policies that prioritize efforts at mitigating effects of agricultural intensification in highly altered, simplified landscapes.

5.2.2. Interactions between landscape alteration and non-native species

Study designs have presented a range of landscape alterations, either categorically (e.g. disturbed or undisturbed) or continuously along a gradient of naturalness, and have assessed invasive species at the level of local plot. In general, the abundance of non-native bees (and plants) is higher in more highly altered landscapes. In Hokkaido, Japan, the non-native European bumble bee *Bombus terrestris* predominates in disturbed deforested areas where it has displaced native *Bombus* spp. to forest habitats (Ishii *et al.*, 2008).

5.2.3. Interactions between pathogens and managed bees

The focus of this body of work has been on pathogen spillover to wild bees from managed honeybees and greenhouse bumble bees. The managed bees in question are often non-native, especially in the case of the globally distributed European honeybee *Apis mellifera* and bumble bee *Bombus terrestris*. In North America, however, native bumble bees exist that have been pressed into service as managed species. When managed *Bombus* spp. occur near wild conspecifics, there is a risk of pathogen spillover to their wild cousins.

From 1877, the European honeybee *Apis mellifera* was repeatedly imported into Japan within the natural range of the Asiatic honeybee *A. cerana japonica*. By 2011, positive cases of the *A. mellifera* viruses 'deformed wing virus' DWV) and 'black queen cell virus' and of the *A. mellifera* parasitic mite *Acarapis woodi*, were found in native *A. cerana* (Kojima *et al.*, 2011). In Belgium, native bumble bees show infection with the *A. mellifera* 'Varroa destructor* Macula-like virus' (VdMLV), with higher prevalence when in proximity to managed honeybee apiaries (Parmentier *et al.*, 2016). In Brazil, viruses of non-native *A. mellifera* have infected the native bumble bee *Bombus atratus*, although this evidence is restricted to commercial conditions where *B. atratus* itself is cultured as a managed greenhouse species (Reynaldi *et al.*, 2013). However, wild populations of *B. atratus* are infected with honeybee viruses and microsporidia in Colombia (Gamboa *et al.*, 2015). Virus spillover from *A. mellifera* has been documented in 11 non-*Apis* hymenopteran species including four native *Bombus* spp. and four species of solitary bees in Pennsylvania, New York and Illinois, USA, suggesting that these viruses freely disseminate via shared contact at flowers (Singh *et al.*, 2010). Evidence for pathogen spillover from non-native *A. mellifera* to native stingless bees has been shown in Australia (Purkiss and Lach, 2019) and in Brazil (Guimarães-Cestaro *et al.*, 2020).

Most work in pathogen spillover has focused on the interface between managed greenhouse bumble bees and wild bumble bees. Beginning in the late 19th century, queens of *Bombus hortorum*, *B. terrestris*, *B. subterraneus* and *B. ruderatus* were exported from the UK to New Zealand where those species survive to this day. By the late 20th century, year-round rearing of *B. terrestris* was commercially practical, and for over 30 years now, colonies of *B. terrestris* have been shipped from Europe to Chile, Japan, Korea and elsewhere for use in pollination of greenhouse vegetables, primarily tomato (Goulson, 2010). These introductions and others have been associated with the movement of parasites from Europe to native bees in New Zealand (Donovan, 1980) and Japan (Yoneda *et al.*, 2008).

In North America, studies have implicated an interaction between pathogens and managed *Bombus* spp. The issue here is management, not exotic species. *B. terrestris* is not legally imported into Canada or the USA; instead, mass-production and domestic movement are concentrated on two native species – *B. impatiens* in eastern North America and *B. occidentalis* in the west, although permissive laws have allowed shipment of *B. impatiens* westward outside its natural range, thrusting *B. impatiens* into the role of an alien species on its own continent. Commercially produced colonies of *Bombus* spp. have higher levels of parasites and pathogens than their conspecifics in the wild (Colla *et al.*, 2006). Thus, an imported commercial box of *B. impatiens* poses an environmental hazard to the wild colonies of *B. impatiens* that live outside the walls of the greenhouse, a risk supported by the fact that greenhouse bumble bees do in fact make their way to forage on plants outside the greenhouse (Whittington *et al.*, 2004; Trillo *et al.*, 2019). In southern Ontario, there is a higher incidence of parasites in wild *Bombus* spp. colonies near to greenhouses than in wild colonies further from them (Colla *et al.*, 2006; Otterstatter and Thomson, 2008).

Szabo *et al.* (2012) scaled up this conversation with an appraisal of the pathogen spillover hypothesis – the idea that pathogen spillover from managed bees in intensified greenhouse agriculture is contributing to the decline of bumble bee species across North America. The team assembled a data set resolving greenhouse density to the level of county, or equivalent municipality in Canada, and superimposed upon it over 65,000 dated records of bumble bee collections from museums and private collections. For two species, *Bombus terricola* and *B. pensylvanicus*, there were reductions in occurrence as greenhouse density increased. These data are the first of their kind linking pathogen spillover to specific bee species declines.

5.2.4. Interactions between artefacts of agricultural intensification

The impact of any stressor on pollinators is expected to be more severe in highly altered landscapes and intensified agriculture owing to a cluster of stressors that follows these conditions. Much of this literature is biased toward honeybees owing to their use in intensified agriculture and their broad research base in toxicology and pathology.

5.2.4.1. Nutrient stress

Landscape intensification almost always reduces a landscape from floral cornucopia to near monoculture. Sometimes those monocultures provide a veritable explosion of nectar and pollen for a few days or weeks of the year, but otherwise the landscape is a food desert. A direct positive relationship has been shown in honeybees between the competence of their immune systems and the diversity of proteins (pollens) in their diets, an effect over and above mere protein quantity (Alaux *et al.*, 2010). An insightful analysis was performed by Naug (2009) who looked at the relationship between declining numbers of managed beehives in the USA and changing land use patterns between 1982–2003, collapsing all land categories into one of four mutually exclusive classes: crop, pasture, range or urban. While declining bee numbers were universal to all classes, the only statistically significant predictor was decreasing rangeland. Naug points out that of all the land classes, rangeland, consisting of non-managed flowering forbs and shrubs, is expected to be the most rich and diverse

in bee forage. Another team, using a spatial habitat model and national land cover data, showed a decline in wild bee abundance across 23% of US land area between 2008–2013 (Koh *et al.*, 2016). The decline was associated with the diversion of unaltered habitats into row cropping. The model identified 139 counties where low bee abundance coincided with 39% of the nation's pollinator-dependent crop area.

A restriction of natural food plants is particularly stressful to long-tongued bumble bee species who rely disproportionately on plants in the family Fabaceae (Goulson *et al.*, 2005; Biesmeijer *et al.*, 2006). Compared to short-tongued *Bombus* species, their long-tongued counterparts exhibit narrower diet breadth for pollen sources, and as a rule rarer species tend to visit fewer flower species (Goulson *et al.*, 2008), making them more vulnerable to local plant perturbations.

Poor nutrition is a potent actor in negative synergies. Nutrient stress interacts with sublethal doses of neonicotinoid insecticides to accelerate honeybee mortality, decrease food consumption and lower haemolymph sugars (Tosi *et al.*, 2017). Nutrient stress also hastens honeybee mortality when bees are artificially inoculated with microsporidian (*Vairimorpha* spp., formerly *Nosema* spp.) parasites. Longevity of infected bees was highest in groups fed either high-quality monofloral pollens or polyfloral pollen mixes (Di Pasquale *et al.*, 2013), demonstrating both that pollens vary in their nutrient quality and that pollen diversity can partially resist the deleterious effects of the parasite. Nutrient stress is shown to exacerbate honeybee morbidity from 'Israeli acute paralysis virus' (IAPV). In laboratory studies, bees fed high-quality diets, i.e. high-quality monofloral pollens or polyfloral mixes, were able to withstand IAPV infections that were lethal to bees under poor diet regimes. Moreover, field colonies that were infected with IAPV and simultaneously deprived of pollen, expressed early worker foraging, a stress response by which young bees precociously jump to the latest and riskiest age task category (Dolezal *et al.*, 2019). Unchecked, this phenomenon can trigger a downward spiral of colony reduction and death (Perry *et al.*, 2015).

5.2.4.2. Pesticides and other agrochemicals

A hallmark of 21st-century intensified agriculture is a heavy reliance on synthetic chemical inputs – insecticides, herbicides, fungicides, fertilizers, antibacterials, plant growth regulators, and accessory compounds such as wetting agents, synergists and carriers. The quantities of these agrochemicals and their potential interactions, one with another or with other stressors, defies description, given the diversity and volume of active ingredients encountered by bees during their normal foraging activities. This single stressor category is significant enough to warrant its own dedicated chapter in Volume II of this work.

For now, readers should be aware that the literature on bees and pesticides is not an equivocal indictment on pesticides. One of the most common problems is that toxicities from 'field-realistic doses' delivered in the laboratory cannot always be replicated in the field, a phenomenon that suggests that either artificial conditions in the laboratory bias dosage effects to a greater degree than experimenters realize, or that nature provides bounteous and largely unknown mitigating buffers. A second and related observation is that the laboratory/field disconnect is especially pronounced in social bees. It appears that many buffers of colonial life give social species a survival advantage over their solitary counterparts. Finally, there are extraordinary references in the literature to what appear to be *beneficial* effects of pesticide exposure on bees. These results must be held as highly conditional, if not idiosyncratic. In my opinion, the overwhelming weight of evidence indicts chronic pesticide exposure as an insidious and serious threat to bees everywhere. Agrochemicals should never be given the benefit of the doubt. All of these issues are more fully developed in Volume II of this work.

5.2.4.3. Pathogen on pathogen interactions

One artefact of highly altered landscapes is a disruption of equilibria among pathogen communities, potentially elevating their virulence to bee pollinator hosts (Meeus *et al.*, 2018). For our purposes here, parasites and pathogens are pooled into one 'pathogen' stressor.

The richest body of evidence for pathogen on pathogen synergies draws from the highly managed honeybee *Apis mellifera*. There is a strong association between this bee's non-natural mite ectoparasite *Varroa destructor* and numerous virus pathogens. Colony levels of 'acute Kashmir Israeli virus' (AKI) and DWV align with the typical mite population increase in colonies over a growing season (Francis *et al.*, 2013). The relationship

between DWV and *Varroa destructor* is so direct that the two are now understood to be in mutualistic symbiosis: the mite vectors the virus, while virus-induced immunosuppression in the bee benefits mite feeding and reproduction (Di Prisco *et al.*, 2016). Infection with spores of the microsporidian *Vairimorpha ceranae* (formerly *Nosema ceranae*) can accelerate DWV replication in bees in a dose-dependent relationship. When bees are supplemented with pollen diets, however, higher doses of *Vairimorpha ceranae* spores are required to induce DWV replication (Zheng *et al.*, 2015). Thus, we have a three-factor interaction – microsporidia, virus and nutrition – a rarity in the literature on bee stressors. In contrast to these damaging synergies, a test for interactions between two arthropod pests, *Varroa destructor* and the hive scavenging beetle *Aethina tumida*, failed to show a harmful synergy between the two on honeybee colony strength (Delaplane *et al.*, 2010).

In bumble bees, there is evidence for virulent pathogen on pathogen interactions between DWV and the neogregarine *Apicystis bombi* (Graystock *et al.*, 2016b).

5.2.4.4. Direct effects of agricultural intensification on bee pathogens

The weight of evidence indicates higher bee pathogen burdens in contexts of increasing agricultural intensification. Pathogen counts tend to be higher in commercially produced *Bombus* spp. colonies than in wild colonies of their conspecifics (Colla *et al.*, 2006). However, a comparative survey of feral versus managed honeybee colonies in the UK showed similar prevalence and quantity of most pathogens, except for DWV which was higher in wild colonies (Thompson *et al.*, 2014).

The question of the impact of intensification is most often raised in the context of migratory beekeeping, a feature of the honeybee industries of Australia, Brazil, South Africa and the USA. It seems intuitive that the strain of mechanized transport by truck across hundreds or thousands of miles and the frequent reorientation to new habitats would leave its mark in the form of measurable stress. However, a direct comparison between migratory and stationary apiaries of African honeybees in Brazil failed to show significant differences in colony loads of *Paenibacillus larvae*, *Varroa destructor*, *Vairimorpha apis*, and *Vairimorpha ceranae* (Cestaro *et al.*, 2017). However, Alger *et al.*

(2018) showed with European honeybees in the USA, migratory colonies returned home (North Carolina) from pollinating California almond with fewer bees and higher loads of black queen cell virus. Stationary colonies exposed to the migratory bees after they returned home experienced a greater increase of DWV compared to isolated stationary colonies. Curiously, 1 month after returning home, migratory colonies had fewer *Varroa* spp. mites than stationary colonies, an effect the authors attributed to a legacy of reduced bee populations in the migratory colonies (Alger *et al.*, 2018). Another study tracked occurrence, prevalence and abundance of black queen cell virus, Lake Sinai virus, Sacbrood virus, *Vairimorpha ceranae* and trypanosomatids over winter and early spring 2013–2014 in three Montana-based migratory operations serving California almonds. There were significant differences across operations and seasons, but pathogen prevalence was highest immediately after almond pollination (Cavigli *et al.*, 2016).

5.2.5. Interactions between climate change, landscape alteration and agricultural intensification

Landscape alteration is expected to interact with climate change to result in mismatches in time and space between the pollinator and its target crop. This was experimentally tested by Parsche *et al.* (2011) who simulated: (i) land use change = pollinator habitat loss, by potting test plants close to, or distant from, favourable bee nesting habitat; (ii) climate change = phenological flowering shift, by manipulating test mustard plants to flower early or normally; and (iii) increased temperature and carbon dioxide (CO_2) = increased vegetative growth/flower height, by keeping potted test plants at ground level or elevating them by 0.5 m. Elevating experimental flowers diminished fly visitation but did not affect bees. Precocious flowering reduced visitation of both pollinators and herbivores but increased seed production, showing that losing synchronized pollinators was compensated for by escaping synchronized herbivores. During the natural flowering interval, more seeds were produced near to favourable nest sites, especially in plants of natural height. The overarching result of the study, however, was that seed production was robust across a range of environmental changes, suggesting that general predictions are difficult. Climate and landscape perturbations may decrease activity of a coevolved mutualist (pollinator) but simultaneously decrease activity of a coevolved antagonist (herbivore).

There are direct effects of climate change on natural ranges of bumble bees, with species in the northern hemisphere expanding their ranges north (Martinet *et al.*, 2015; Biella *et al.*, 2020) or becoming restricted to higher-elevation refugia (Pyke *et al.*, 2016; Biella *et al.*, 2017; Jacobson *et al.*, 2018), sometimes with desynchronization of bee activity with flowering schedules.

5.3. Modelled Predictions of Bee Decline

Over the last decade, the limits of single- or at most two- or three-factor studies in the field have been recognized, as the complexity and geographic breadth of the bee health problem has unfolded. The ready-made science (Stillman *et al.*, 2015) of *in silico* analysis – computer modelling – has made impressive inroads into synthesizing the diverse and complex data from the field for honeybees and bumble bees. Interested readers are encouraged to consult Henry *et al.* (2017) for an excellent review and guidance on developing and understanding predictive models for bee health.

One team has developed an online predictive model, BEEHAVE (Becher *et al.*, 2014; www.beehave-model.net, accessed 6 January 2021) that integrates natural honeybee colony growth dynamics while exploring interactive agents contributing to colony failure. The viewer can test outcomes of different scenarios by manipulating model inputs on parameters such as presence or absence of parasites, food quality, forage distance from hive, and presence or absence of pesticides. A companion model, *Bumble*-BEEHAVE, was subsequently developed for bumble bees and made downloadable from the same website above (Becher *et al.*, 2018).

5.4. Bee Decline and Impacts on Pollination

So far, I have focused on bee decline – whether it is happening, how severely it is happening, where it is happening, and why it is happening. Ultimately though, this is relevant to our purposes only if bee problems are limiting crop pollination either by bee absence or by morbidity-induced pollination inefficiencies. It is a fair question to ask whether the loss of all these pollinators translates to lost pollination and reduced crop yields.

This means that we should linger a while on the concept of *pollination deficit* (Wilcock and Neiland, 2002) which may occur as pollination *limitation*, a level of stigmatic pollen delivery insufficient in quantity or quality for optimum seed-set or fruit-set to occur, or pollination *failure*, a wholesale absence of pollen dispersal to stigmas of a plant dependent on crossing. Although pollination deficits can happen from causes 'on the plant side' such as predispersal failures of physiology, chromosomal aberrations, and mechanical loss, or postdispersal failures from incompatibilities of pollen and stigmatic surface, we are concerned here with deficits owing to impaired performance or inadequate numbers of bee pollinators.

5.4.1. Pollination deficit from sick bees

Let us begin with the first scenario – the idea that sick bees are impaired pollinators. The question was addressed indirectly by Anderson and Giacon (1992) who showed that honeybee colonies infected with Sacbrood virus or the microsporidian *Vairimorpha apis* collected significantly less pollen than colonies fed only a control sucrose solution. Much later, Lach *et al.* (2015) inoculated honeybees with spores of *V. apis* and analysed incoming bees at hive entrances for spore counts and loads of target pollen. There was a significant negative correlation between a bee's spore count and pollen load, leading the team to conclude that microsporidian infection, even at relatively low doses, can negatively impact the pollinating efficacy of individual foragers.

However, to my knowledge the question has been addressed directly only by my PhD student, Amanda Ellis (Ellis and Delaplane, 2008), who manipulated honeybee colonies to have different levels of the parasitic mite *Varroa destructor* or the scavenging beetle *Aethina tumida*. She tented infected and non-infected control colonies with flowering plants of oilseed rape or rabbiteye blueberry, and measured resulting fruit-set. On the basis of single-bee flower visits, fruit-set was lower in blueberry tented with bees from *Varroa*-parasitized colonies. However, when pollination effects were measured at the level of colony, there were no differences in blueberry fruit-set, number of blueberry pollen grains deposited on the stigma, or pod-set in oilseed rape among colonies across the pest treatments or controls. We

concluded that individual inefficiencies of *Varroa*-compromised foragers are compensated for by multiple flower visits fielded by colonies of this social species.

Two overarching conclusions draw from these admittedly limited studies. First, evidence exists that reduced pollination efficacy is a feature of foragers whose health is compromised, whether from direct infections of microbial pathogens or from indirect infections of a colony-level parasite like *Varroa destructor*. The Ellis and Delaplane (2008) study suggests, however, that social species can provide a measure of compensation in the form of repeat flower visits. This is an attractive idea, but certainly warrants substantiation with other social species, pathogens and cropping systems.

5.4.2. Pollination deficit from bee shortage

By far, more attention has been paid to the scenario of pollination deficit from bee shortage. Do bee declines of the kind shown in Table 5.1 translate into negative effects on crop yield? The question is not a trivial one, given the significant sums of public research money that continue to be funnelled toward pollinator health and conservation with an eye on food security.

In an influential paper published in *Science*, Biesmeijer *et al.* (2006) enumerated the criteria necessary for establishing a connection between declining pollinators and costs to pollination. One must be able to show: (i) overall decline in pollinator density; or, (ii) reductions in species diversity or species composition, so that loss of some pollinators has not been compensated for by the increase of functionally equivalent species; and (iii) declines in reproductive success or abundance of pollinator-dependent plants. These authors proceeded to collect almost 1 million records of bee and hoverfly observations in the UK and The Netherlands, concentrating on 10×10 km geographic cells rich in observations, before and after 1980. Significant reductions in bee species richness were shown for 52% of British cells and 67% of Dutch cells. Shifts in hoverfly species richness were inconsistent across the same period, with no directional trend in the UK and no change to slight increase in The Netherlands. In the absence of data on shifts in pollinator flower visits or pollen deposition, Biesmeijer *et al.* (2006) instead examined shifts in plant species distributions between the two countries and showed a general decline in bee-dependent plant species, an increase in abiotically-pollinated species, and a slight decline and intermediate position for self-pollinated species. This study, however short it may fall of its own criteria, remains a high watermark for arguing a causal connection between pollinator fate and plant reproductive success.

Translating such discoveries from ecosystem to agricultural contexts requires essentially the same criteria as those named by Biesmeijer *et al.* (2006). However, a 2008 workshop of the Food and Agriculture Organization (FAO) 'International Initiative for the Conservation and Sustainable Use of Pollinators' simplified the matter by stating, 'Crop pollination deficit refers to inadequate pollen receipt that limits agricultural output' (Vaissière *et al.*, 2011, p.2). This generalization is appropriate, given that plant reproductive success is ancillary to the purposes of crop production where priority is spread more broadly across yield, palatability, nutrient content and consumer acceptance of seeds and fruits.

With the value of pollination largely uncoupled from plant reproductive success, in agriculture we can think of pollination like we think of other inputs that vary continuously and have end points of diminishing return. To the extent that output improves with the addition of the input – be it fertilizer, pesticide, irrigation or pollinators – the input has value.[i] Just as irrigating during a rainy season is probably a waste of money, so is inputting more pollinators when crop stigmas are already being saturatively covered with viable pollen. Figure 5.1, adapted from the discussion by Reilly *et al.* (2020) summarizes the various scenarios that may confront

a farmer contemplating a pollination deficit in a pollinator-responsive crop.

Moreover, pollination interacts with other agronomic inputs to reveal what may be a hierarchy of limiters. Garratt *et al.* (2018) showed that oilseed rape can partially compensate for suboptimal fertilizer and pollination regimes by producing larger seeds and more pods; increasing pollination inputs improved yield only after fertilizer minima were satisfied. Once basic agronomic needs are met, the benefits of pollination are comparable to other limiters of economic importance. Such was the case in a study of hybrid leek seed production in Italy and France where investigators (Fijen *et al.*, 2018) showed that pollination was at least as important to marketable seed yield as were size and vigour of the plants. The interacting effects of pollinators against other agronomic and horticultural limiters is a new and long-neglected area of research, promising to more efficiently integrate pollination into large-scale food production.

Most evidence for crop pollination deficits has come in the form of short-term, localized studies that contrast fruit-set achieved under ambient conditions, e.g. open plots, against fruit-set achieved under artificially saturative pollination, e.g. pollination by hand or by tenting plants with bees. With some exceptions (Petersen *et al.*, 2014), such methods almost always show higher pollination rates in the artificially saturative conditions. My PhD student Selim Dedej found 68.9% fruit-set in open-pollinated blueberry compared to 79% in blueberry tented with honeybees (Dedej and Delaplane, 2003), which suggests that this particular orchard was underpollinated by about

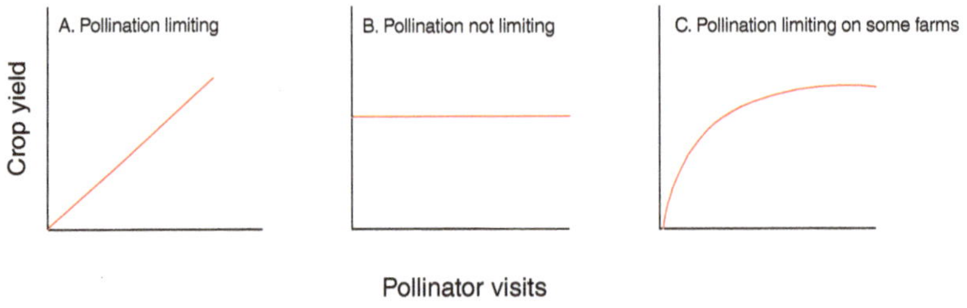

Fig. 5.1. Appraising crop pollination deficits, assuming a crop is pollinator-responsive. (A) As long as yield increases with increasing pollinator visitation, we can assume that pollination is limiting on that farm. (B) If yield is unaffected by an increase in pollinator visits, then pollination is non-limiting, and any pollination inputs by the grower are superfluous. (C) Variable pollination deficits will be seen across farms under two general scenarios: (i) pollinator flower visitation rates are variable across farms; or (ii) agronomic limiters such as water, fertilizer or pests are variously managed, obscuring pollination deficits even if they exist. A corollary of this appraisal is that commercial farms, typically well managed for agronomic inputs, will be especially sensitive to pollination deficits. Adapted from Reilly *et al.* (2020).

10% of its physiological capacity. Similar effects of augmentation over ambient pollination have been shown in apple (Garratt *et al.*, 2014b), to name another example among many. In fact, a review of such 'pollen supplementation experiments' has shown conclusions tipping in favour of pollination limitation in 62–73% of species or cases, a finding that begs for explanation given that sexual selection theory predicts that female resources rather than male resources should limit the number of seeds produced (Knight *et al.*, 2006). This incongruence between theory and real-world data could be an artefact of bias against publishing negative results (studies that fail to show differences between saturative and ambient pollination do not get published) (Knight *et al.*, 2006), or it could be an authentic signal that plant populations and their pollinators are not in equilibrium (Ashman *et al.*, 2004).

I lean toward the latter explanation. For one thing, it aligns with regional ecosystem trends of the kind shown by Biesmeijer *et al.* (2006). Second, the possibility of disequilibrium between plants and their pollinators seems nowhere more likely than in the gross distortions in species distributions that occur in agricultural landscapes.

However, feasibility is not evidence. The truth is, it has proven difficult to experimentally accommodate the criteria of Biesmeijer *et al.* (2006) and directly show a connection between pollinator loss and cost to pollinator-dependent cropping systems at scales larger than a farm.

One natural experiment comes close (Roubik, 2002). The self-pollinating coffee plant *Coffea arabica*, was exported from its native Africa to South America, Central America, the Caribbean and Indonesia. Across its range it was considered non-responsive to pollinator visitation. However, beginning in the 1980s the East African honeybee *Apis mellifera scutellata* began colonizing the coffee growing regions of South and Central America, establishing dense feral populations, after which coffee production dramatically increased (bee pollination has since been confirmed to help coffee fruit-set) (Klein *et al.*, 2003). Similar yield increases did not occur in coffee growing regions of Africa where widespread agricultural intensification had eliminated feral bee habitats, nor in Indonesia, nor in the Caribbean region that remained uncolonized by *A. m. scutellata*. It is now generally understood that coffee culture was operating under a chronic and unrecognized pollination deficit prior to the arrival of *A. m. scutellata*. By inverting this case

history, we may infer what happens in the situation of wide scale pollinator loss in a pollinator-responsive cropping system.

Perhaps the most comprehensive controlled experiment on this matter comes to us from Reilly *et al.* (2020) who associated yield with on-ground bee visitation rates for seven major pollinator-responsive crops in their primary areas of production in the USA and Canada. For each crop/state (province) combination, the investigators modelled and tested the three scenarios shown in Fig. 5.1 and determined that pollination is limiting for apple in Michigan and Pennsylvania, sour cherry in Michigan, sweet cherry in Washington, and for highbush blueberry in Michigan, Oregon and British Columbia. For watermelon (CA, FL), pumpkin (PA) and almond (CA), there were no signals of pollination limitation. For Florida blueberry, the investigators did a pollination supplementation experiment (see discussion above) including hand-pollinated controls, and showed that pollination limitation is generally not a problem in this crop and is limited to farms with low bee visitation rates.

The practical difficulties in showing direct effects of pollinator loss at regional or national scales is the reason we have relied so heavily on inferences based on historic changes in relative cropping areas and the pollinator dependence ratios (D) of those crops (see Chapter 3, this volume; O'Grady, 1987; Melathopoulos *et al.*, 2015). This method is blind to changes in bee numbers, but it is reasonably good at pointing out vulnerabilities to cropping systems under the assumption that pollinators are limiting.

In an influential analysis of trends in worldwide cultivation of pollinator-dependent crops, Aizen *et al.* (2008) reported that between 1961–2006, global crop yields (metric tonnes/hectare) increased at an annual rate of about 1.5%, a rate roughly the same for pollinator-dependent as for non-dependent crops. If pollinators were a limiting factor during that interval, one should have detected slower rates of increase in pollinator-dependent crops, which did not occur.

In an accompanying study, Aizen *et al.* (2009) categorized the world's principle pollinator-dependent crops according to D and estimated that in the total absence of animal pollinators, the expected direct reduction in global agricultural production would range between 3–8%.

In what is surely a demonstration of species disequilibria in intensified agriculture, Aizen and Harder (2009) showed with FAO data for 1961–2007 that global stocks of managed honeybees increased by 45% during that period, highlighting along the way that the much publicized honeybee decline is limited largely to the USA and former Soviet bloc. However, during those same decades, the fraction of world crops dependent on animal pollination increased by 300%, raising the possibility that global pollination demand is out-pacing pollinator availability.

Gallai *et al.* (2009) analysed the world's crops with a *vulnerability ratio* – the ratio of the value of insect pollination for the focal crop relative to its total economic value. Under a scenario of total pollinator loss their model predicted a rate of vulnerability of global agricultural production for human food at 9.5%. This average obscures what for some regions and crops are catastrophic vulnerabilities, ranging from 22% (nuts in East Africa) to 94% (stimulant crops in Oceania).

In reading these papers, one detects notes of cautious watchfulness, not claxons of emergency. The fraction of the world's crops dependent on pollinators is increasing. Local bee losses and pollination deficits are demonstrable and no doubt contribute to lost revenues for farmers at the local scale. Agricultural pollination deficits at global scales, however, are within the range of 8–10% of crop values which, albeit obscuring some regionally acute vulnerabilities, still fall short of a 'global pollination crisis'.

Why are we *not* in a crisis? The question seems fair given the libraries of papers written, the very penning of which seems to imply a backdrop of ecological peril. For one thing, our species could survive without bee-pollinated crops. As noted in Chapter 4 (this volume), 60% of global food supply derives from plants that are wind-pollinated or passively self-pollinated for whom pollinators, their health and abundance are of little or no consequence.

Second, there is a robustness and unpredictability to nature which often belie the gloomy predictions of human prognosticators. The loss of one pollinator is made up for at the community level by another (Hallett *et al.*, 2017). Pollinators living in social colonies compensate for those of their number who are sick and inefficient pollinators. Plants and pollinators labelled by ecologists as 'specialists' are found to be less obligately narrow than formerly thought (Chittka *et al.*, 1999). In the case of the persistent winter losses for honeybees enumerated in Table 5.1, it is helpful to remember that the average 12-month mortality of a new honeybee swarm in temperate latitudes is around 75% – a statistic all the more remarkable for the fact that it was discovered in a time and place before the ravages of the pan-global *Varroa* mite (Seeley, 1985, p.46). High winter mortality in first-year swarms is the rule, not the exception, for temperate-evolved *Apis mellifera*.

Third, our species of short-lived hominins is not good at discerning crises from ebbs in demography and species persistence. Plants may not be evolved for saturative pollination and 100% fruit-set – the unstated gold standard for agricultural pollination management. Consider long-lived perennials for which increased fecundity may be purchased only at the cost of a long-term hold on their habitats. When the perennial European spring pea *Lathyrus vernus* was experimentally saturatively pollinated, its seed production increased threefold; however, flowering was reduced the following year (Ehrlen and Eriksson, 1995).

Finally, in the face of chronic pollinator limitation, plants are expected to evolve compensatory mechanisms such as increased rates of autonomous selfing (Thomann *et al.*, 2013).

In short, there are many buffers and mitigating dynamics that sustain fruit, vegetable and seed availability from pollinator-responsive agricultural crops, even while pollinator declines register locally across many parts of the globe.

What do we do with this state of knowledge? We are at a point in our species's history where we understand the value of pollinator-dependent crops, and more and more of the whole human family wants a share in their benefits. We understand the connection between food security and healthy, diverse and abundant pollinator populations, and we know the drivers that are pushing local and regional bee declines. Bees have charisma, and increasingly in all quarters of the world we find people newly aware of the plight and value of bees and other pollinators and willing to help. These are knowledge tools and social assets, not for purchasing us the leisure of complacency, but for spurring us to develop pollinator health and conservation initiatives that are science-based, humanitarian, practical and profitable. We are poised for *preventing* a pollination crisis.

Note

[i] In reality, a farmer would weigh the expected income-increasing potential of the input against the input's cost. Here, we are simplistically considering only yield-limiting factors.

6 Applied Bee Conservation

Since the beginning of the 18th century, there has been a steady increase in the worldwide coopting of wild lands into agricultural production. Between 1700 and 2007, cropland and pasturelands increased fivefold from ~3 to 15 and ~5 to 27 million km², respectively. Significant technical advances began making their mark in the 1930s. Plant breeders developed varieties with enhanced yields and resistance to pests, disease and drought. The Haber-Bosch process enabled the synthesis of nitrogen fertilizer from virtually limitless atmospheric nitrogen. Synthetic herbicides, fungicides and insecticides became plentiful and widely used. Cheap fossil fuels paved the way for mechanization. The net effect of these developments has been described as nothing less than revolutionary – the highest per capita calorie production levels in world history. Even though average rates of land clearing have slowed since the 1950s, productivity shows little sign of slowing down (Ramankutty *et al.*, 2018). Between 1961 and 2006, global crop yields (metric tonnes/hectare) increased at an annual rate of about 1.5% (Aizen *et al.*, 2008).

These gains have come at great environmental cost. One outcome of this centuries-long process has been the widespread elimination of natural habitats and their replacement with radically simplified ecosystems consisting at the macroscale of little more than the monoculture and a handful of 'weeds'. The abundance and diversity of wild bees drops precipitously in these simplified systems, ironically at a period of history when pollinator-dependent crop acreage is increasing and unprecedented demands being made on our pollinators (Aizen *et al.*, 2008; Koh *et al.*, 2016). Studies consistently show strong association between the nearness of patches of suitable bee forage and nesting sites, and the abundance and species richness of wild bee visitors to the crop with corresponding improvements to pollination outcomes (see section 10.2, this volume). It follows, therefore, that re-diversification of conventional farmlands is an essential part of redressing the

excesses of earlier farming paradigms and integrating farming habitats into neighbouring ecological networks. The expected outcome is improved stability of community populations, assisting not only pollinators but predators and parasites that constrain herbivorous crop pests. Some of the most exciting work in this area has shown that such landscape improvements, far from sacrificing tillable acres to 'idle' uses, instead increase net profits (Morandin and Winston, 2006; Gurr *et al.*, 2016).

Habitat conservation and restoration are among the most cost-effective ways to increase pollinator abundance and diversity on a farm. The changes brought about by habitat diversification programmes tend to be long lasting, spreading the cost of their installation over many years. One analysis showed that a US$4000 installation cost for a 300 m hedgerow field edge would be recovered on the basis of insecticide savings within 16 years. If pollination services of wild bees are included, the recovery interval is reduced to 7 years (Morandin *et al.*, 2016). The benefits are also long lasting since permanent nesting sites and pastures encourage large, locally recurring bee populations.

A growing literature is parsing out the characteristics of natural and restored habitats that promote increasing bee abundance and species richness. A self-sustaining bee population requires long-lasting, undisturbed nesting sites and plants that annually produce nectar and pollen during bee nesting seasons. The habitat can be literally conserved – by setting aside natural land near the crop and sparing it from cultivation and disturbance, or actively restored – by installing hedgerows or flower strips of attractive bee plants along field margins.

Principles and theory of wild bee conservation are addressed in Chapters 2 and 10 (this volume), as well as in sections 3.2, 8.2, 9.1.1 and 9.2.1 (this volume), leaving us in this chapter to focus more on its applied aspects.

© Keith S. Delaplane 2021. *Crop Pollination by Bees,* 2nd Edition, Volume 1 (K.S. Delaplane)
DOI: 10.1079/9781786393494.0006

6.1. Natural Bee Habitats

Osborne *et al.* (1991) ranked some natural habitat types in central and Atlantic Europe for their suitability for bees (Table 6.1). Their ranking is based on the nesting sites and plant types available in those habitats and illustrates some general features of habitats that have universal significance in bee conservation.

From Table 6.1 we can see that bee activity and reproduction are optimized in open, sunny habitats with an abundance and diversity of food plants. This is in contrast to flower-poor, shaded woodlands. Habitat conservation efforts for bees, therefore, places a premium on sunny, open undisturbed meadows, field margins, sun-drenched undisturbed patches of semi-bare soil (Fig. 2.7, this volume), roadsides (Fig. 2.6, this volume), ditch banks, hedgerows and woodland edges. Habitat must also provide nesting materials (mud, leaves, etc.) appropriate to the bee species. A shortage of mud could be a limiting factor with orchard mason bees (*Osmia* spp.), a group that uses mud in nest construction (section 9.3, this volume). Bumble bees need grassy thatch or abandoned rodent burrows in which to build their nests. If a bee sanctuary meets the general standards discussed here, appropriate nesting materials

Table 6.1. Ranking of some European habitats for their suitability for bee forage and nesting sites. From Osborne *et al.*, 1991.

Ranking	Atlantic Europe	Central Europe
1 (best)	calcareous (limed) grasslands heaths[a]	wet meadows
2	fens[b] hedges[c] wastelands neutral grasslands woodland edges	fens heaths fresh meadows
3	bogs[d] marshes wastelands	
4	oak woods ash woods moorlands[e]	oak woods alder woods some beech woods
5 (worst)	beech woods conifer woods	beech woods conifer woods

[a]Open, barren land with acidic, poorly drained soil and shrubby plants
[b]Lowlands covered wholly or partly in water, unless drained
[c]Dense shrubbery at field margins
[d]Wet, spongy, acidic soil with a characteristic flora
[e]Open, rolling, boggy lands dominated by grasses and sedges

will probably follow suit, but the matter should not be disregarded. Maintaining areas such as these in an undisturbed state will increase the abundance of bee nesting sites and diversity of flowering plant species on farms. The presence of such bee sanctuaries in the agricultural landscape of western Poland is one reason Banaszak (1992) gave for the constancy of bee diversity recorded there over 40 years.

6.2. Restored Bee Habitats

6.2.1. Plant lists

Bee conservation can go beyond habitat preservation, which is essentially a passive process, to active habitat improvement through the establishment of restorative plantings, typically hedgerows or flower strips. The goal is to provide a reliable source of quality nutrition which will serve to increase bee numbers, either by attracting bees to the area, increasing the number nesting in the area, or by increasing their reproductive output. Benefits of perennial pastures can be long lasting since wild bees tend to nest near to where they were reared the previous year (Butler, 1965; Osborne *et al.*, 1991).

The literature contains voluminous references to bee and flower associations. Much of it is intended to identify promising honey plants for honeybee keepers (Pellett, 1976; Crane *et al.*, 1984; Ayers *et al.*, 1987; Williams *et al.*, 1993; Villanueva-Gutiérrez, 1994). However, an increasing number of regionalized plant lists have been developed for bees, often focusing on native plant species (Williams *et al.*, 2015; Rowe *et al.*, 2018). The Xerces Society publishes recommended plant lists on a regionalized basis (available at: www.xerces.org/pollinator-conservation/pollinator-friendly-plant-lists, accessed 6 January 2021).

Conservationists are primarily interested in identifying assemblies of plants that are nutritionally useful to bees, easy to grow, cost-effective, non-invasive and bloom over a long period of time. Some researchers have identified bee forage plants for particular cropping systems such as apple (Heller *et al.*, 2019), blueberry (Blaauw and Isaacs, 2014), cranberry (Patten *et al.*, 1993), field bean and oilseed rape (Fussell and Corbet, 1991).

6.2.2. Importance of season-long bloom

The value of restorative bee plantings to local bee populations is optimized in those pastures that have a season-long succession of bloom. This principle is best illustrated with bumble bees. With an annual life cycle, the top priority of a bumble bee colony is to produce a new crop of mated queens for the following season. Each colony has only a few weeks to start a nest (as a solitary queen), rear a foraging force of workers, and collect enough food to produce new queens and males. The number of queens a colony can produce depends largely on the number of workers it can produce in the weeks leading up to the queen production period (Heinrich, 1979). Producing workers requires energy, so a colony's reproductive success ultimately hinges on season-long availability of food. The link between good nutrition and high queen output was underscored by Bowers (1986) who showed that new queens appear earlier in those meadows with the richest flower densities.

A dearth of mid-summer nectar can be disastrous. Again, bumble bees illustrate the magnitude of this problem. Unlike honeybees that store large surpluses of food, bumble bees store enough nectar for only a few days at most. This makes them vulnerable to nectar dearths. Worker bumble bees stop incubating brood and respond lethargically to invading predators and parasites in colonies that are experimentally deprived of nectar for even 1 day (Cartar and Dill, 1991).

6.2.3. Importance of native perennials as bee pasture plants

Although some annuals provide rapid and relatively abundant bee forage, native perennial herbs and shrubs are generally superior forages and preferred by native bees (Fussell and Corbet, 1992; Petanidou and Smets, 1995; Morandin and Kremen, 2013). Compared to annuals, perennials are richer nectar sources owing to their ability to store and secrete sugars from the previous season. Perennials provide bee populations with a dependable food source year after year and encourage repeated bee nesting in the area. This is important for explaining why the number of bee and plant species tends to increase together over time in undisturbed meadows. A focus on native perennials has the collateral benefit of helping conserve native plants and associated fauna (Fiedler et al., 2008). Considering the repeated labour and inputs required for annuals, perennials are a cost-effective, low-maintenance choice for growers installing restorative plantings (Carreck and Williams, 2002).

6.2.4. Importance of age and diversity of restorative plantings

The maturity of a bee habitat is expected to have a strong effect on the diversity of the foraging bee community and consequently on its pollination performance on near crops. Kremen et al. (2018) studied this matter directly with a long-term study in the Central Valley of California, an area rich in highly diverse agriculture ranging from conventional row crops to vineyards. The investigators chose five farm edges for restoration and ten to leave alone as control edges. They planted hedgerows, each about 350 × 3–6 m wide, in 2007 and 2008 with native woody shrubs and trees. In 2008 they planted nine 1×8 m flower strips with mixtures of nine native annual and perennial forbs. Pollinators were systematically sampled at all sites from 1 year pre-restoration until 2014; nine field seasons inclusive. In this 9-year data set it was floral diversity that was the direct driver, not the age of the restoration project. As hedgerows aged, they did accumulate floral species, and bee abundance responded positively to these changes, but increases in bee abundance levelled off as hedgerows matured, suggesting a saturation effect. Curiously, bee species *evenness* – the similarity in number of individuals of each species – decreased as flower diversity increased. This is thought to be an effect of accumulating rare bee species. The relationships among bee species diversity, increasing plant diversity and hedgerow age were nuanced. The expected positive relationship between plant and bee diversity did occur, but only in the early years when hedgerows were young and maturing. Once hedgerows are mature and saturated with plant species, it does not necessarily follow that the preferred plant species are the most abundant; nor does it follow that total flower numbers track with hedgerow age. Indeed, this is consistent with earlier observations that bee visitation tends to be higher in mid-succession plant communities (Dramstad and Fry, 1995).

Albrecht et al. (2020) performed a meta-analysis with 35 data sets from North America, Europe and New Zealand to appraise the relative effects of flower strips and hedgerows on pollination and pest control in adjacent crops. Flower strips and their harbourage of parasites and predators improved pest control in adjacent fields by 16% compared to control fields without flower strips. Pollinator performance was improved near flower strips or hedgerows and decreased exponentially with increasing distance from them; however, there was no overarching difference in pollination outcome between fields with or without restorative plantings. Plant species richness in the restorative plantings was positively associated with crop pollination performance but not with pest control. Crop pollination services tended to increase with time since establishment of the adjacent flower strip, but there was also a plateau effect as discussed above, such that the greatest gains in pollination performance were accomplished in the first 2 years. Ultimately, there were no significant differences in yield between crops with or without restorative plantings.

The studies highlighted here do not show a clear and consistent benefit from bee conservation efforts, yet they do treat the question at the largest scales, one at a scale of time (Kremen *et al.*, 2018) and the other space (Albrecht *et al.*, 2020). The strong results of Albrecht *et al.* (2020) for enhanced pest control with restorative plantings join other studies of similar scale (Gurr *et al.*, 2016) to support

the importance of taking steps to re-diversify intensified farmlands. However, the effects of restorative plantings on pollination performance are inconsistent when viewed categorically – that is, fields with or without restorative plantings. It is also true, however, that within that data cloud are components that are consistent. Pollination performance decreases exponentially with increasing distance from restored habitats. Plant species diversity is associated with increasing bee species richness and improving pollination performance. Increasing age of restorative plantings is associated with increasing plant diversity and improving bee pollination performance. There seems to be special value to restorative plantings at mid-stages of ecological plant succession. In sum, there are optimum returns on restorative plantings that are plant species rich, at mid-levels of plant succession, and very 'edgy' relative to area of focal crop. Albrecht *et al.* (2020) propose a dense spatial network of relatively small restorative plantings rather than a few large ones.

7

Honeybees: Their Biology, Culture and Management for Pollination

The western honeybee (*Apis mellifera*, family Apidae), a highly eusocial and adaptable species and the famous maker of honey, is also the most common managed pollinating bee in the world (Fig. 2.13, this volume). It forms large perennial colonies in hollow trees or other cavities, and it readily accepts artificial hives as domiciles. The honeybee's association with humans is not only ancient; it is *sympatric*, the two lineages having evolved in Africa together during synchronous epochs of natural history, beginning a little over 5 million years ago (Leonard, 2002; Kotthoff *et al.*, 2013; Cridland *et al.*, 2017). Each had its expansion out of Africa: *A. mellifera* beginning around 300,000 years ago (Wallberg *et al.*, 2014) into Europe, the Middle East and western Asia; and *Homo sapiens* by around 130,000 years ago (Reyes-Centeno *et al.*, 2014) ultimately into every habitable corner of Earth. In their earliest interactions, the relationship between the two was one of predator and prey, human on bees; I personally think there's no better extant evidence for our sympatry than the highly effective mass-stinging reaction of a disturbed bee colony. No lesser reaction would do, it seems, against the world's most dangerous predator. In time, however, predation by humans transitioned into husbandry by humans, the earliest signs of which are industrial scale use of beeswax in Israel by around 3500 BCE (Kritsky, 2015) and textual allusions to beekeeping from the First Dynasty of Egypt *c*.3150–2613 BCE (Ransome, 1937; Crane, 1999).

7.1. Bee Colony and Beekeeper Demographics

Today, according to data from the Food and Agriculture Organization (FAO) for 2018, the number of managed beehives in the world is 101,438,884 which, if one applies a modest multiplier of 30,000 as a 12-month average colony population, leads us to the staggering number of over three trillion managed honeybees on Earth. At a superficial read, it seems difficult to reconcile such a number against the gloomy prognostications of honeybee decline from many parts of the world. Yet, the overwintering losses given in Table 5.1 (this volume) are for managed bees, and one can be excused for thinking that these numbers should be better. When I was a boy growing up it was axiomatic that winter losses should be no higher than 10%. It is true that beekeepers can quickly make up winter losses by dividing surviving colonies, but this is not without cost as it requires inputs of new queens, feed and labour. It is this steady drumbeat of attrition that stresses beekeeping operations and raises alarm about the long-term sustainability of the beekeeping industry (see Chapter 5, this volume).

Official censuses of beekeepers are harder to come by, but their numbers are reported at 125,000–150,000 for the USA (Hoff and Willett, 1994), 35,000 for the UK (Carreck and Williams, 1998) and 7000 for Canada (CHC, 2020). In Europe there has been a downward shift in the numbers of beekeepers: whereas numbers were stable or even growing between 1965–1985, there was universal decline in the number of practitioners across all regions reporting between 1985–2005 (Potts *et al.*, 2010). Structural change is also happening in the US beekeeping industry, mirroring that for other agricultural sectors (Daberkow *et al.*, 2009). The number of beekeeping operations is declining while the size of remaining operations is increasing. The largest number of beekeepers is found in Appalachia, the Corn Belt, and North-east, whereas the Pacific, Northern Plains and Mountain states harbour the largest numbers of colonies.

DOI: 10.1079/9781786393494.0007

7.2. Honeybee Biology

In nature, western honeybees nest in rock crevices, hollow trees or other similar dry, hollow places. Worker bees secrete scales of beeswax from glands on the underside of their abdomens. They shape the wax scales into repeating, hexagonal cells (Fig. 7.1) that collectively make a comb. It is in these cells that food is stored and immature bees are reared. A natural bee nest contains up to ten combs.

The life cycle of a honeybee colony revolves around surviving winter or a similar dearth period and producing one or more new colonies early enough in spring to give the new colony time to collect food for next winter. Bees cluster together in a tight ball during winter to conserve heat, and the queen starts laying eggs in the centre of the nest in mid-winter when days start getting longer. Colony populations grow rapidly once natural *nectar flows* and *pollen flows* begin. By early spring, colonies are crowded with bees, and these congested colonies split and form new colonies by a process called *swarming*. A crowded colony rears several daughter queens, then the mother queen flies away from the colony accompanied by up to 60% of the workers. This *swarm* (Fig. 7.2) eventually occupies a new nest site, usually a hollow tree or similar void. Back at the original nest one of the daughter queens kills her rival sisters and inherits the colony. After the swarming season, the bees concentrate on storing honey and pollen for winter. By late summer, a colony has a central area full of brood and situated below layers of honey and pollen.

Due to this perennial life cycle a honeybee colony is potentially immortal, and nest sites tend to be occupied year after year. A colony may occasionally abandon a nest, a process called *absconding*, during a severe food dearth or if the colony is continually harassed by predators. However, even these abandoned nest sites are quickly reoccupied by new colonies because the odour of old nests attracts swarms.

7.3. Honeybees as Pollinators

Honeybees are generalists that visit a wide assortment of blooming plants during a season. They are manageable, movable, well known and effective pollinators for many crops; hence they are the standard against which all other bee pollinators are measured. However, because they are generalists, honeybees are not the most efficient pollinator for every crop. Unlike some solitary bees whose life cycles and behaviours are synchronized with a coevolved crop, honeybees play

Fig. 7.1. Hexagonal beeswax cells are built contiguously to form combs.

Table 7.1. Top 20 countries ranked by numbers of managed honeybee hives for the year 2018. From Food and Agriculture Organization (FAO), www.fao.org/faostat/ (accessed 6 January 2021).

Country	No. managed beehives	Country	No. managed beehives
India	13,048,275	Mexico	2,172,107
China	9,048,546	Korea	2,165,616
Turkey	7,947,687	Central African Republic	1,679,762
Iran	6,601,394	Romania	1,602,453
Ethiopia	6,018,223	Poland	1,586,063
Russian Federation	3,182,399	Greece	1,556,404
Argentina	3,020,370	Kenya	1,533,668
Tanzania	3,019,784	Angola	1,153,618
Spain	2,965,557	Brazil	1,017,506
USA	2,803,000	Serbia	914,134

Fig. 7.2. A reproductive swarm in process. The clustering mass of bees is slightly visible in the shadow of the foliage.

the field for the richest reward (Westerkamp, 1991). Their highly celebrated recruitment dance language, a marvel of evolution allowing a colony to exhaustively plumb the floral resources within its flight range, is nevertheless the bane of crop growers when that communication web permits a colony's foragers to abandon a focal crop in preference to a flowering weed 0.5 km away. Individual honeybee foragers tend to specialize on floral resources (Free, 1963), whether pollen, nectar or plant oils. This means that in the case of hybrid seed production which employs male-fertile and male-sterile lines, honeybee pollen specialists may rarely if ever visit male-sterile flowers, thus dramatically reducing their effectiveness at delivering cross-pollen.

There is evidence that honeybees, a social species, are able to compensate for the comparative inefficiency of their individual flower visits by the fact that a single colony fields hundreds or even thousands of foragers who effect multiple flower visits. This was shown, in principle, with a field study in which flowering blueberry bushes tented with bee colonies received increasing rates of flower visitation and corresponding increase in fruit-set from 25 to 79% as bee populations increased from 400 to 6400 bees (Dedej and Delaplane, 2003). In another context, honeybee colonies were inoculated with the parasitic mite *Varroa destructor* while another group was treated to be nearly mite-free (Ellis and Delaplane, 2008). Both groups of colonies were tented with flowering blueberry bushes. Based on single-bee flower visits, individual bees from mite-infested colonies were severely disadvantaged as pollinators (2.2% probability of setting fruit) compared to individual bees from mite-free colonies (33.2% probability of setting fruit); however, this difference disappeared when fruit-set was averaged by tent (colony). Fruit-set was 54.5% with mite-infested colonies and 43.9% in mite-free colonies. The conclusion was that mite-compromised foragers can still accomplish acceptable fruit-set as long as their colony fields multiple flower visitors.

A subsequent meta-analysis (Rollin and Garibaldi, 2019) of published research on the effects of honeybee density on crop yield elaborated on a recurrent recommendation: that importing high densities of pollinators, usually honeybees, is the answer to yield-limiting pollinator deficits. The investigators noted the diversity of measures employed in the literature for describing honeybee density: colonies per hectare, number of bees per tree, number of bees per tree per minute, number of bee visits per flower, number of bee flower visits per minute, number of bees per 100 flowers, or number of bees per 100 m². Despite this patent non-standardization, the investigators were able to detect signals that both colony density and visitation rates increase all productivity measures. The effect size is greater for visitation rates; however, the effect of visitation rates is non-linear, plateauing to a point beyond which subsequent visits are either unhelpful or even damaging to yield. The average optimum range of flower visits appears to be 8–10 visits per flower. This is one of the most interesting take-away messages of this study: the possibility for *too much* flower visitation – visitation optima beyond which yield decreases. Finally, the effect size of visitation rate is greater for crops with unisexual, rather than hermaphrodite flowers which is intuitive, given that pollen vectoring demand is greater when floral sexes are separated. These meta-results highlight that one-size-fits-all recommendations for industrial scale pollination are bound to miss important crop-specific idiosyncrasies of flower type and optimum visitation rates.

Honeybees can be practically bred for selected characteristics, including foraging behaviours. The availability of instrumental insemination brings this to a high pitch of precision, but such technology is expensive, difficult to learn and not widely adopted by beekeepers. It is possible to select for honeybees that preferentially collect a certain type of pollen, as shown in alfalfa (lucerne) (Nye and Mackensen, 1970). However, beehives used for pollination are normally cycled through many crops in one season, so it makes more sense to select for high pollen-hoarding strains rather than ones that prefer particular crops. Pollen foragers are generally more effective pollinators than nectar foragers (Vansell and Todd, 1946), probably because honeybee pollen foragers prefer inflorescences with relatively greater numbers of both male and female flowers (Gonzalez *et al.*, 1995). Fortunately, honeybees can be selected for high pollen-hoarding behaviour (Hellmich *et al.*, 1985; Gordon *et al.*, 1995).

The literature on honeybee pollination efficacy is far from equivocal. There are cropping systems and regions for which the presence of managed honeybees has been shown superfluous as long as abundant or species-rich assemblies of wild bees are present. This is the case for studies on apple in Wisconsin (Mallinger and Gratton, 2015), highbush blueberries in Michigan (Isaacs and Kirk, 2010), sweet cherries in Hesse, Germany (Holzschuh *et al.*, 2012), and tomato and pepper in New Jersey and Pennsylvania (Winfree *et al.*, 2008), to name a few.

The most comprehensive appraisal of honeybee pollination efficacy, however, was a highly coordinated global study with 600 sites and 41 cropping systems (Garibaldi *et al.*, 2013). Positive associations between fruit-set and wild insect flower visitation were universal in all 41 crop systems studied, compared to positive associations between fruit-set and honeybee flower visitation in only 14% of the systems studied. The effects of flower visitation by wild insects and honeybees appeared to act independently, suggesting that pollination by honeybees supplements rather than substitutes for wild insect pollination (but see section 7.3.1).

This seems the place to acknowledge that a kind of partisanship is detectable in the bee pollination scientific literature, and honeybees (and to a lesser extent, all managed bees) represent one side of it (Aebi *et al.*, 2012; Ollerton *et al.*, 2012). It is part of a necessary correction to a hegemony of honeybees that existed in 20th-century crop pollination science and praxis, especially in the USA where in 1991 the venerable Cornell professor of entomology, Roger Morse (1991), could state of honeybees, '[It's] pollination by brute force, and it works.' We have come a long way from acquiescing to such received orthodoxies, yet it is limited progress to trade one bias for another. The new literature slant is not so much detectable in the *evidence* presented as it is in the *titles*, where authors have opportunity to insert value-laden language that pitches the conclusions in a certain direction. creating and sustaining this counterproductive boundary. I object to this; it is antithetical to the purposes of science to carve up nature into domains and assign one value over another. Second, like all polarizing energies it reduces complex systems whose understanding requires time, nuance and context into simplistic caricatures of themselves. Finally, it is damaging to the pursuit of sustainable crop pollination for the 21st century – trafficking as it does in the fiction that one pollinator group is less desirable than another when the overwhelming sum of evidence shows that *large and taxonomically diverse local admixtures of pollinators are the key to optimized crop pollination*. What was said in the first edition of this volume (2000) remains true for today: we need all the pollinators we can get.

Let us admit that the main advantage of honeybees is their sheer numbers, manageability and diversified revenue streams. Their supreme manageability means that a grower can increase the number of pollinators in their orchard by orders of magnitude with one telephone call and overnight delivery. Not only does the bees' social structure permit a degree of compensation for their individual inefficiencies, but the presence of other bee species can also compensate for their inefficiencies (see section 7.3.1) and synergize with the honeybees to create highly effective pollination outcomes. Millions of inefficient pollinators are not such a bad thing. The fact that honeybees make honey adds a buffering income stream for beekeepers that helps stabilize the industry. For introducing crop pollination benefits at a farm scale in poorer parts of the world, it is often honey production that is the initial economic incentive.

7.3.1. Synergies with other bee species

Studies have shown cases in which honeybee pollination is augmented in the simultaneous company of other species of flower visitors. In California almond, one team of investigators (Brittain *et al.*, 2013) studied the flower visiting behaviour of honeybees (the dominant pollinator) in orchards with honeybees only and in orchards with honeybees plus naturally occurring non-*Apis* spp. bees. In orchards with non-*Apis* spp. bees, honeybees responded by switching almond rows more frequently, the effect being a higher rate of cross-pollen transfer among

varieties[i] and higher yield. The row-switching behaviour is believed to result from floral resource depletion or odour marks left by other species that provoke foraging honeybees to prolong their foraging flights and increase likelihood of crossing rows.

In commercial apple plantations in Upper Galilee, it was shown that honeybee pollination is enhanced when orchards are supplemented with hives of the European bumble bee *Bombus terrestris* (Sapir *et al.*, 2017). The benefit was most pronounced in the variety Gala. Overall, the benefit of supplemental *Bombus* spp. colonies was twofold: (i) the *B. terrestris* colonies foraged earlier in the morning and in more inclement weather conditions than their *Apis* spp. counterparts; and (ii) mirroring the results reported above for almond, the presence of *Bombus* spp. altered foraging behaviour of the honeybees, increasing their movement across rows of pollenizers and increasing their frequency of performing effective top-working[ii] flower visits, both of which can be interpreted as responses to increasing competition for floral resources.

In another example with apples, a team in Brazil found that orchard hive stocking rates of 12 hives for stingless bees per hectare and seven *Apis mellifera scutellata* hives per hectare provided higher fruit production than stocking rates of seven *A. m. scutellata* hives per hectare alone (Viana *et al.*, 2014). In sweet cherry in Belgium, increases in *Bombus* spp. abundance and species richness stimulated honeybees to increase their flower visitation rate and frequency of changing rows (Eeraerts *et al.*, 2020). In Maine, honeybee pollination efficacy in wild blueberry *Vaccinium angustifolium* was improved if the flower had been previously visited by bumble bees (Drummond, 2016). Apparently, the sonicating activity of *Bombus* spp. freed pollen for subsequent pickup by the non-sonicating honeybee.

With hybrid sunflower seed production in the Central Valley of northern California, a common problem with honeybees is that pollen specialists predominate at male flowers and nectar specialists predominate at female flowers. Neither situation promotes good cross-pollination. However, when wild non-*Apis* spp. bees are present in the plantation they increase the pollination efficiency of honeybees up to fivefold by modifying their behaviour (Greenleaf and Kremen, 2006b). When a honeybee on a male flower encounters a wild bee, she has a 20% chance of relocating to a female flower, whereas if she encounters another honeybee she expresses only a 7% likelihood of switching.

Apparently, the disturbance of a non-conspecific encounter disrupts the pollen specialist's orderly pursuit of exclusively male flowers. The investigators calculated that, remarkably, the indirect benefits of wild bees at promoting honeybee behavioural modifications were more important than the wild bees' direct contributions to pollination. The overarching conclusion was that interspecific mixes of pollinators are not only a good bid for direct pollination but a good strategy for improving the efficiency of the most abundant industrial scale pollinator.

7.3.2. Africanized honeybees and pollination

All New World honeybees are descendants of honeybees brought to North and South America by European colonizers beginning in the 1600s. Bees imported from Europe flourished in temperate areas of the New World, and within three centuries there were large sustainable populations of European honeybees in North America and temperate South America. However, European bees are not well adapted to tropical conditions, and to this day European honeybees do not prosper in the neotropics unless they are intensively managed by beekeepers.

Researchers imported queens of the African subspecies *Apis mellifera scutellata* from Africa into Brazil in 1956 in an effort to improve beekeeping profitability in the New World tropics. In contrast to their European predecessors, these African bees were well adapted to the tropical conditions of Brazil and began colonizing South America, hybridizing with and displacing European subspecies. Compared to the relatively gentle European bees, Africanized honeybees are very defensive. Large numbers of them may sting people and livestock with little provocation. They began spreading northward, and today most of South America, all of Central America, Mexico, and parts of the southern USA have established populations of Africanized honeybees.

The first naturally arrived colony of Africanized bees in the USA was found near Hidalgo, Texas in October 1990. By 1996, Africanized bees were present in parts of Arizona, California, New Mexico, Puerto Rico, St Croix and Texas, and by 2011 that list had expanded to include Arkansas, Florida, Louisiana, Nevada, Oklahoma and Utah (Szalanski and Tripodi, 2014). Their arrival in Georgia was marked with a tragic human stinging fatality (Berry, 2011), but aggressive monitoring

and trapping by the Georgia Department of Agriculture seems to have eradicated the offending colonies, and today this state is apparently free of established feral populations.

However, rather than being due to human heroics, it is more likely that the slowdown in Georgia is part of a wider trend noted elsewhere – natural ecological limits to range expansion. Consistent with its tropical legacy, *A. m. scutellata* does not seem to saturatively colonize temperate regions. There is a kind of symmetry between its southernmost limit in Argentina (34° S latitude; Visscher *et al.*, 1997) and its emerging limit in North America (35.5° N latitude; Kono and Kohn, 2015). Rather than encountering monolithic Africanized populations of the kind in South and Central America, US beekeepers are more likely to encounter an occasional 'hot' colony, a problem that can be cleared up by replacing its queen.

Nevertheless, to the extent the defensive phenotype expresses itself, these bees offer challenging situations compared to European honeybees. They exhibit extreme defensive (stinging) behaviour. Colonies may stay defensive for days after they are worked, endangering livestock, farm workers and non-involved bystanders (Danka and Rinderer, 1986). Africanized colonies have smaller forager populations (Danka *et al.*, 1986b); thus, on a per-colony basis, Africanized colonies field fewer potential pollinators. They do not retain large populations after being relocated. In Venezuela, 15 Africanized colonies and 15 European colonies were moved to six different crop sites over 2 months. Debilitating losses of adult bee populations were over twice as high in the Africanized colonies (Danka *et al.*, 1987). Africanized bees forage more closely to their nests (Danka *et al.*, 1993). Thus, Africanized colonies must be distributed nearer the focal crop (increasing sting hazard for farm workers) and more uniformly throughout the crop (increasing handling costs).

The tropical adaptations of *A. m. scutellata* in some situations can be advantageous relative to European bees, however. Africanized colonies field a higher percentage of foragers (Danka and Rinderer, 1986). Flower handling time is equivalent, as shown in cotton trials in southern Mexico (Loper and Danka, 1991). In sesame, they forage slightly more rapidly which may improve the rate of pollen dispersal (Danka *et al.*, 1990). Africanized bees are less susceptible to the insecticides azinphosmethyl, methyl parathion and permethrin (Ambush® or Pounce®) (Danka et al., 1986a). As they forage more closely to their hives, Africanized bees may be satisfactory pollinators in small restricted areas such as isolated fields used for growing hybrid seed (Danka et al., 1993). Africanized bees, especially in feral populations, have proven to be extremely effective and valuable pollinators of coffee in South and Central America where European honeybees have been ecologically unable to gain a foothold (Roubik, 2002). Africanized bees collect significantly more target pollen in hybrid sunflower plantations, as shown in a study from Argentina (Basualdo et al., 2000).

From the perspective of their impacts on native bees, the introduction and spread of A. m. scutellata cannot be considered an ecological success story, but neither has it been a catastrophe. As early as 1978 in French Guiana, it was noted that the advance of Africanized honeybees was associated with a decline in abundance of native meliponine stingless bees (Roubik, 1978), and subsequent studies experimentally demonstrated resource competition between the two groups (Roubik, 1980). However, independent studies have also shown equivocal results (Roubik, 1983), and a 17-year bee trapping regimen on Barro Colorado Island, Panama, including 10 years following the arrival of A. m. scutellata, detected no change in annual abundances of 15 native bee species (Roubik and Wolda, 2001).

When it comes to the effects of exotic A. m. scutellata as a pollinator, one approaches the question expecting evolutionary mismatches between native plant and exotic pollinator that render the exotic less efficient in this ancient transaction between angiosperm and bee. As with the literature on effects of Africanized bees on native bees, the record is mixed. Compared to the native meliponine Trigona nigra, A. m. scutellata in Mexico transferred 2.5× less pollen on a per bee visit to flowers of the native Arizona poppy Kallstroemia grandiflora; however, they also visited flowers 2.65× more frequently than T. nigra, leading investigators to conclude that the exotic bee's pollination performance was equivalent to that of the native (Osorio-Beristain et al., 1997). Turning to crop plants, investigators studied the pollination performance of A. m. scutellata on the indigenous tomato (Solanum lycopersicum) and habanero pepper (Capsicum chinense) in Yucatán, Mexico and revealed superior pollination performance in native bees Exomalopsis spp. (Apidae) and Augochloropsis spp. (Halictidae) based on fruit weight, number of seeds and fruit-set (Macías-Macías et al., 2009). Both native bees are solitary and capable of sonicating flowers (see Chapter 3, this volume), a flower-handling character state important for efficacious pollination in Solanaceous crops and others.

7.4. Simplified Beekeeping for Pollination

In some cases, a grower may decide to buy and maintain honeybee colonies for the sole purpose of pollination. This section explains basic beekeeping equipment, how to start colonies and some minimum required management practices necessary to keep colonies strong and healthy for pollination.

A grower's first step is to determine the density of beehives recommended for a particular crop (see Volume II of this work) and then decide if keeping one's own beehives is cost-effective. It is important to remember that the recommended density of beehives is high for commercial plantings, but this need not be a serious deterrent. By housing bees in good equipment, treating bees for parasites and feeding bees as needed, an average person can keep honeybees alive and reasonably productive with little other special attention. The concept of grower-owned honeybee colonies is not new. Commercial lowbush blueberry growers in Nova Scotia own and operate large apiaries to ensure honeybee availability, and increasingly the same is true for rabbiteye blueberry in the southeastern USA.

7.4.1. Basic hive parts and configuration

A honeybee *colony* is any single nest of bees containing combs, a queen and a supporting population of workers; the term can apply to both wild and managed bee nests. A bee*hive* is a man-made structure that contains a colony of bees (Fig. 7.3). In a standard Langstroth configuration, a beehive is made up of stacked boxes called *supers*. Each super contains 8–10 removable combs. Supers come in three common sizes – a deep super or *hive body* that is 24.1 cm tall, a *medium super* that is 16.8 cm tall and a *shallow super* that is 14.6 cm tall (US dimensions). The heavy hive body (60+ lb when full of bees and honey) promotes good production of *brood* (young developing bees), but the

Fig. 7.3. A typical Langstroth US-style beehive. The bottom box is a hive body and contains the queen and brood. The beekeeper is pulling a frame of honey out of a shallow super. Between the hive body and super (not visible) is a queen excluder, a kind of screen that restricts the queen to the lower hive body. The smoker behind the beekeeper is an important tool for calming the bees. The hive behind the beekeeper has a Boardman entrance feeder with a jar for feeding syrup. Photo courtesy of Pilar Delaplane.

Fig. 7.4. The frame in the top image is newly installed with black rigid plastic brood foundation. The inset shows detail. The foundation is embossed with the shape of the hexagonal cells. The bees 'draw out' the foundation into their three-dimensional combs. The bottom image shows a fully drawn comb covered with brood and bees.

two smaller sizes are much easier for an average person to handle. A hive has a *bottom board* on which are placed the supers. It is important to set the bottom board on concrete blocks or rails of steel or preserved wood in order to resist wood decay. Six or seven medium supers per hive should provide enough year-round space for bees, brood, honey and pollen.

The combs inside supers are each made of a wooden frame in which is inserted *foundation*, a sheet of beeswax or plastic embossed with the shape of hexagonal cells (Fig. 7.4). Bees use foundation as a framework on which they construct their combs. The wooden frame makes the comb strong, movable and interchangeable.

A lid goes on the topmost super. The simplest of these is a flat *migratory lid*. These are inexpensive and practical if one plans to pack hives tightly on trailers during moves (Fig. 7.5). All-weather laminate board is good material for migratory lids as it warps very little.

Hive equipment within countries tends to be standardized and generally interchangeable. Hives are not inspected as intensely with simplified beekeeping for pollination as they are for honey production, and this justifies increased attention to good initial hive construction. Exterior parts should be assembled using wood glue in the joints and galvanized nails or woodscrews as fasteners. By drilling pilot holes for nails or screws one greatly reduces the chance of splitting wood. Finally, all exterior surfaces (but not the interior surfaces) should be covered with a good quality exterior-grade paint.

Fig. 7.5. Each of these single hive body hives has a flat migratory lid. A beekeeper can use a circle saw to cut a hole in the lid to accommodate a feeding jar of syrup with a perforated lid. This is a fast and effective way to feed large numbers of colonies.

7.4.2. Other required beekeeping equipment

Other tools one will need for keeping honeybees include:

- *Smoker* (Fig. 7.3): the most valuable tool for working bees. A smoker calms bees and reduces stinging behaviour. Pine straw, dry grass and burlap make good smoker fuel.
- *Hive tool*: ideally shaped for prying apart supers and frames.
- *Veil, gloves and bee suit* (Fig. 7.6): protect the body from stings.
- *Feeders*: hold sugar syrup that is fed to bees when natural food supplies become low. Several types of in-hive feeders are available from online catalogues. Among the easiest to use is a simple quart or half-gallon jar with a perforated lid. A circle saw is used to cut an accommodating hole in the hive lid (Fig. 7.5).

7.4.3. Buying colonies

The simplest, and sometimes most economical, way to get started keeping bees is to buy established colonies from a reputable beekeeper. It is advisable to arrange to inspect the colonies before buying and to ask the seller to provide a recent certificate of health inspection. Buying bee colonies can be intimidating to the uninitiated, and one way to boost one's confidence and gain a measure of protection is to invite a government bee inspector or trusted and qualified acquaintance to inspect the colonies and offer an expert opinion on the colonies' condition. It is important to buy bees that are housed in standard equipment and to reserve caution over bees that are housed in equipment that is shabby and deteriorated. The quality of the woodenware may be a good indicator of the quality of care the bees have received.

Once the colony is opened, the bees should be calm and numerous enough that they fill most of the spaces between combs. Each super should have at least nine frames of comb. If the adult bee population appears adequate, then it is time to inspect the brood. Capped brood is tan-brown in colour. Young, uncapped brood is glistening, pearly white (Fig. 7.7). A good quality queen will have produced at least five or six combs of brood by mid-spring, and she lays eggs in a solid pattern so that there are few skipped cells.

Fig. 7.6. These members of the National Bee Unit of the Animal and Plant Health Agency, UK, are demonstrating full-coverage bee suits and working the UK national hive – similar to its US counterpart but square.

Fig. 7.7. Healthy young brood is glistening and pearly white.

If there is brood with perforated cappings, or if the larvae are tan, brown or black, the colony may have 'American foulbrood disease'. The deciding factor is the 'ropy test' which is done by taking a small stick or toothpick, inserting it in a suspicious cell of brood, mixing it up and withdrawing it. If the brood is dead from American foulbrood disease it will be stringy and 'rope out' up to 2.5 cm. This disease is very serious, highly contagious to other colonies and very difficult to control. It is also a reason to walk away from the potential purchase.

The parasitic mite *Varroa destructor* is widely considered the most serious honeybee parasite in the world. Regular miticide treatments are now a normal part of beekeeping, and beekeepers who are negligent on this count will have relatively stressed and unhealthy colonies. It is a good idea to enquire of a seller's mite treatment regimen. In warmer latitudes the most recent miticide treatment should have been within the last 6 months. This interval may be longer in colder latitudes. Optimum mite treatment intervals vary considerably across regions, and this is a matter where an outside expert's opinion could be useful.

Beehives are easiest to move during winter, when they weigh less and the bee populations are low. Moving hives requires at least two persons. It is best to move them at night when all the bees are in the hive. Hive entrances are usually closed with a piece of folded window screen, and supers are fastened to each other and to the bottom board with a strap or hive staples. Hives are then lifted on to a truck bed or a trailer, and securely tied down with strapping or rope. It is important to remember to open hive entrances after the hives are relocated. If the temperature is very cold and bees are completely inactive there is no need to delay the move until night; however, extra care must be taken to avoid dropping or jostling hives in freezing temperatures because the cold, brittle combs may shatter and the bees may not be able to reform a tight cluster.

7.4.4. Installing package bees

Another way to start keeping bees is to buy package bees and install them in new hives. This method is more costly initially, but one has the assurance of healthy bees with a known history and the assurance of equipment built to one's own standards. Bees are routinely shipped in 2–5 lb (0.9–2.3 kg) packages of about 9000 to 22,000 bees. Detailed instructions for installing package bees are available from government extension services and online videos, bee supply companies, and local and state bee associations.

7.4.5. Minimum hive management

Bee colonies used for pollination require a minimum necessary amount of care, or they simply will not survive. Beekeepers must monitor colonies closely to prevent starvation during times when nectar is not available. The colony's weight relative to others in the apiary can be estimated by tipping each hive from the back side to get an idea of its weight. Colonies probably need supplemental feed if they weigh under 50 lb (22.7 kg). It is best to feed the colony with a heavy syrup made of two-parts sugar to one-part water.

Protecting colonies from parasitic *Varroa* spp. mites (Fig. 7.8) is a top priority, even more so with the minimal management of the kind suggested here, where pollination, not honey production, is the goal. If one wants to keep colonies as free as possible from the depredation of the mites and the secondary viral diseases they vector, then two miticide treatments a year are recommended, one in late winter/early spring and another in late summer/early autumn. Miticides and their availability vary across time and government jurisdiction. All beehive medications and miticides are regulated by government agriculture or health agencies. It is important to check with your extension service or government inspectorate to make sure whether a

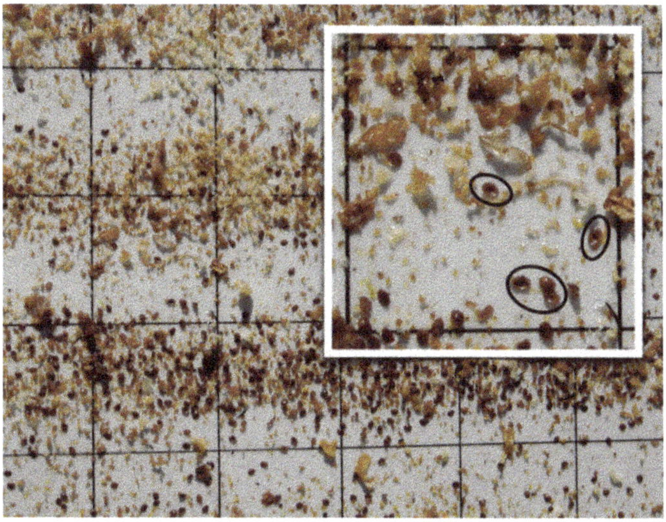

Fig. 7.8. Beekeepers check for *Varroa* spp. mites by placing a sticky sampling sheet on hive floors, beneath a screen to protect the bees. The insert shows five mites encircled, each of which is about the size of a sesame seed. This bottom screen has trapped hundreds of mites, a very severe infestation level for this colony. *Varroa* mites are widely considered the most damaging biotic threat to honeybees in the world today.

particular chemical is legal to use in your area. Disease and pest control technology changes rapidly, and it is helpful to join a beekeeping association or subscribe to at least one beekeeping publication to stay abreast of the most current management recommendations.

7.5. Managing Honeybees for Pollination

Crop growers who need beehives for pollination and the beekeepers who rent them have different agendas. Growers want bees on the crop during the critical pollination window, but they also want them removed soon after bloom is finished so they will not interfere with other tasks. Beekeepers want income from colonies, but they also are concerned about sting liability, insecticide exposure and keeping colonies strong for other uses. It is common for colonies to decline in strength while they are on crops that are poor sources of nectar and pollen. These motives underlie much of the negotiating when beekeepers and growers are making pollination agreements. Education, understanding and the use of contracts can help bridge these differences.

7.5.1. A good pollinating hive

Pollination brokers, state and provincial extension services, and government departments of agriculture have established minimum colony strength standards, but the enforcement of those standards ebbs and flows with the vagaries of government budgets. Nevertheless, published strength standards provide a good starting point for pollination contracts. The standards given here are consistent with most published standards, including those for the western USA (Sagili and Burgett, 2011).

In well-developed pollination markets, such as California almond, contracts and fee structures can be scaled to colony strength (Goodrich and Goodhue, 2016). Either party, the grower or beekeeper, may hire a third-party inspector who will sample 10–25% of the delivered colonies and determine the average 'number of frames fully covered by bees' – a positive indicator of colony health and strength, with eight being a common benchmark. With a fixed compensation contract, a grower may want to hire an inspector to verify that the benchmark was met. With incentivized contracts, the beekeeper is committed to a benchmark frame count, but for every frame above the benchmark she receives a bonus (usually capped), and correspondingly for every frame below the

benchmark she is penalized. With an incentivized contract, the beekeeper may wish to hire an inspector to document that the superior service was rendered.

In practice, the type of contract used is complicated by the unpredictability of honeybee overwintering mortality. As contracts are made months in advance, the beekeeper needs to anticipate overwintering mortality levels. Almond bloom is early in the season (Feb–Mar in the northern hemisphere), with little or no time to make up losses. For such contingencies, there is a third type of contract for 'field run' colonies, representing the bottom end of the fee range, stipulating neither strength standards nor inspections.

Nowhere is the honeybee health crisis more acute than with this tension of early season demand for top-strength colonies. It is true that beekeepers can split colonies to make up for overwintering losses of the kind shown in Table 5.1 (this volume), but it is a sore challenge to do so in time for California almond, arguably the most important revenue opportunity of the year.

It is normal for beekeepers to add supers to hives to accommodate growing bee populations and honey stores. Thus, a very tall hive is probably strong. A grower should not rely on external appearances, however. It is reasonable for the grower to ask the beekeeper to open a few hives for random inspection, but the grower must know what to look for. When the hive lid is removed, bees should immediately 'boil over' and blanket the tops of 6–10 frames (Fig. 7.9). Bees do not well up so dramatically in colonies with small populations (Fig. 7.10). There should be enough adult bees to fully cover 6–10 combs; 4–6 of those should be well filled with brood. When brood are young, it is possible to see glistening white larvae in their cells. Older brood are covered with cardboard-coloured wax cappings. Bees are best motivated to collect pollen, and hence are more efficient pollinators, when they have young, uncapped brood. A colony in two hive bodies usually meets these minimum strength criteria. Single-storey hives can make good pollinating units, but with singles it is important to make sure they have enough bees and brood. Very strong colonies are superior pollinators, and the beekeepers who provide them should get premium rental fees.

7.5.2. Moving hives

Beehives are moved at night when the bees are not flying and temperatures are cooler. Smaller operators

Fig. 7.9. A strong colony in a ten-frame hive body with bees blanketing the tops of almost all the combs. Photo courtesy of Jack Garrison.

Fig. 7.10. A comparatively weaker colony than that shown in Fig. 7.9. Photo courtesy of Jack Garrison.

will screen hive entrances individually and manually load hives on and off a trailer (Fig. 7.11). In large-scale operations, beehives are palletized (Fig. 7.12) and loaded with a forklift on to a trailer or flatbed truck; in these cases, the beekeeper may net the entire truckload to contain flying bees. In either case, the grower should be prepared for night-time arrival and arrange details in advance with the beekeeper about field access and hive placement.

7.5.3. Timing

It is best to use honeybees that are inexperienced at foraging in the area around the crop of interest. That way, upon delivery to the orchard the bees will immediately begin working on the crop because they have not yet discovered other more attractive plants blooming in the area.

To gain the benefit of inexperienced bees, it is necessary to move hives into the crop after it has already begun flowering a little. If the colonies arrive before the crop starts blooming, there is a

strong likelihood that the bees will learn to forage on non-target plants such as dandelion. Once bees are trained to such competing flowers, they may ignore the crop when it blooms. In Israel, honeybee pollination of Red Delicious apples was significantly increased when half the hive stocking rate (2.5 hives/ha) was delivered at 10% bloom and the remainder (another 2.5 hives/ha for a total of 5 hives/ha) at full bloom (Stern *et al.*, 2001). Volume II of this work has information on timing recommendations for specific crops.

7.5.4. Irrigation and bee activity

Overhead irrigation decreases the bee foraging rate for nectar and, possibly, for pollen (Teuber and Thorp, 1987). Open flower designs, like those found in melons and cucumbers, can fill up with water and lose their attractiveness to bees. As much as possible, growers should avoid irrigating during bloom times and during daylight hours when bees are pollinating (Fig. 7.13).

Fig. 7.11. Sixteen trailered colonies pollinating a fruit orchard. Photo courtesy of Dewey Caron.

7.5.5. Recommended bee densities

Research confirms that the probability of having honeybees in one's crop is correlated with the presence of managed hives on the farm (Mallinger and Gratton, 2015). The ideal number of hives per hectare, however, depends on the attractiveness of the crop, the population density of non-managed bees, the abundance of competing natural nectar sources, strength and location of beehives, weather and the grower's experience. The goal is to use the minimum hive density that provides a maximum crop yield. Generally, any existing factor that reduces overall pollination efficiency (unattractive crop, few non-managed bees, many competing nectar sources, poor weather, etc.) calls for an increased rate of imported pollinators to compensate for the natural deficiency. A grower can generally consider 2.5 hives/ha as a starting point in the decision making process, then use more or fewer according to the advice of crop consultants or extension specialists. Chapters in Volume II of this work provide recommended bee densities for specific crops.

7.5.6. Hive placement

It is helpful to orientate hives so that entrances are exposed to early morning sun. This stimulates bees to visit flowers early, and pollination occurring early in the day is important in many crops. Hives should be placed on knolls or high ground, and never in low areas which are prone to accumulate cool, damp air. Bees in chilly, shaded conditions are comparatively less motivated to fly. Hives should be protected from strong winds with bales of straw, hedges or similar types of wind barriers. As much as possible, hives should be located away from farm workers, pedestrians and livestock. Likewise, hives should not be placed near dwellings nor

Fig. 7.12. Inspecting palletized colonies for pollinating sunflowers. Photo courtesy of Dewey Caron.

irrigation valves. It is important that the bees have a source of water, especially during summer drought conditions. The beekeeper can provide this by placing open containers of water near the hives. Floating wood or polystyrene chips in the containers will help prevent bees from drowning.

It is in the grower's best interest to have hives placed as close as possible to the focal crop. The number of bees visiting the crop, amount of pollen collected and crop yields predictably increase the closer the hives are to the crop, as shown in alfalfa (lucerne) (Bohart, 1957), almond (Cunningham *et al.*, 2016), faba bean (Cunningham and Le Feuvre, 2013), oilseed rape (canola; Manning and Wallis, 2005) and red clover (Bohart, 1957; Peterson *et al.*, 1960; Alpatov, 1984), to name a few.

Foraging distances honeybees will undertake depend on richness of floral resources and complexity of landscape. Here and elsewhere, 'simple' landscape means visual monotony with large patch sizes and small proportions of semi-natural habitats, while 'complex' landscape means smaller patches, more varied land types and consequently more transition zones between habitat types. In Lower Saxony, Germany, average foraging distance for pollen foragers was longer in structurally simple landscapes (1743 m) over structurally complex landscapes (1543 m) (Steffan-Dewenter and Kuhn, 2003). Pollen forager distances also decreased during months when rich floral resources were available, a trend noted by others (Beekman and Ratnieks, 2000). The principles at work here are that the behaviour of pollen foragers is of special interest given their general superiority as pollinators, and short foraging distances – desirable energetically for the bees and conducive to crop visitation – are optimized when hives are in complex habitats with rich forages, hopefully representing the crop itself.

Using much smaller optimum foraging distances, Levin (1986) applied these principles to make

Fig. 7.13. Palletized bee colonies underneath drip irrigation in carrot. Photo courtesy of Dewey Caron.

recommendations for placing hives at a focal field. Assuming that honeybees prefer to work within 92 m of their colony, Levin proposed that by putting hives at 153-m intervals throughout a field, one can place a whole field within ordinary bee foraging range (Fig. 7.14); however, this is not always practical. If the interior of a field is inaccessible, one can group hives around the edges. In these cases, the centre of the field is less likely to be visited by bees, but the beekeeper can offset this problem by putting more colonies in central groups along the field edge; this increases competition and forces bees to forage deeper into the field (Fig. 7.15).

Hive dispersal can be a point of difference between growers and beekeepers. For the beekeeper, dispersing hives throughout a field is labour-intensive, but it is desirable for the grower to have well-dispersed hives. It is possible that a hive's microclimate is more important than its location in the field. Hives are better located in sunny, wind-protected sites along

field edges than in low, cool spots in field interiors. The degree of hive dispersal, and who will do it, should be clearly worked out in a contract with the rental fee adjusted accordingly. Sometimes beekeepers simply deliver hives at a central location and it is the grower who distributes them with her own equipment.

7.5.7. Non-crop or 'competing' bloom

The idea of removing wild plants whose bloom coincides with the crop is a discredited vestige of the 'honeybees forever' paradigm (Morse, 1991). There are records where honeybees are indeed diverted from the focal crop by nearby bloom. In one study, honeybees ignored apple blossoms and instead concentrated on dandelion that was growing in the same orchard (Mayer and Lunden, 1991). Honeybee activity in alfalfa (lucerne) and subsequent seed-set were higher after nearby

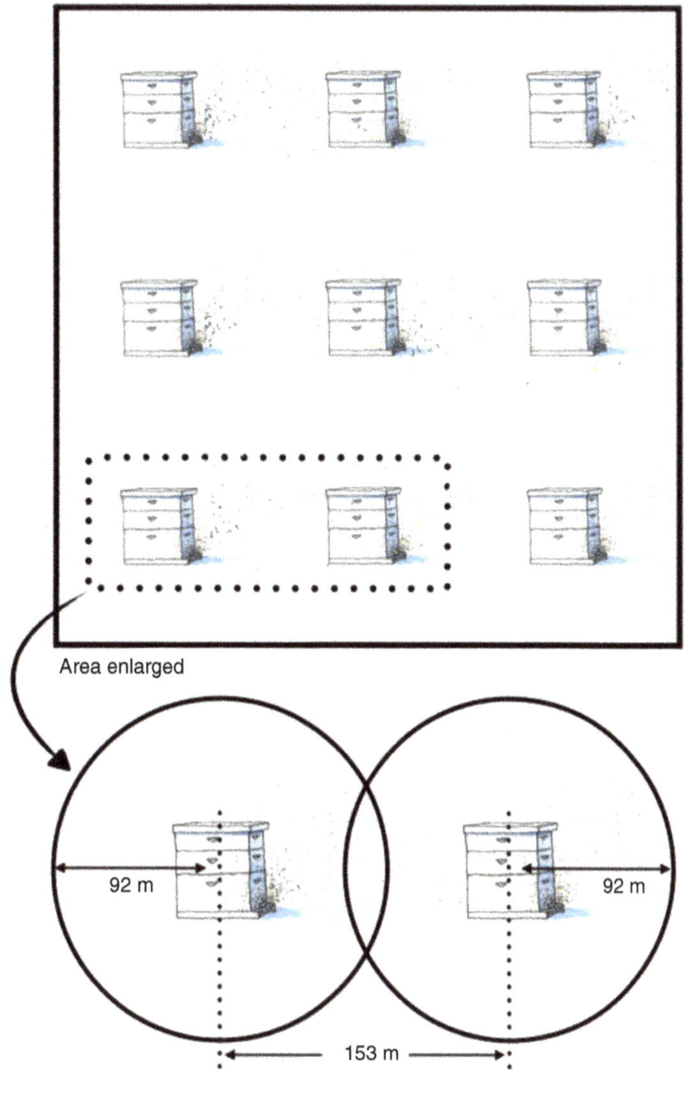

Fig. 7.14. A saturation situation in which hives are spaced within the field at 153-m intervals so that their 92 m optimum foraging radii overlap. Redrawn from Levin, 1986.

flowering mustard was mowed (Linsley and McSwain, 1947). Honeybee foraging on a focal watermelon crop in Georgia, USA, nearly ceased when peripheral sunflower began blooming on the same farm (Ellis and Delaplane, 2009). However, an abundance of flowering plants can also attract foraging bees to an area, encourage a taxonomically diverse bee community near the crop, and improve the reproductive output of bumble bee colonies (see Chapter 6, this volume). Bee visitation

(including wild bees) and fruit-set in sweet cherry was unaffected by flowering ground vegetation (Holzschuh *et al.*, 2012), and visitation rates by honeybees to focal oilseed rape was higher in landscapes rich in alternative foraging habitat, refuting the notion of 'competing bloom' altogether (Woodcock *et al.*, 2013). Recent thinking is clearly in favour of enhancing, not eliminating, alternate bee forage in the interest of building up diverse populations of pollinators.

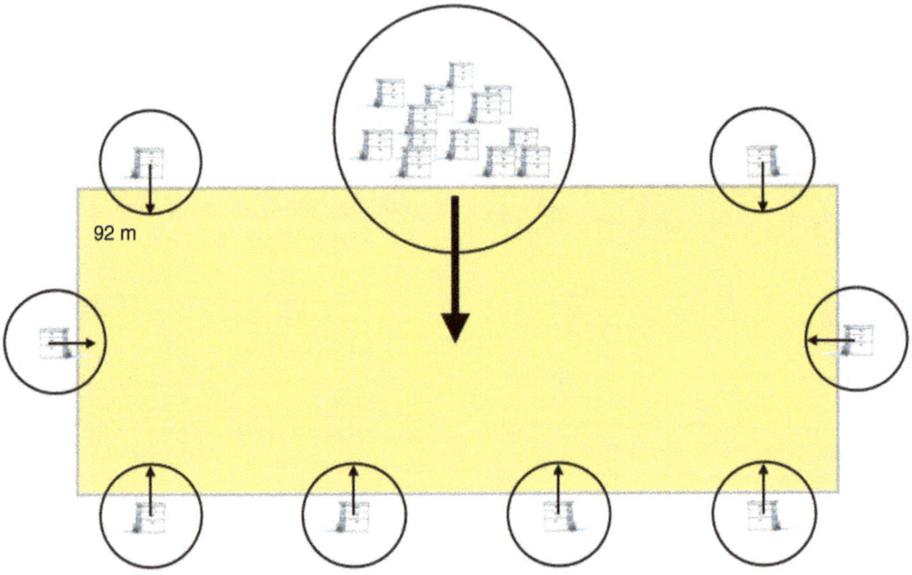

Fig. 7.15. It is always preferable to disperse bee hives throughout a crop. But if this is not practical, a beekeeper can space hives around field edges, including one or more clusters of a larger number of hives near the middle of the field edge. Competition is expected to force bees from this cluster (centre) to forage further from their hives and pollinate the inaccessible field interior. Redrawn from Levin, 1986.

7.5.8. Pollen or biocontrol dispensers

Pollen dispensers are devices that fit at the entrance of beehives and hold pollen of desirable pollenizer varieties in such a way that the bees dust themselves with the pollen as they leave the hive (Hatjina, 1998). Dispensers possibly stimulate bees to forage for the pollen type in the insert (Lotter, 1960) which, if true, would certainly be an argument in their favour. Part of the rationale for using hive entrance pollen dispensers is the fact that bees inadvertently exchange pollen as they jostle against each other inside the hive. Departing foragers carry a variety of pollen grains on their bodies, not all of which is from the plant species they are currently working (DeGrandi-Hoffman *et al.*, 1984). The concentration of target pollen could conceivably be increased if departing foragers are forced to wade through target pollen, collecting it on their bodies as they leave to visit the crop.

In one study with apricot in France, hives fitted with entrance pollen dispensers failed to cause a comparative increase in fruit-set over controls. Moreover, the investigators noted that departing foragers coated with pollen after wading through the dispenser tended to immediately pack it into their corbiculae for food, rendering it unavailable for pollination (Vaissière *et al.*, 2001).

Pollen dispensers may be most effective when orchard layouts are not well designed for providing compatible cross-pollen with pollenizer varieties. Inserts are considered advisable with old solid-block orchards that have no pollenizers planted nearby (Mayer and Johansen, 1988). A study in California almond (Dag *et al.*, 2000) found that pollen dispensers supplied with pollenizer pollen increased fruit-set in an orchard with suboptimal pollenizer bloom overlap with the main variety and a 1:2:1 planting design (pollenizer on either side of two rows of Nonpariel), compared to an orchard with good bloom overlap and a 1:1:1 planting design. Interestingly, the benefit of dispensers was greatest in the suboptimal orchard on the two Nonpariel row sides facing each other.

Hive inserts may also open the possibility for using bees to deliver biological pest control agents in crops (Fig. 7.16), an emerging technology called *entomovectoring* (Maccagnani *et al.*, 2020). In this manner, honeybees have been used to deliver bacteria antagonistic to 'fireblight disease' in apple and pear (Thomson *et al.*, 1992); a beneficial virus to

Fig. 7.16. A hive entrance biocontrol agent dispenser. This device contains a bacterium antagonistic to mummy berry disease in blueberry, diluted with talcum powder. Bees pick up the agent on their bodies and deliver it directly to the blueberry flower. Photo courtesy of Selim Dedej.

control defoliating caterpillars in clover (Gross *et al.*, 1994); fungi antagonistic to *Botrytis* disease in strawberry (Kovach *et al.*, 2000); and bacteria antagonistic to 'mummy berry disease' in blueberry (Dedej *et al.*, 2004). Efforts are under way to expand this technology to other bee vectors such as bumble bees and *Osmia* spp. (Maccagnani *et al.*, 2020).

7.5.9. Pollen traps

Pollen traps are devices attached to the entrances of beehives that are used to harvest pollen loads off foraging bees. It has been thought that pollen traps may induce a pollen deficit in the colony, thus increasing the proportion of bees foraging for pollen and thereby increasing the effectiveness of the colony as a pollination unit. Webster *et al.* (1985) tested this hypothesis with honeybee colonies in almond and plum orchards and determined that colonies with pollen traps had higher proportions of foraging bees with pollen loads than did colonies without traps. This potential benefit was partly offset by a decrease in brood production in colonies continuously fitted with traps, but this is not a repeatable effect. Pollen traps have been shown in other situations not to decrease brood production (Goodman, 1974).

7.5.10. Honeybee attractants

The use of pheromone-based honeybee attractants experienced a heyday in the 1990s, but promising results in experiments could not be translated into consistent benefits for growers. The companies innovating with the technology two decades ago have since been bought out, and their product lines are no longer available. Honeybee attractants are designed to increase bee visitation to treated crops with the goal of increasing pollination, fruit-set, yield and ultimately profits. These attractants are mixed with water and applied to crops with conventional spray equipment.

Several attractants have been marketed, and most have had doubtful performance records. The best were formulated around synthetic *pheromones* – 'external hormones' that insects secrete to regulate the behaviour and physiology of other individual insects. Honeybees have a rich battery of pheromones.

In general, attractants are warranted only when conditions are suboptimal for pollination or when the crop is not attractive to bees. The idea is to focus bees away from competing bloom, improve their efficiency when foraging conditions are poor, or to improve their efficiency when their numbers are low relative to the amount of bloom needing pollination.

Bee attractants encourage bee *visitation*, not necessarily bee pollination. If the flowers are not appealing to bees, no chemical attractant will make bees work them. Likewise, if there are no bees in the area, an attractant will not draw them in from great distances. A grower's first priority must be the bees themselves.

The most promising avenue for bee attractants was focused on synthetic honeybee *queen mandibular pheromone* (QMP). Although the existence of QMP was known since the 1960s (Butler and Fairey, 1964), all of its components were not characterized until over 25 years later (Slessor *et al.*, 1988, 1990; Kaminski *et al.*, 1990). Researchers and industry synthesized and developed QMP into commercialized products, one of which was named Fruit Boost®. The mechanism by which QMP enhanced crop pollination was by stimulating greater bee recruitment to treated plots and by stimulating individual foragers to stay in treated plots longer and visit more flowers (Higo *et al.*, 1995).

QMP-based bee attractant increased honeybee visitation in pear varieties Anjou and Bartlett in Washington and British Columbia. The attractant also increased bee visitation in Red Delicious apples in British Columbia, but did not affect yield nor fruit quality (weight and diameter).

However, the attractant increased fruit diameter in pear which translated to a US$1055/ha increase in farmgate revenue (Currie *et al.*, 1992b). In a later study, QMP-based attractant increased fruit size in Anjou pear by 7% which translated to a US$400/ha increase in farmgate revenue; however, the attractant did not increase bee visits, fruit-set, nor fruit size in Bing sweet cherry (Naumann *et al.*, 1994).

QMP-based attractant increased bee visitation in cranberry varieties Crowley and Stevens and in the highbush blueberry Bluecrop (Currie *et al.*, 1992a). Maximum attractiveness to bees was achieved in cranberry with a concentration about ten times less than that for blueberry or for the apple and pear data given by Currie *et al.* (1992b). This suggests that more attractant is needed for a three-dimensional surface (i.e. bush or tree crops) than for a flat surface (i.e. a cranberry bog). Bee flight conditions were poor in the first year of the cranberry study and the attractant increased yield 41% and farmgate revenue by US$8804/ha. In the second year, weather conditions for pollination were excellent and the attractant did not improve yield nor revenue. In two out of 3 years of highbush blueberry trials (Currie *et al.*, 1992a), QMP attractant increased fruit yield by at least 6% and farmgate revenue an average of US$900/ha.

In 2009 my student Amanda Ellis (Ellis and Delaplane, 2009) tested Fruit Boost® under conditions that would seem to epitomize the rationale for using honeybee attractants – a situation of floral competition. We were testing whether Fruit Boost® would improve honeybee flower visitation and harvest parameters in seedless Sugar Heart watermelon when several plots of sunflower on the same farm began blooming simultaneously. Over the 3-week course of field observations, the sunflowers began blooming in the second week, provoking a rapid decline in honeybee visits to watermelon. Fruit Boost® was not able to stop the exodus: numbers of honeybee visits to female flowers – potentially effecting fruit-set – were identical between plots treated, or not treated, with Fruit Boost®. Accordingly, there were no significant differences in fruit-set nor melon weight between plots treated, or not treated, with Fruit Boost®.

Notes

[i] Almonds are planted with alternating rows of compatible pollinizer varieties to provide cross-pollen. Bees tend to forage at successive plants along a row, but cross-pollination is maximized when bees cross rows.

[ii] Honeybees are notorious for 'side-working' apple flowers – accessing the floral nectaries from the side of the flower while avoiding contact with the sexual column, which reduces cross-pollinating potential of the flower visit.

8 Bumble Bees: Their Biology, Culture and Management for Pollination

The non-parasitic bumble bees (*Bombus* spp., family Apidae) are large, hairy bees whose species express simple eusociality in which a single mated female overwinters and emerges in spring to found a new nest, relinquishing foraging duties only after she has accumulated a critical mass of workers. The species of *Bombus* occupy a vast range in geography and habitat from alpine meadows at 5600 m in the Himalayas to lowland tropical forests in Amazonia (Williams, 1985). Although the *Bombus* spp. are geographically diverse as a group, their ecological abundance is concentrated in the northern temperate latitudes where physiological adaptations for cold tolerance have enabled them to exploit niches too extreme for other bees (Heinrich, 1993).

8.1. The Genus *Bombus*

Bombus has at least 54 recognized species in North and Central America (Michener *et al.*, 1994), over 250 species worldwide, and 38 subgenera (Williams, 1998). The species of one of these subgenera, *Bombus* (*Psithyrus*), are 'cuckoos', or social parasites of other bumble bees. *Bombus* (*Psithyrus*) spp. have lost their worker caste and produce only reproductive males and females. A female will invade a nest of social *Bombus*, kill or subdue the resident queen, and trick the colony's workers into caring for her and her own brood. *Bombus* (*Psithyrus*) spp. are considered pests in commercial bumble bee production for pollination.

Another subgenus, *Bombus* (*Bombus*), is distinctive for representing four of the five bumble bee species used worldwide for commercial pollination (Velthuis and Van Doorn, 2006). One of its members, *B.* (*B.*) *terrestris*, has been intentionally exported outside its native range into other countries including Argentina, Australia, Chile, UK, Japan, Jordan, Korea, Mexico, Morocco, New Zealand, Saudi Arabia, South Africa, Taiwan, Tunisia and Uruguay

(Donovan, 1980; Yoneda *et al.*, 2008; Dafni *et al.*, 2010; Goulson, 2010). In North America, where bumble bee declines have tracked similar reductions in the UK, the heaviest declines seem to be concentrated in members of the subgenus *Bombus* (*Bombus*), a situation possibly connected to the transport between 1992–1994 of North American bumble bees to European rearing facilities, where the bees were housed alongside *B.* (*B.*) *terrestris*, potentially exposed to European pathogens, and their colonies shipped back to North America and released (Flanders *et al.*, 2003; Winter *et al.*, 2006).

Bumble bees today occupy a niche in agricultural pollination similar to that of honeybees – each occurs as highly managed cultured colonies and as wild colonies in the field. Bumble bees are distinctive, however, for their pre-eminence as pollinators in the greenhouse environment.

8.2. Bumble Bee Biology

Bumble bees are social bees with an annual colony life cycle (Fig. 8.1). They first pass through a solitary phase (as single queens) before reproducing into a colony of numerous, social individuals. This is contrasted with the perennial honeybee colonies that never have a solitary phase.

The bumble bee life cycle begins with a young, mated queen that overwinters in isolation in some dry, safe harbourage in the ground or under loose tree bark. In spring she becomes active, foraging for early nectar and pollen by which she builds up energy reserves for brood production. She seeks out a nest site in such places as grass thatch, piles of hay or abandoned rodent nests. She must choose a dry, well-drained site that will be safe from flooding.

The queen builds a thimble-shaped beeswax *honey pot* in which she stores nectar. Nearby she forms a lump of field-collected pollen, excavates a depression in it, and lays one or more eggs in the depression, covering it with wax. The eggs hatch

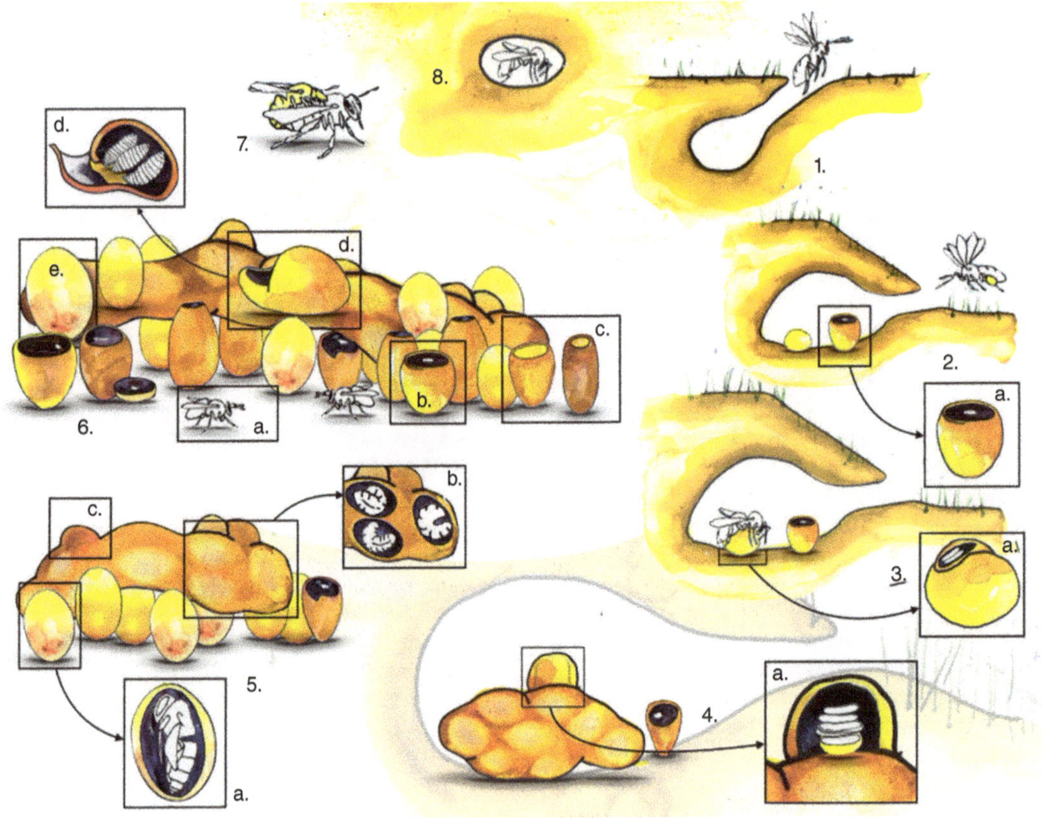

Fig. 8.1. Generalized colony life cycle of *Bombus* spp. The overwintered queen emerges from her hibernaculum in early spring (1) to forage and search for a nest site, often an abandoned rodent burrow in thatch (2) in which she makes a ball of pollen, dampening it with nectar. She builds a wax honey pot near the entrance (2a) in which she stores nectar. The queen deposits a few eggs on top of the pollen ball (3), covers it with a wax seal (3a) and incubates it with her body heat. The larvae consume the pollen ball and the queen alternates time between incubating and foraging for more pollen, enlarging the brood/pollen mass as the larvae grow (4). She starts another brood clump on top of the existing brood clump, making another pollen ball, laying eggs on it and covering it in wax (4a). Soon, three cycles of immature forms are developing simultaneously, together producing an irregular comb (5), with the first clutch entering the pupal phase (5a), the second still larvae (5b), and the third represented by one or more new brood clumps of eggs (5c). Once workers start emerging (6a), the colony expands rapidly. Larvae can be oriented in any direction, but pupation is usually in a vertical posture. Once larvae spin their silken cocoons, workers remove wax from the cocoon to use it elsewhere. After adults emerge from their cocoons, the empty cells are repurposed for storing nectar (6b) or pollen (6c). The so-called 'pollen-storer' species store pollen in pots (6c) until it is fed to larvae; workers open the wax seal of a brood clump, insert the pollen and close up the wax cover. The 'pocket-makers' build open pockets on the sides of brood clumps into which they push pollen where the larvae can access it. Over successive brood cycles the colony may achieve a population of a few tens to hundreds of individuals. By mid- to late summer, the colony begins increasing feed to a few select female larvae who grow larger (6e) and are destined to be next year's queens. The colony mother stops producing workers and the colony begins a slow decline. Males are produced about this time and mate with the young queens (7). Eventually, the old queen dies along with the workers and males, and newly fertilized females burrow into some shelter to pass the winter in solitary hibernation (8). The only connection to the next generation is the newly fertilized young queen and the sperm in her spermatheca.

and the young larvae feed on their bed of pollen, and as they grow the queen opens the wax covering and adds more pollen and nectar. When she is not foraging, the queen perches on this *brood clump*, incubating the larvae to speed their development. As the larvae mature, each spins a cocoon of silk in which it pupates and completes development into an adult. After new workers emerge, their empty cocoons are used as storage pots for honey or pollen. More pollen lumps with eggs are deposited alongside or on top of the old ones, and thus the irregular comb grows (Fig. 8.2). Eventually, there are enough workers to do the foraging and housekeeping tasks so that the queen can concentrate on laying eggs. Colony population peaks at a few hundred individuals (Sladen, 1912; Heinrich, 1979).

Bumble bee species fall roughly into two camps depending on the mode in which pollen is stored and fed to larvae (Sladen, 1912). The *pollen-storers* collect pollen into repurposed cocoons (Fig. 8.1, 6c) and regularly feed larvae from their stores by opening up the wax covering of the brood clump, inserting pollen and closing the wax cover again. Their feeding strategy is therefore called *progressive*-provisioning, as workers adjust food constantly according to the growth stage of larvae. In contrast to this we have the *pocket-makers* who build one or more pockets on the side of a brood clump (Fig. 8.1, 6d) into which they push large masses of pollen on to

Fig. 8.2. *In situ* natural colony of *Bombus auricomus*, a common species of eastern North America. The downy nest material is evident, as is an open honey pot (bottom left). Most of the wax has been removed from the vertical pupal cells, revealing their white silken cocoons. Photo courtesy of Sydney Cameron.

the cell floor upon which the larvae directly feed. This mode is called *mass*-provisioning. The greatest size divergences between workers and queens is found in the pollen-storers, which has been thought to show that workers regulate caste differentiation by directly controlling the amount of food given to larvae (Pendrel and Plowright, 1981). However, later work seems to indicate that caste fate is determined early in a young queen's life and that subsequent worker feeding regimens are a result, not the cause, of this differentiation (Pereboom *et al.*, 2003). Finally, there is evidence that a larva's location in the nest dictates the amount of food and attention it receives, with larger larvae in the nest centre and smaller larvae at the periphery; thus location may serve as a first condition for fixing an individual's caste fate (Couvillon and Dornhaus, 2009). In any case, queen-destined larvae are fed more frequently in both pocket-makers and pollen-storers. More immediately to our purposes, the majority of bumble bee species that have been successfully cultured for commercial pollination are pollen-storers (Evans *et al.*, 2017) and comparatively short-tongued (see section 8.3; Carnell *et al.*, 2019).

The colony switches from producing workers to producing males and new queens sometime in mid- to late summer. Males leave the nest a few days after emerging. The new queens linger in the nest a while longer, eating food collected by workers, foraging for their own food and occasionally foraging for the colony. The new queens mate and seek out a suitable overwintering site. Males, workers and the old queen die before winter.

The entire season's activities – nest founding, worker production and food collection – are all aimed at colony reproductive success – the mid- to late-summer production of next year's queens. Although parasite intensity plays a role (Schmid-Hempel and Durrer, 1991; Goulson *et al.*, 2018), the main driver of reproductive success is thought to be season-long constancy of floral resources (Bowers, 1986). Constant bloom fuels production of workers who in turn produce queens, an association that hearkens to the famous analogy of Oster's and Wilson's (1979) that the social insect nest can be thought of as a factory with workers as the machinery and reproductives the product. The relationship between habitat resources and reproductive success, however, can be nuanced and indirect. Williams *et al.* (2012) in their analysis of

39 mixed agricultural/natural landscapes in northern California found a positive and direct effect of floral resources on numbers of workers and males, but not on queens. Instead, queen output was positively associated with worker numbers, so the effect of floral resources was indirect. This peculiar dynamic may be explained by the fact that of the 39 landscapes analysed by Williams *et al.* (2012), not one of them had high floral resources in both early and late season. Perhaps a large forager workforce increases food stores in the nest, providing a partial buffer against late-season dearths. As likely as this sounds, there appears to be a limit to this buffering benefit, as suggested by Westphal *et al.* (2009) who found that even high worker numbers could not always compensate against a late-season dearth.

Even thinking narrowly in terms of crop pollination alone, however, floral-rich landscapes can be expected to directly promote large numbers of flower visitors – worker foragers – in the focal crop. The effects of season-long bloom constancy on queen production and long-term pollinator population stability can only be positive, even if only by supplementing the food reserves procured by an early large retinue of foragers.

The relative effects of urban versus agricultural landscapes on *Bombus* spp. reproductive success are similarly equivocal. In one study from Surrey, UK, colony reproductive success was found higher in an urban-leaning vector, suggesting that urban landscapes provide an ecological refuge to bumble bees against the pesticides and floral dietary homogeneity of intensified agriculture (Samuelson *et al.*, 2018). In a similar study from Michigan, USA, published that same year, however, Vaidya *et al.* (2018) could detect no effect of urbanization; instead, reproductive success there was heavily driven by the depredations of larvae of the scavenging wax moth *Vitula edmandsae*. These authors made the important point that reproductive success in *Bombus* spp. is prompted by drivers that are both bottom-up (resource-dependent) and top-down (natural enemies).

The unpredictability of such top-down drivers on colony survival, and by extension the production of new queens, was reinforced by the data of Goulson *et al.* (2018) who monitored 47 bumble bee nests over 2 years and found that 71% produced queens in 2010 while only 21% produced queens in 2011. This effect was largely driven by parasitism of the trypanosomatid *Crithidia bombi*,

but the investigators also noted a steady procession of hopeful predators at the nest entrances including birds and small mammals, leading the authors (Goulson *et al.*, 2018, p.168) to conclude that: 'Bumble bee nests are at the heart of a rich web of interactions between many different predator and parasite species.'

Finally, mortality of young queens is naturally high, a fact that may contribute to large and stochastic year to year differences in *Bombus* spp. reproductive success in any locality. Of the hundred or so queens that one colony produces, on average only one survives to produce another generation of queens (Heinrich, 1979).

8.3. Bumble Bees as Pollinators

Like honeybees, bumble bees are generalists and visit an assortment of flowering plants. However, because of their different morphology and behaviours bumble bees are superior pollinators for certain crops, especially in greenhouses.

Species of bumble bees vary in tongue length, and they segregate themselves among crops so that long-tongued species predominate at crops with longer corollas and vice versa (Ranta and Tiainen, 1982; Plowright and Plowright, 1997). This is an evolutionary strategy for avoiding foraging inefficiencies represented by visiting flowers depleted by previous visitors. Optimal foraging models (Rodríguez-Gironés and Santamaría, 2006) show that when nectar resources are abundant, long- and short-tongued bees visit flowers in a landscape indiscriminately, but as nectar begins decreasing bees segregate themselves by tongue and flower length. This segregation is associated with improved foraging efficiency of the respective bee. In lavender (corolla depth=7 mm), long-tongued bumble bees (tongue=7.8–8.9 mm) not only outnumbered honeybees (tongue length=6.6 mm) by a relative abundance of 92% vs 8%, but the bumble bees probed almost twice as many flowers (Balfour *et al.*, 2013). In short-corolla white clover, Plowright and Plowright (1997) showed that short-tongued bumble bees outperformed long-tongued bumble bees at the speed and efficiency at which they pollinated the crop. Thus, the pairing of tongue length and corolla depth in *Bombus* spp. may track with the pollination efficiency of the respective bee visitor (but see King *et al.*, 2013).

With reflection, it is easy to see that this pairing dynamic is not ecologically symmetrical. Short-tongued bees will stop visiting a deep corolla as soon as long-tongued bees have drained its nectar beyond their reach, but a long-tongued bee can keep exploiting shallow flowers as long as they have any nectar at all. Thus, nectar competition must be intense before the morphological segregation by flower type is complete. Drivers such as these may explain how in ecological time it is generally long-tongued bees that act as floral generalists and short-tongued bees as specialists (Stang et al., 2006).

Two corollaries of this discussion warrant emphasis. First, these particulars of bumble bee foraging ecology reinforce the importance of taxonomic diversity of bee crop visitors, bumble bee or otherwise. The breadth of bee morphologies so represented stands to accommodate the floral morphologies of any focal crop. Second, this logic thread supports the importance of crop plants that not only produce nectar but produce it copiously. Copious nectar-bearing flowers stand to accommodate a broad range of bees and bee morphologies and the speed at which they vector pollen from flower to flower.

Bumble bee workers exhibit a range of body size, even within colony, and larger workers are shown to outperform smaller workers at depositing pollen on stigmas (Willmer and Finlayson, 2014). The general largeness of bumble bees is thought to explain why they pollinate male-sterile cotton more efficiently than honeybees (Berger et al., 1988), and increase fruit-set in cucumbers more effectively than honeybees when compared at an equal number of flower visits (Stanghellini et al., 1997). Large pollinators also tend to forage earlier and later in the day, thus expanding the range of their pollinating activity (Willmer and Finlayson, 2014). Bumble bees have long been noted for their foraging tenacity during rain, strong wind and cool temperatures (Corbet et al., 1993; Drummond, 2016). Curiously, however, foraging temperature sensitivity is independent of worker body size. Worker bumble bees of all sizes foraged indiscriminately within an ambient temperature range of 16°–36°C (Couvillon et al., 2010).

Bumble bees are not only among the largest, they are among the hairiest bees, and hairiness (technically, *pilosity*) is positively associated with pollinator efficacy (Stavert et al., 2016; Phillips et al., 2018).

Some flower designs impede good pollination by honeybees but are no problem for bumble bees. For example, alfalfa (lucerne) flowers must be 'tripped'

to expose the sexual parts to a visiting insect. Bees automatically trip the flower with their body weight as they enter it from the front, causing the pistil and anthers to snap up and strike the bee on the head. Honeybees learn to avoid this bad experience after an initial encounter, switching on subsequent visits to nectar thievery – probing flowers for nectar from the side and bypassing the sexual parts altogether (Heinrich, 1979). Bumble bees are less intimidated by this insult from the flower and keep foraging on alfalfa legitimately (Brunet and Stewart, 2010; Brunet et al., 2019).

Bumble bees are among the 74 of 508 recognized bee genera (58% of all bee species) (Cardinal et al., 2018) capable of sonicating flowers (see Chapter 3, this volume). This makes them valuable pollinators of crops with poricidal anthers such as blueberry, aubergine, seed potato and tomato (Plowright and Laverty, 1987; Cane and Payne, 1990; Nunes-Silva et al., 2013a). One of the most important contexts for bumble bee sonication, if not bumble bee pollination altogether, is the pollination of tomatoes. For this crop, the value of bumble bees as sonicators eclipses their value as 'large' bees (see discussion above). Thoracic weight (a reliable proxy for bee size) did not correlate with bee removal of tomato pollen; instead, bees adjusted time spent sonicating a flower according to the amount of pollen remaining on it. Yet, even one flower visit was usually sufficient to fully pollinate the tomato flower (Nunes-Silva et al., 2013a).

Bumble bees are the shining stars of greenhouse pollination and they do not fly against windows like honeybees. Although they are less likely to forage outside the greenhouse when given the opportunity, managed bumble bees do often make their way outdoors (Whittington et al., 2004) – a point of concern over the spread of pathogens from managed to wild populations (Colla et al., 2006; Otterstatter and Thomson, 2008). In the main, however, year-round availability of cultured bumble bees in tandem with greenhouse technology has been a boon to year-round availability of vegetable crops.

It was long thought that bumble bee foragers lack anything like the honeybees' famous recruitment dance that communicates resource quality and location to nestmates. Each bumble bee was imagined a solitary agent making foraging decisions on her own (Heinrich, 1979). This landscape naïvety was interpreted as an advantage of bumble bee pollinators over honeybees, as it was thought that habitat scouting and recruitment contribute to honeybee infidelity to a focal crop. This model has been revised with the discovery in *Bombus terrestris* that workers can and do inform nestmates of the scent and general availability of a novel food resource. A successful forager, upon returning to her nest, makes excited runs punctuated with bouts of wing fanning by which she dissipates an airborne pheromone product of the *tergal gland* corresponding, if not homologous to, the honeybee tergal gland that produces the 'Nasanov orientation pheromone' in that species (Dornhaus *et al.*, 2003). Previously inactive nestmates respond to this 'dance' by increasing activity and eventually flying out of the nest to search for food, albeit without any specific directional knowledge, as far as we know, beyond the floral scents they pick up from the scouts (Dornhaus and Chittka, 1999, 2001). The end result is a colony-level response of 'more foraging', and the advantage of bumble bee landscape naïvety is sustained only to the extent that the floral scent they are searching for matches the focal crop. In general, the bumble bee nest functions as an 'information centre' in which candidate foragers appraise food needs by assessing colony food stores. From their successful forager sisters they learn the scent of profitable sites and receive excitatory pheromone stimulants (Dornhaus and Chittka, 2004a). Their recruitment is not as sophisticated as that of honeybees, but neither are bumble bee foragers wholly independent actors.

A possible application of this knowledge exists in the context of bumble bee greenhouse pollination. Producers of bumble beehives provision each hive with a reservoir of sugar syrup to sustain the bees until they can forage on their own. This resource satiation, however, can disincentivize bees to forage once they are in service (Dornhaus and Chittka, 2005). Molet *et al.* (2009) applied the volatile components of *B. terrestris* recruitment pheromone (eucalyptol + farnesol + ocimene; Mena Granero *et al.*, 2005) in a vial with a cotton wick inside bumble beehives and were able to increase foraging rate by 1.5 to 3.6 times.

8.4. Conserving Wild Bumble Bees

Chapter 6 (this volume) covers bee conservation in more detail. However, for wild bumble bees there are two principles that warrant repeating here. First, bumble bees nest in grassy thatch and abandoned rodent nests. One can conserve nesting sites by simply leaving unmown hedgerows and grassy margins around fields and orchards. It is important that these areas are never disturbed by heavy machinery compaction, herbicides, insecticides or ploughing. Second, of all the wild bees, bumble bees are the most dependent on a season-long succession of blooming plants. Production of queens and males in mid- to late summer depends largely on the quantity of food plants available in the preceding weeks. A mid-summer dearth can reduce a colony's output of reproductives. One way to provide an unbroken succession of bloom is to install bee pasture, preferably with native perennial plants. Government agencies publicize regionalized plant lists for sustaining native pollinators. For the US and Canada, the Xerces Society website (www.xerces.org, accessed 6 January 2021) is rich in recommendations for supporting pollinators at a farm scale.

8.5. Rearing Bumble Bees

There are three approaches to producing hived bumble bees for use in pollination: (i) hiving natural colonies collected from the field; (ii) setting out artificial nest boxes to attract wild queens; and (iii) inducing bumble bees to nest year-round in captivity. The first two approaches appear simple, but they give the grower minimum control over colony growth or synchrony with crop bloom; moreover, their success rate is poor. The third approach is labour-intensive but lets one produce full-size bumble bee colonies at any time of the year; it constitutes the full domestication of the bumble bee. Methods for year-round production of bumble bees are not generally available because they are held as proprietary secrets by business interests that sell bumble beehives for pollination. Nevertheless, a body of information exists and the field is ripe for development.

Since the first edition of this volume, two additional resources have been published that offer practical guidance on rearing bumble bees (Strange, 2009; Evans *et al.*, 2017), but neither provides guidance on continuing domestic production to second-generation queens and beyond. Neither do the instructions here, although the successes and failures are frankly disclosed and an attempt is made to update the methodology with tips gleaned from the more recent literature. Collectively, these published methods are adequate for providing an interested party with incipient colonies for use in pollination, research or for personal interest and pleasure.

8.5.1. Hiving colonies from the field

Natural bumble bee nests are not easy to find because they are usually located underground or in thick thatch. if a natural colony is found, however, it is possible to hive it using a few simple tools including a bee veil, gloves, a sheer insect-collecting net, one or more quart jars with lids, a shovel and some kind of hive for the colony. Any kind of weather-resistant box with a removable lid and small entrance hole can serve as a hive. The hive's entrance must be temporarily stapled shut with window screen or fine hardware cloth.

After one has donned protective clothing and before the nest is exposed it is possible to catch a large number of bees with a net as they fly to and from the entrance. Bees can be transferred from the net to a quart jar by scooping them up in the jar as they walk up the sides of the net that has been inverted so that the net opening faces down. After a while one can then begin excavating the nest, catching bees simultaneously. For nests near the surface, it is relatively easy to pull aside grass to expose the comb, carefully pick up the comb and put it in the hive. A shovel will be necessary for underground nests. Most bees will have been caught by the time the comb is reached. After the comb is transferred to the hive and all the bees are in the jar, the bees can be briskly shaken into the hive and the lid quickly closed. The new hive should be relocated at least 1 mile (1.6 km) away to discourage foragers from flying back to their original nest site. The temporary closure across the entrance can now be removed.

An alternative method is to transfer the comb and bees at night with the help of a red light (bees have difficulty seeing red). That way, one can capture more of the workers since most of them will be in the nest for the night. They are also less likely to fly in the dark.

Relocated bumble bee colonies are stressed and one way to help them survive the ordeal is to temporarily provide them with food. They can be fed sugar syrup (1-part sugar:1-part water) or diluted honey in a gravity feeder that is inserted through the wall of the hive.

Berger *et al.* (1988) transferred field-caught and hived bumble bee nests to a field of male-sterile and male-fertile cotton in Texas. They set out eight nests at ~15-m intervals along the 5-ha field. All colonies survived the transfer, resumed foraging and continued rearing brood. Although only eight bumble bees were seen foraging on the cotton in 4 weeks of observation, up to 21% of the pollen reserves in the nests was cotton pollen. All colonies were parasitized with cuckoo bees by late August which seriously reduced the likelihood of the bumble bee colonies producing queens for the next season.

Field bumble bee colonies are susceptible to the small hive beetle *Aethina tumida*, a transglobal pest of honeybees since the late 1990s when it was inadvertently spread from its native range in Africa. A sap beetle of the family Nitidulidae, *A. tumida* is attracted to pollen, honey, bee brood and virtually any of the natural odours emanating from a honeybee colony. Once inside the nest, adults and larvae feed on honey, pollen, and in the most severe infestations the bee brood itself. Unchecked populations can cause the loss of formerly strong and robust *Apis mellifera* colonies. Spiewok and Neumann (2006) demonstrated in Florida that *Aethina tumida* can parasitize and successfully reproduce in field colonies of the eastern North American bumble bee *Bombus impatiens*. The beetle's larvae pupate in soil outside infested bee nests, but the predatory soil nematodes *Steinernema riobrave* and *Heterorhabditis indica*, when applied to soil outside beehives, can induce 88–100% mortality in beetle larvae (Ellis *et al.*, 2010). These nematodes are available from organic farming supply companies for *A. tumida* biocontrol, a use that can be expected to apply equally well to cultured bumble bee colonies in the field. I am not aware of any evidence of significant *A. tumida* depredation on bumble bees in greenhouse environments, which by extension restrict access of free-flying adult beetles and provide marginal habitat for soil-dependent beetle pupation.

8.5.2. Providing artificial nesting sites in the field

The rationale for using artificial nest boxes is to encourage local bumble bee populations by supplementing available nesting sites. This presupposes that nesting sites are a limiting factor in the local habitat. Queens become active and start seeking nest sites on warm days in spring when the first pollen sources begin to bloom. This is the time to set out artificial nest boxes.

Hobbs *et al.* (1960) and Hobbs (1967) attracted nesting queens in Alberta, Canada by setting out cubical plywood boxes roughly 15.2 cm on all sides (Fig. 8.3). The plywood was 1.9 cm thick. The entrance holes were 1.6 cm in diameter which

Fig. 8.3. An above-ground artificial bumble bee nest box after the design of Hobbs *et al.* (1960) and Hobbs (1967). The lid is open to reveal the upholsterers' cotton within. The box is attached to the ground with a loop of heavy wire running through two eyelets and into the ground. A long entrance tube is used to create the illusion of an underground burrow.

Fig. 8.4a. Setting up a false underground bumble bee box after Hobbs (1967).

Fig. 8.4b. The false underground box is completed by covering the midsection of the plastic pipe with soil or sod. This creates the illusion that the exposed end of the plastic pipe is leading to a subterranean cavity.

permitted entry to bees but excluded mice. The boxes had hinged lids and were filled with upholsterers' cotton which the bees used to line their nests. Boxes were fastened to posts with wire to keep skunks from turning them over. Hobbs (1967) favoured a 'false underground' nest, a nest box set on the ground but modified by using a 30.5-cm section of plastic pipe as an entrance tunnel and placing a piece of sod over the pipe in such a way as to leave the pipe entrance exposed (Fig. 8.4a,b). Presumably the sod gives the illusion that the pipe tunnel leads to an underground nest. Acceptance by queens was highest when the boxes were placed in fallow backyard gardens, beside fence posts in a prairie and along thickets of small aspen trees. Bumble bee occupancy over 6 years averaged 44±23%. Working in New Zealand, Macfarlane *et al.* (1983) achieved about 30–60% bee occupancy with a slightly different box design. Frison (1926), working near Urbana, Illinois, USA, reported 48% occupancy in his underground artificial domiciles made of wood and tin.

To my knowledge, these records constitute the literature's high-water marks for artificial domicile occupancy. In a 2-year unpublished trial that I conducted in Georgia in 1996–1997, my 2-year occupancy rate (of 50 domiciles) with the Hobbs false underground nest was zero. Occupancy rates have fared somewhat better in trials from New Zealand with an above-ground domicile (mean 4-year occupancy of 84 nests = 13%) (Barron *et al.*, 2000);

from Ontario with above-ground, underground and false underground nests (occupancy 17 of 346 nests = 4.9%) (Johnson *et al.*, 2019); and from southern England and central Scotland with a variety of above-ground, subterranean and semi-subterranean domiciles (23 of 736 nests = 3.1%) (Lye *et al.*, 2011). The effect size of artificial domiciles is presumably greatest in areas where a shortage of natural nest sites is a limiting factor to bumble bee populations, but this has not been experimentally confirmed.

In order for bumble bee nest boxes to be useful for crop pollination, it is important to set out boxes near the crop of interest or to relocate occupied boxes near to the crop. However, moving nests causes a loss of foragers (presumably because some spend the night in the field) and provokes queens to revert to foraging behaviour which increases the chance of their loss. Colonies that are moved produce fewer queens. Nevertheless, if one can achieve a reasonably good occupancy rate, then this practice may be able to increase bumble bee numbers at a crop. Relocating bumble bee boxes should be limited to crops that bloom relatively late in the season because this allows more time for colonies to recover an adequate foraging force.

One could conceivably increase the population of bumble bees in an area over time by repeatedly introducing bumble bee colonies which release new queens at the season's end. With similar objectives, Clifford (1973) increased peak densities of local bumble bee populations by importing and releasing 100 queens each spring for 3 years; however, bee densities returned to their previous levels when queen importations stopped.

8.5.3. Rearing bumble bees year-round

The keys to rearing bumble bees year-round are: (i) inducing queens and males to mate in captivity; (ii) bypassing or abbreviating the queen's natural diapause interval; (iii) inducing queens to rear a brood in captivity; (iv) growing colonies so that they produce a foraging force of workers and subsequently another generation of reproductives; and (v) retaining some queens and males in order to start the cycle over again. A few companies rear bumble bees efficiently on a commercial scale, but their methods are proprietary secrets. Nevertheless, the scientific literature gives guidance on bumble beekeeping from which the following summary is made, supplemented by personal experiences (Plowright and Jay, 1966; Heinrich, 1979; Pomeroy and Plowright, 1980; Röseler, 1985; Griffin *et al.*, 1990; Van den Eijnde *et al.*, 1990; Tasei, 1994; Tasei and Aupinel, 1994; Ribeiro *et al.*, 1996; Strange, 2009; Evans *et al.*, 2017; Carnell *et al.*, 2019).

8.5.3.1. Honeybees as a source of pollen and surrogate workers

It is useful to have on hand one or two colonies of honeybees (see Chapter 7, this volume) with entrance pollen traps. Honey beehives thus fitted can provide fresh bee-collected pollen for feeding the bumble bees. Pollen must be collected daily from the trap and immediately frozen. It is important to collect and freeze enough pollen to sustain the operation during winter if the intention is to grow bumble bees year-round.

Young queens of the European species *Bombus terrestris* and North American *B. appositus*, *B. bifarius* and *B. centralis* can be stimulated to begin brooding when they are housed with young honeybees; in these cases, honeybee hives are necessary to provide these surrogate workers.

Surrogate honeybee workers should be used only when they are less than 12 h old. To harvest these young workers, one must open a honeybee hive and look for a comb of emerging worker bees. Combs of emerging bees contain a large quantity of capped brood and upon close examination one can see the ragged edges of recently opened cells. Light-coloured young bees (<12 h old) will be walking on the comb, and one may see bees emerging from cells. When such a comb is identified, it is brushed free of all adhering bees and placed inside a white plastic bag to be stored overnight at comfortably warm temperatures. The next morning there will be numerous young bees walking around in the bag. As long as the bag does not overheat and contains some honey from the comb, the bees can remain there until they are added to the bumble bee queen starter boxes within the next few hours.

8.5.3.2. The queen starter box

Queen starter boxes are small boxes in which mated queens are placed and induced to begin nesting. Reported dimensions are variable and probably do not matter greatly. One design is a box of ~22.9×11.4×5 cm with two chambers – a nest chamber and a feeding/defaecating chamber (Figs 8.5 and 8.6). This design keeps the box interior dark, but the nest chamber can be easily opened and bees observed through the clear inner cover (light bothers some bumble bee species; others do not seem to mind). Evans *et al.* (2017) present photographs and dimensions for starter boxes along with downloadable plans at www.befriendingbumblebees.com (accessed 6 January 2021).

The floor of the defaecating chamber can be made of small-mesh hardware cloth or can be lined with a square of corrugated cardboard that is replaced as needed. In one design, there is no permanent floor in the defaecating chamber and instead a piece of cardboard or heavy blotting paper is taped to the bottom of the box and replaced as required. Bees can pass between the nesting and defaecating chambers through a small circular opening. A defaecating area that is separate from the nesting area helps maintain general nest sanitation.

Some authors use single-chamber starter boxes. One such design is a box of ~12×5.5×11 cm with transparent acrylic side walls, and another is a box of ~11.3×4.5×4.3 cm with an acrylic lid.

Regardless of the design chosen, it is helpful to provide each box with a plastic weighing cup in the nest chamber, on which pollen will be provided and the queen will build her brood clump. By encouraging the queen to nest on a plastic weighing cup, it is a relatively simple matter to move the comb later when the incipient colony is graduating up to a finisher box. It is also helpful to attach a plastic honeybee queen grafting cup to the floor of the plastic weighing cup with melted beeswax; these cups simulate the shape and size of a natural bumble bee honey pot and may encourage nesting (Fig. 8.7).

A small wad of upholsterers' cotton should be placed in the nest chamber; the queen will fashion the material into a fibrous shell around her comb (Fig. 8.8). Some authors recommend chip foam

Fig. 8.5. A two-chamber queen starter box opened to reveal its components. A partition separates the chambers but has a passageway in order to allow bees to move between chambers. The chamber at the left is the feeding/defaecating chamber. There is a hole in the lid to accommodate a vial of syrup or honey water. The floor of the box is lined with disposable corrugated cardboard. A plastic weighing cup on the floor catches drips. The nest chamber on the right has upholsterers' cotton and a plastic weighing cup with a ball of pollen dough. A transparent acrylic inner cover allows the beekeeper to inspect the nest chamber with minimal disturbance to the bees. Photo courtesy of Nancy B. Evelyn.

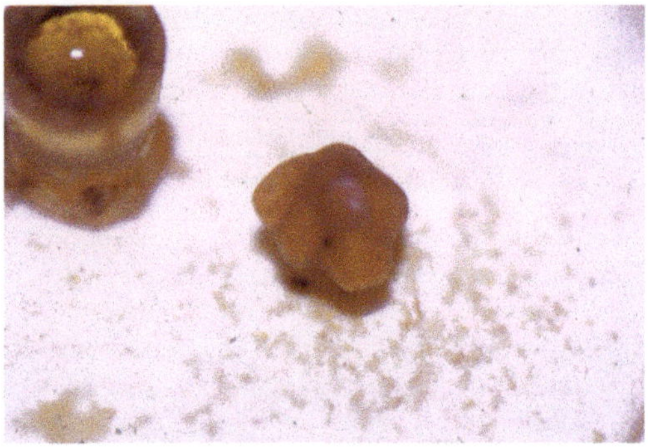

Fig. 8.7. A plastic honeybee queen grafting cup (top left) simulates the size of a natural bumble bee honey pot and may encourage nesting. The brood clump (centre) has five brood cells that look like bulges around the central core of pollen. Under each bulge are developing eggs or larvae.

Fig. 8.6. An assembled two-chamber queen starter box. A piece of hardboard covers the acrylic inner cover of the nest chamber. Photo courtesy of Nancy B. Evelyn.

Fig. 8.8. The queen has used the upholsterer's cotton in this nest chamber to form a fibrous shell around her brood mass. One emerged worker is barely visible behind her in the centre. An open cocoon is visible. The queen is in a typical defensive posture in response to the opening of her nest – she is on her back and her stinger pointed straight at the intruder.

instead of upholsterers' cotton; it is easier to handle and replace than upholsterers' cotton. By cutting out holes in layers of chip foam, one can make a cavity for the pollen ball that is enclosed yet easily accessible (Fig. 8.9).

Some kind of gravity feeder for dispensing syrup or diluted honey should be inserted through the lid or wall of the feeding/defaecating chamber. The feeder can be a pipette with one end sealed off or a small inverted vial with a tiny perforation in its lid from which bees can drink. Another feeder design is a block of solid plastic with numerous feeding wells drilled into it. The wells are filled with syrup and the block placed in the feeding/defaecating chamber. Feeders should be cleaned at least after every third filling.

8.5.3.3. The finisher box

Queens that successfully rear brood in starter boxes must soon be transferred to larger finisher boxes in which the colony can grow to maturity. Conceivably, one could bypass the starter boxes, but queens seem to initiate nesting better in small-volume domiciles. Like starter boxes, published designs for finisher boxes are variable. Outside dimensions for one design are ~30×21×17.5 cm. There are separate chambers, but in the finisher box the nesting chamber is larger than the feeding/defaecating chamber. Ventilation is provided by holes drilled in the walls of the feeding/defaecating chamber and covered with hardware cloth (Fig. 8.10). The floor of the defaecating chamber can be made of metal mesh or lined with some kind of disposable absorbent material. The feeding/defaecating chamber must have a gravity syrup feeder. See also the finisher box design of Evans *et al.* (2017).

It is ideal if the floor of the nest chamber is sloped upward from the centre, like an inverted cone, at an angle of 35°. This design conforms to the natural shape of a growing bumble bee comb. One author achieves this effect by moulding nests of the correct shape out of porous concrete. Another way to do this is to lay several layers of chip foam inside the nest chamber. A hole is cut out of the centre of each layer of foam, each hole successively larger than the one below (Fig. 8.11).

8.5.3.4. Ambient rearing conditions

It is important to provide favourable climate conditions for starter boxes and finisher boxes. Some authors maintain rearing rooms at ~28–30°C, 50–65% relative humidity and in total darkness. Some light-sensitive species do better if the area is illuminated with red light while the operator is feeding or inspecting the young colonies. Species that are not light sensitive do not need dark rearing rooms, but even with these species it is advisable to use boxes that are designed to keep interiors dark.

One option to humidifying the room is to humidify each starter box. One can do this by placing damp filter paper just under the lid of the nest chamber. This works best in starter boxes provisioned with chip foam instead of upholsterers' cotton (Fig. 8.12).

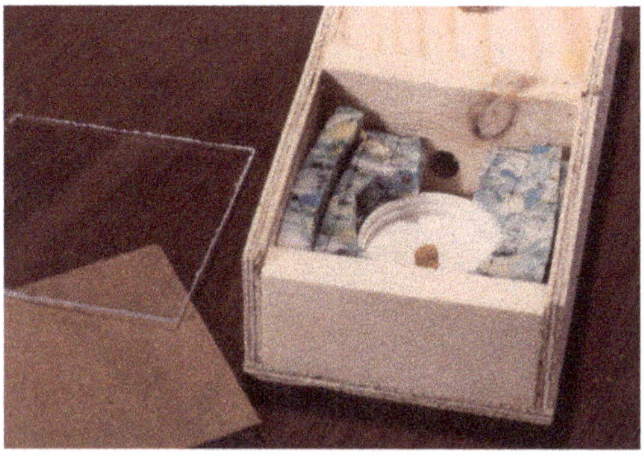

Fig. 8.9. A nest chamber with chip foam instead of upholsterers' cotton. By cutting out holes in successive layers of chip foam, one can make a cavity for the pollen ball that is enclosed yet easily accessible. Photo courtesy of Nancy B. Evelyn.

Fig. 8.10. A two-chamber finisher box designed to accommodate a queen and her growing colony. There are ventilation holes covered with hardware cloth in the side walls of the feeding/defaecating chamber and a flight entrance, closed in this photograph with hardware cloth and thumb tacks. Photo courtesy of Nancy B. Evelyn.

Fig. 8.11. A finisher box opened to reveal the floor of the nest chamber. A hole is cut out of the centre of each layer of chip foam, each hole successively larger than the one below, in order to create a cavity suitably shaped for a growing brood comb. Photo courtesy of Nancy B. Evelyn.

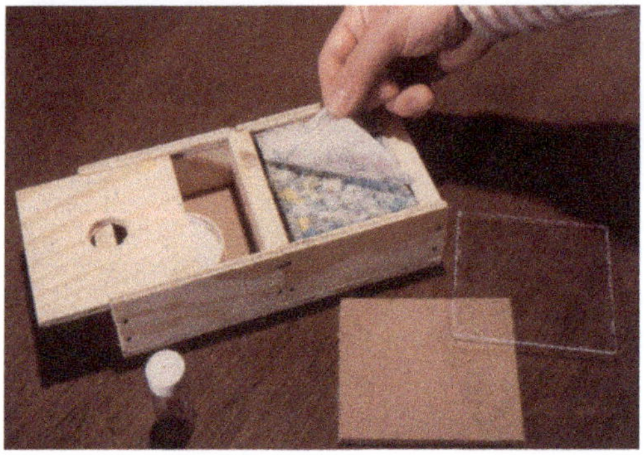

Fig. 8.12. A damp filter paper just under the lid of the nest chamber is one way to humidify a starter box. It must be changed daily to avoid mould. Photo courtesy of Nancy B. Evelyn.

The damp paper must be changed daily to avoid mould problems.

8.5.3.5. Feeding colonies in captivity

Fresh food must be prepared at the same time the incipient colonies are being started. A syrup made of one-part sugar to one-part water is prepared and placed in the gravity feeders. One alternative to sugar is a syrup made of one-part honey to one-part water. Honey syrup ferments more rapidly than sugar syrup and must be replaced more frequently (at least every 2 days), but bees can locate it more easily because of its attractive odour.

Fresh bee-collected pollen must be cleaned of visible debris, ground into a fine consistency and added to 50% (v/v) sugar syrup until it achieves a slightly sticky dough-like consistency. Pollen must be kept fresh-frozen in its bee-collected pellet form until it is needed. Queens produced from colonies that are fed fresh-frozen pollen are larger, have lower mortality rate and produce larger colonies than do queens produced from colonies fed dried-frozen pollen (Ribeiro *et al.*, 1996).

Evans *et al.* (2017) suggest coating pollen balls in a thin layer of melted beeswax. The beeswax helps prevent the pollen from drying out and seems to encourage the queen to lay eggs and initiate a brood clump. Clean beeswax is melted to the lowest possible melting point (not so hot that it will damage pollen nutrient quality), a ball of pollen is speared with a toothpick and the ball is dipped into the beeswax.

8.5.3.6. Catching queens and initiating nests

Queen bumble bees become active and fly on warm days in spring when early pollen sources become available. It is best to catch them while they are still searching for a nest site and have not started a nest; one can find such queens flying low to the ground in a zigzag pattern obviously exhibiting searching behaviour. Queens with pollen on their legs have already started a nest in the field, are unlikely to resume nesting in captivity and should be left free to live naturally.

Queens can be caught with an insect net. They should be transported individually in small jars padded inside with paper towelling and with lids loosened to provide ventilation and prevent overheating. The queens should be transferred to starter boxes as quickly as possible.

Success rate (bumble bee queens that begin brooding and produce a clutch of workers) in starter boxes generally ranges from 20–50%, and you should start at least twice as many queens/starter cages as you hope to end up with. In the case of *Bombus terrestris*, each queen can be given three or four young honeybees as surrogate workers. In the case of *B. pascuorum*, one team (Carnell *et al.*, 2019) reports success by pairing queens in the same starter box and waiting for one to show dominance – interpreted as the one who spends the most time in the centre of the box, standing on the pollen. Submissive individuals are more active and spend their time roaming the edges of the box.

These individuals are re-paired with another sub-missive in a new box and once again allowed to resort themselves, and so forth. The end result is either a starter box with a dominant singleton or a box of cooperative pairs. This pairing technique appears to be species-specific; the method did not work for *B. hortorum* whose queens were highly aggressive to each other.

Using the North American species, *Bombus appositus*, *B. bifarius* and *B. centralis*, Strange (2010) achieved 59.1% nest initiation success with conspecific queen pairings, 33.3% success pairing the queen with two young honeybee workers, and 16.7% success when the queen was by herself.

A bean-sized lump of sticky pollen dough is placed on the plastic weighing cup in the nest chamber and close to the artificial honey pot. Gravity feeders and the artificial honey pot are filled with syrup. Starter boxes are placed in proper temperature and humidity conditions and left undisturbed for at least 24 h after which the queens are checked for nesting activity. Promising signs are (more or less in this order): wax deposits on the floor; wax deposits on the plastic honey pot; a natural wax honey pot; natural honey pots filled with syrup; excavations on the pollen lump; a brood cell in the pollen lump; and the queen incubating the brood clump (Fig. 8.13). A brood cell looks like a bulge on the side of the pollen lump. The queen excavates a depression in the pollen lump, lays one or more eggs in it and covers the eggs with a dome of wax (Figs 8.1 and 8.7). Under this dome each egg may be partitioned from its neighbour or all eggs may be grouped

Fig. 8.13. A queen *Bombus impatiens* incubating her brood clump. Photo courtesy of Elaine Evans.

together, depending on the species. The pollen lump can now be referred to as a brood clump.

If after 24 h there are no signs of nesting activity, the pollen lump should be replaced and the queen left alone for another 24 h. The presence of faeces in the defecation chamber indicates that the queen is eating pollen and may eventually lay eggs. Once the queen starts making brood cells, new pollen dough lumps should be placed next to the brood clump, three times a week. However, overfeeding is a common problem and a queen should not be fed more than she consumes. A good guideline is to feed no more pollen dough than one-third the volume of the brood clump. If the brood clump does not grow, this indicates that the larvae may be dead or that the young colony is being underfed. Any uneaten food must be removed regularly.

Beginning on day 34, Carnell *et al.* (2019) provisioned each nest box of *Bombus pascuorum* or *B. hortorum* queens with two 5-day old cocoons of surrogate worker pupae of *B. terrestris*. This was thought to stimulate egg laying and brood care in the host queens.

8.5.3.7. Graduating incipient colonies to finisher boxes

After the first worker bumble bees emerge, the queen, workers and brood clump can be transferred into a finisher box. The brood clump, plastic weighing cup base and all, is placed in the centre of the nest chamber floor of the finisher box (Fig. 8.14). The growing colony continues receiving syrup and pollen lumps, and the quantity is gradually increased as the colony grows. At this stage, some species will readily accept natural pollen pellets as collected from the pollen trap; this saves the labour of making pollen dough. The pollen pellets can be simply sprinkled on top of the brood clump. It remains important to avoid overfeeding and to cut back on food if uneaten pollen accumulates in the nest.

8.5.3.8. Graduating colonies into pollination units

A finisher box colony is ready to go outside for free flight and pollination as soon as it has around 80 worker bumble bees. It is helpful to continue feeding colonies pollen and syrup for a few days to help the bees through this transition period. Colonies can be placed in the field or orchard in a manner similar to that for honeybees. Colonies should rest on concrete

Fig. 8.14. Using pliers with flat opposing surfaces, the entire young colony – plastic cup, brood cells, bees and fibrous shell – is transferred into the nest chamber of the finisher box. Photo courtesy of Nancy B. Evelyn.

blocks to keep them off the ground, and they should be secured to tree trunks or posts so that they cannot be overturned by skunks or other predators. It is important to shade the colonies against direct sun.

8.5.3.9. Mating queens and inducing incubation

To maintain year-round production, one must induce new queens and males from cultured colonies to mate in captivity. In appearance, males have longer antennae, more barrel-shaped abdomens and more variable body coloration compared to workers (Fig. 8.15). New queens are noticeably larger than workers and more brightly coloured (Fig. 8.16). Queens and males are produced after the colony has reached maturity.

To induce mating, queens and males from two or more colonies are released into a flight room or cage. The flight room should be ~3.7×5.8 m in size, and warm and sunny. A sunny porch or similar screened enclosure can suffice. Some bumble bee species will mate in small cages covered with gauze or wire mesh. In these cases, the mating cage should be ~70×70×70 cm. Working with *Bombus impatiens*, Evans *et al.* (2017) used a 1×1×1 m mating cage. Whether using a cage or a flight room, the enclosure must be provisioned with shallow pans of syrup and pollen, and a shallow pile of moistened peat moss or soil must be deposited on the floor and covered with a layer of raw cotton, straw or dry leaves. If a flight room is used, bees seem to find the food more easily if it is placed on a windowsill. Release about twice as many males as

Fig. 8.15. A male *Bombus vagans* of eastern North America. Notice the non-modified hind tibia (white arrow). Photo courtesy of Brooke Alexander via Flickr, USGS Bee Inventory and Monitoring Lab.

Fig. 8.16. A female *Bombus vagans* of eastern North America. The outer surface of the hind tibia (white arrow) of all females is broadened into a corbiculum used for transporting pollen in the field. See also Fig. 2.2 (this volume). Photo courtesy of Brooke Alexander via Flickr, USGS Bee Inventory and Monitoring Lab.

females into the flight cage from different colonies to avoid inbreeding. Bees should be released in the morning. After the queens mate, they invariably bury themselves in the soil on the flight room floor. The beekeeper can then dig out these queens, place each one individually in a vial half-full of damp peat moss (Fig. 8.17), and store them at normal

Fig. 8.17. Queens are placed individually in vials half-full of damp peat moss and refrigerated for up to 8 months.

Fig. 8.18. Queens are narcotized with carbon dioxide (CO_2) gas in their individual vials to stimulate egg laying and brooding.

refrigeration temperatures (~5°C). The vials and peat moss should first be heat-sterilized to reduce mould problems. Handling queens is easiest if you use forceps or pliers with flat opposing surfaces. The flat surfaces let you pick up the bee by the legs with ease and without causing injury.

8.5.3.10. Activating second-generation queens

The length of the hibernation period, or lack thereof, represents a flexible stage in bumble bee culture that permits year-round availability of colonies. The queen's hibernation period can be eliminated or truncated according to the demand for colonies. A necessary tool for this step is a canister of carbon dioxide (CO_2) gas with a regulator, connected to a few feet of rubber tubing. CO_2 acts as an anaesthesia and egg-laying stimulant in bees.

If you wish to resume rearing colonies immediately, the hibernation period must be circumvented. Queens are not refrigerated after they dig into the soil in the flight room, but instead 1 day after mating are placed individually in glass jars, each of which is treated with CO_2 (narcotized) until she is immobilized. The queens are then returned to the flight room or mating cage and provided with syrup and pollen pellets. The CO_2 treatment is repeated after 24 h. Two to 4 days after the last CO_2 treatment, each queen is placed in a starter box as before.

If a period of hibernation is desired, then queens can be kept refrigerated in their peat-filled vials for up to 8 months. To activate refrigerated queens, CO_2 is injected directly into each vial for 10 s (Fig. 8.18), then the queens are moved to a slightly elevated

temperature (15°C) overnight. In the morning the queen vials are moved into the flight room or mating cage at room temperature. Two or 3 h later, queens are released so that they can fly, and the room is provisioned with syrup and pollen pellets. After several minutes of flight, each queen is placed individually in a jar and treated with CO_2. Queens are returned to the flight room and the CO_2 treatment is repeated the next day. Three to 6 days after the last CO_2 treatment, each queen is placed in a starter box as before.

It is very important that activated queens fly and eat pollen and syrup. If they are sluggish in the flight cages and do not visit feeders, it may help to pair them with a conspecific queen of the same generation or with young honeybee workers as described in section 8.5.3.6. At least once a day each queen should be encouraged to fly in the flight cage until she shows brooding behaviour in the starter box, at which point she should remain confined to the starter box.

8.6. Managing Hived Bumble Bees for Pollination

8.6.1. Managing bumble bees in the field

Bumble bees are capable of foraging long distances from their nests, with some astonishing records at a scale of kilometres (Table 8.1). A perusal of Table 8.1 shows that there are species-specific categories of short-, mid- and long-range foragers. *Bombus pascuorum*, *B. sylvarum* and *B. muscorum* are, for example, called 'doorstep foragers' for their proclivity to fly only a few hundred metres from their nests, while *B. terrestris* and *B. vosnesenskii*

Table 8.1. Some values for worker foraging distance from the nest for *Bombus* spp.

Species	Location	Distance (m or km)	Reference
B. balteatus	alpine Colorado, USA	≤221 m, 85.4±15 m	Geib *et al.* (2015)
B. bifarius	alpine Colorado, USA	≤221 m, 110.3±41.7 m	Geib *et al.* (2015)
B. distinguendus	Outer and Inner Hebrides and Orkney Islands, Scotland, UK	mode 391 m, 95%≤955 m	Charman *et al.* (2010)
B. flavifrons	alpine Colorado, USA	≤203 m, 23.8±10.1	Geib *et al.* (2015)
B. hortorum	Buckinghamshire, UK	273±20 m	Redhead *et al.* (2016)
B. lapidarius	Buckinghamshire, UK	536±16 m	Redhead *et al.* (2016)
	Hertfordshire, UK	≤450 m	Knight *et al.* (2005)
	Hesse, Germany	≤1500 m, 78%≤500 m, 9% 1001–1500 m	Walther-Hellwig and Frankl (2000)
B. muscorum	Hesse, Germany	≤650 m, 100%≤500 m	Walther-Hellwig and Frankl (2000)
B. pascuorum	Buckinghamshire, UK	337±20 m	Redhead *et al.* (2016)
	Hertfordshire, UK	≤449 m	Knight *et al.* (2005)
	Hesse, Germany	≤800 m	Kreyer *et al.* (2004)
	Southampton, UK	≤312 m	Darvill *et al.* (2004)
B. pratorum	Hertfordshire, UK	≤674 m	Knight *et al.* (2005)
B. ruderatus	Buckinghamshire, UK	502±34 m	Redhead *et al.* (2016)
B. silvicola	alpine Colorado, USA	≤291 m, 74.7±56.3 m	Geib *et al.* (2015)
B. terrestris	Hesse, Germany	43%≤500 m, 25% 1500–1750 m	Walther-Hellwig and Frankl (2000)
	Hertfordshire, UK	≤1500 m	Osborne *et al.* (2008)
	Hesse, Germany	≤2.2 km	Kreyer *et al.* (2004)
	Halle, Germany	≤800 m, 267±180 m, 40%≤100 m	Wolf and Moritz (2008)
	Buckinghamshire, UK	551±40 m	Redhead *et al.* (2016)
	Hertfordshire, UK	≤758 m	Knight *et al.* (2005)
	Southampton, UK	≤625 m	Darvill *et al.* (2004)
B. vosnesenskii	Oregon, USA	≤11.6 km, 1.3–3.7 km (2007), 3.4–7.3 km (2008)	Rao and Strange (2012)

are long-distance foragers, and *B. lapidarius* is intermediate. These species-specific differences have implications, not only for hive placement for pollination, but also for conservation considerations, as longer-ranging species are presumably less vulnerable to local perturbations of foraging habitat. Across species, however, it is generally true that foraging distance decreases in landscapes with uninterrupted semi-natural vegetation (Redhead *et al.*, 2016).

The principle at work here is that bees make energetic decisions constantly as they forage, weighing the calories gained against the calories spent foraging. Energetic models (Cresswell *et al.*, 2000) show that bumble bees should rarely forage >1 km from their nest and when they do so it is a concession: either nearer sources are too rare or poor in quality, or distant sources rich enough to justify a long flight in spite of the extra costs. It is far better for the bees, and for the grower, if the focal crop is both near and energetically rich, i.e. a good source of nectar or pollen. Dramstad *et al.* (2003) present experimental evidence and literature examples that challenge this assumption but do not develop a supporting theoretical counter-argument. In my opinion, the weight of theory and field evidence support the placement of hives as near to the focal crop as possible.

Mayer and Lunden (1996), working in apple and pear orchards in Washington, USA, recorded good bumble bee foraging at relatively short distances, within 18 m of commercial hives. In Israel, Zisovich *et al.* (2012) recorded yield-enhancing numbers of bumble bees per pear tree at 50-m distances from hives. If these data are representative of foraging patterns under commercial field pollination conditions, this means that pollination will be optimized when hives are numerous and well dispersed in an orchard. Meeting this goal in field conditions may be difficult because commercial bumble beehives are expensive.

Bumble bees are easily stressed after moving. Workers may get lost when they leave the nest for the first time and queens may abandon the nest. To reduce these problems, one can mark hive entrances with distinguishing designs so that workers can visually orientate to their home hive. To discourage queens from leaving, it is important not to move colonies until there are at least 80 workers or, at the very earliest, the second batch of brood has emerged. It may help to feed the colonies syrup and pollen for 1–2 days. Outdoor colonies can gain some protection from parasitic cuckoo bees (*Psithyrus* spp.) if the beekeeper uses small entrances. Sladen (1912)

used a small entrance 0.7×1.1×1.0 cm to exclude mice, and he 'reduced the size of the hole in the mouse-excluder' (Sladen, 1912, p.97) to exclude cuckoo bees. Colonies can be protected from skunks and other predators by mounting them on elevated stands or wiring them to posts or trees. Hives should also be shaded in all but the coolest climates, as the species used in commercial production are naturally temperate ground nesters and not adapted to high temperatures.

8.6.2. Managing bumble bees in the greenhouse

Bumble bees have distinguished themselves as pollinators of greenhouse plants, especially tomato, and rearing and packaging methods have reached a high pitch of standardization and efficiency, especially with *Bombus terrestris* in Europe, and *B. impatiens* and *B. occidentalis* in North America. Colonies are sold ready-to-use, leaving the grower to only space them in the greenhouse, open them for bee flight, and protect them from climate extremes and pesticides (Fig. 8.19). Hive designs consist of plastic compartments contained in a single corrugated plastic sheath and include a reservoir of sugar syrup to sustain the bees in transit. Once at the greenhouse, colonies should be fed syrup continuously. Pollination is optimized when hives are distributed evenly throughout a greenhouse (Morandin *et al.*, 2001),

Fig. 8.19. Commercial bumble beehives in a greenhouse. Photo courtesy of Arnon Dag.

and one manufacturer recommends stocking a greenhouse crop in a checkerboard fashion 'with a hive in every black square'. Published hive densities include one hive per 625 m² (Torres-Ruiz and Jones, 2012), 1 per 667–1430 m² (Morandin *et al.*, 2001), 1 per 1667 m² (Van Ravestijn and Van Der Sande, 1990), and 1 per 2000–2500 m² (Birmingham and Winston, 2004). Crop-specific densities are given in Volume II of this work.

In a review on insect pollination in greenhouses, Dag (2008) pointed out numerous physical considerations that can affect bumble bee pollination efficacy. Greenhouses are sometimes covered in ultraviolet (UV)-absorbing sheets to manage insect-borne viral diseases, but bumble bees require UV light for navigation, and disrupted navigation has been traced to poor pollination. One offered solution is to place bumble beehives along southern greenhouse walls that receive more illumination. Greenhouses tend to have higher relative humidity than ambient conditions, a problem that subtracts quality of floral reward; nectar is more diluted and pollen grains less available from the anthers. Dag recommends adequate greenhouse ventilation during the pollinating period to increase nectar sugar content and to stimulate bee foraging.

9 Managed Solitary Bees

Manageability is not the monopoly of the social honeybees and bumble bees, however conspicuously high these bees rate in charisma and public appreciation. The solitary bees field at least as many manageable species, some systems for which are highly intensified and fully integrated into commercial scale crop production. This chapter covers three of the most economically important of these taxa – the alfalfa leafcutting bees, alkali bees and orchard mason bees.

9.1. Alfalfa Leafcutting Bees[i]

The alfalfa leafcutting bee (ALCB, *Megachile rotundata*, family Megachilidae) was accidently introduced to North America from Eurasia sometime after the mid-1930s. Today it occupies the northern 75% of the contiguous USA and extends from British Columbia to the Great Lakes region of Canada. It has become the major pollinator of seed alfalfa (lucerne) in the western USA and Canada. The bee is also managed as an alfalfa pollinator in Europe, New Zealand, South Australia and parts of South America.

Corresponding with the arrival of ALCBs in North America, yields of alfalfa seed, traditionally pollinated by honeybees, tripled from a typical 450 kg/ha to 1300 kg/ha (Pitts-Singer and Cane, 2011). Today, ALCBs are fully integrated into commercial alfalfa seed production. No other solitary bee is managed more intensely.

9.1.1. Biology

Megachile is a large genus with over 1400 described species distinctive for their derived habit of lining nests with pieces of leaf (Michener, 2000). *M. rotundata* is solitary and nests in pre-existing holes in wood or other materials. The bees are from 0.5–1 cm long and 0.2–0.4 cm wide (Fig. 9.1). Females are larger than males and are black with short white hair on parts of the body. The abdomen of the female is more pointed than that of the male and has four or five stripes of white hair across the top and a pollen-carrying brush of long white bristles on the ventral side called the *scopa*. Males have buff-coloured hair, two light spots on the terminal abdominal segment and lack the pollen brush. The male bee has mandibles each with a prominent tooth that helps him cut through the leaf plug sealing his cell. The female's mandibles have smaller teeth adapted for cutting pieces of leaf that she uses to line her cells.

Adults emerge from cells in late spring or early summer, depending on temperature. Developing bees need a chilling period to break diapause. In the north-western USA, depending on location and weather, males emerge from early to mid-June and females 1 week later. Males usually outnumber females 2:1, but 1:1 or even female-biased ratios sometimes occur. Males are likely to harass females by repeated attempts at copulating, a hazard which peaks at male-biased ratios of around 3:1 and exacts reproductive costs to females in time lost foraging, as well as dropped and lost foraging loads. When male-biased ratios reach 4:1 or higher the problem abates somewhat as males become more distracted at chasing each other away (Rossi *et al.*, 2010).

Despite the frenzied male competitive behaviour, females mate only once and thereafter forage for nectar and pollen while their eggs develop. Around their second or third day post-emergence they start building cells. Cells are made back to back, linearly in a pre-existing tunnel (Fig. 9.2a). The mother lays female eggs in the innermost cells and male eggs in the outermost. This arrangement, common to many solitary bees, enables the early-emerging males to chew out of their cells and leave the tunnel without trampling female cells.

A female makes a thimble-shaped cell out of alfalfa (lucerne) leaf pieces which she cuts, carries and shapes by chewing their edges and pushing them against the wall of the tunnel. Her chewing of

© Keith S. Delaplane 2021. *Crop Pollination by Bees,* 2nd Edition, Volume 1 (K.S. Delaplane)
DOI: 10.1079/9781786393494.0009

each leaf piece edge renders it sticky so it adheres to previous leaf sections (Trostle and Torchio, 1994). It takes about 15 leaf pieces and between 81 min (Maeta and Adachi, 2005) to 2 h 27 min to make one cell, and an additional 5 h 3 min and 17

Fig. 9.1. A female alfalfa leafcutting bee *Megachile rotundata*. Photo courtesy of Jason Gibbs.

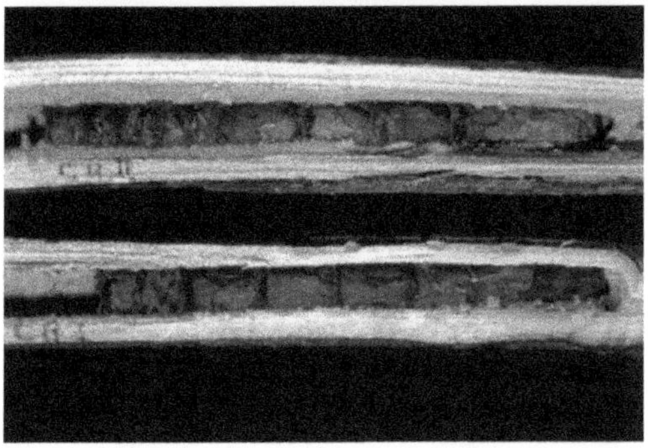

Fig. 9.2a. Natural leaf-lined cells of *Megachile rotundata* in a hollow stem. Photo courtesy of Theresa Pitts-Singer.

foraging trips to provision it (Klostermeyer and Gerber, 1969). Even though a female makes 400–600 leaf collecting trips in her lifetime (Fairey and Lefkovitch, 1994), the economic benefit of her carrying out pollination on her provisioning trips exceeds the cost to photosynthesis caused by her foraging for leaf material.

After a cell is formed, the female gathers nectar and pollen as food for the larva. The average provision mass is 64% nectar and 36% pollen. The female lays an egg on the surface of the nectar–pollen mass and caps the cell with round leaf pieces which form the base of the next cell. She makes, provisions and lays eggs in successive cells, sometimes failing to exploit the entire space available to her (Fig. 9.2b). Lastly, she makes a 0.6-cm entrance plug out of round leaf pieces. A female builds an average of four to seven cells in a tunnel before she plugs it and repeats her labours in another tunnel. Under ideal conditions in a greenhouse, one female can produce two cells per day and achieve lifetime production of 57 cells in 7–8 weeks (Maeta and Kitamura, 2005). In more natural conditions, however, the reproductive output of females is hugely resource-dependent (Peterson and Roitberg, 2006) and is negatively impacted when bee densities are too high for available floral resources (Pitts-Singer and Bosch, 2010). In general, a female will make about 28 cells in her lifetime under controlled conditions and 16 in unmanaged feral populations. Females may take over nests from other females. About 3–5% of all nests contain progeny of more than one female (McCorquodale and Owen, 1994).

Males cluster in nests or other cavities at night and their numbers dwindle after the females begin nesting. Females spend the night in the nest facing inward. They turn and face the entrance as temperatures

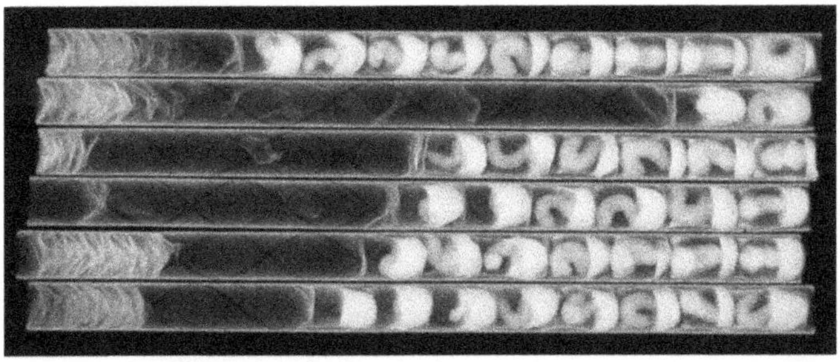

Fig. 9.2b. X-ray images revealing interiors of commercial straws of *Megachile rotundata* showing successive linear cells, each partitioned from the next with a section of leaf and each possessing a bed of pollen and a larva at varying stages of maturity. Photo courtesy of Theresa Pitts-Singer.

rise in the morning, but they rarely fly until temperatures exceed 21°C. The number of nesting females drops sharply 6–7 weeks after their emergence. Males live for 15–23 days and females for about 30 days (Richards, 1984).

The ALCB larva is distinctive for its rapid rate of development, consuming all its pollen provisions in only 3 days and advancing quickly through five larval instars. After spinning a cocoon, the larva morphs into a non-feeding quiescent prepupa. This is the stage in which the ALCB overwinters, with diapause breaking the following spring with increasing temperatures or when the cells are artificially incubated. In some regions including the north-western USA and Nevada, however, 10–20% of prepupae skip diapause and emerge in the same season as their parents. These are called second- or summer-generation adults.

9.1.2. Alfalfa leafcutting bees as pollinators

The ALCB has distinguished itself as an efficient and practical pollinator of alfalfa (lucerne). Alfalfa is largely self-fertile but requires the mechanical release of pollen by a flower visitor. The flower has a sexual column of fused stamens and pistil held under tension within the keel. When a bee lands on the flower, the sexual column is released, often striking the bee with force. This flower 'tripping' is crucial to the performance of any flower visitor. On average, female ALCBs trip 78% of single flowers they visit, and males trip 51%. This is significantly better than single visits of the honeybee which accomplish 22% flower tripping (Cane, 2002). Each female ALCB trips enough flowers in her lifetime to produce up to 0.1 kg of seed. These bees can nest in many kinds of portable artificial and natural cavities above ground and be propagated under a range of climatic conditions. These are significant advantages over another candidate alfalfa pollinator, the native soil-nesting alkali bee *Nomia melanderi* (see section 9.2) whose nests are permanently fixed in localized patches of particular soil types. ALCBs, when actively employed in field shelters, predictably forage in straight transects away from and back to their home shelters, and evidence suggests that 97.4% of pollen moved by ALCBs is delivered within 2 m of the originating plant (Amand *et al.*, 2000).

ALCBs made alfalfa (lucerne) production possible in areas of western North America where it was formerly not profitable. In 1950, Canada was

importing alfalfa seed to meet domestic demand, but by 1988 western Canada was exporting 1.1 million kg of seed (Richards, 1993), a turnaround explained largely by the successful introduction of ALCBs. Pitts-Singer and Cane (2011) explain how today the contribution of ALCBs to crop pollination happens at a commercial scale rivalled only by managed honeybees. In 2004, ALCBs were responsible for 46,000 metric tonnes of alfalfa seed produced in North America, representing two-thirds of world production for that year. Yields can exceed 1100 kg/ha.

Despite its name, the ALCB has shown promise as a pollinator of crops beside alfalfa (lucerne), namely red clover (Fairey *et al.*, 1989), lowbush blueberry (Stubbs and Drummond, 1997; Javorek *et al.*, 2002), alsike clover, birdsfoot trefoil, cicer milkvetch, crown vetch, sainfoin, white clover, white sweet clover, yellow sweet clover (Richards, 1990, 2020), hybrid canola seed (Pitts-Singer and Cane, 2011) and native legumes grown for wildland restoration (Cane, 2006).

ALCBs also perform well in caged environments, as shown in studies with hybrid carrot (Davidson *et al.*, 2010), hybrid canola (Soroka *et al.*, 2001), male-sterile soybean (Yang *et al.*, 2005) and numerous perennial legumes (Richards, 2020). They are generally reliable foragers under greenhouse conditions (Pitts-Singer and Cane, 2011), and their potential as fruit and vegetable pollinators under glass seems underexplored.

ALCBs do not perform well in all crops or contexts. Although they visited lowbush blueberry in Maine at low temperatures between 13.5–23°C, they also experienced lethal low temperatures while pollinating the same crop in the Canadian maritimes, highlighting that ALCBs are essentially a summertime pollinator (Sheffield, 2008). In field contexts they are found to be easily distracted to non-focal crops; ALCBs are, in fact, moderately catholic in taste and known to visit flowers of 21 species of plants representing 14 genera and seven families (Small *et al.*, 1997).

9.1.3. Recommended bee densities

Target ALCB densities for alfalfa (lucerne) in Canada are 61,000–74,000 bees per hectare and in the USA 100,000–150,000 bees per hectare (Pitts-Singer and Cane, 2011). Growers favour high bee densities because they result in rapid pollination; computer simulations do support that pollination is completed more rapidly the more bees are introduced (Strickler, 1996, 1997). These simulations are consistent with field evidence that individual bees do not forage widely on a single foraging trip. A focal *M. rotundata* forager has only a 56% probability of leaving a given raceme on a given plant on a given foraging bout (Strickler and Vinson, 2000), a liability that large numbers of bees can be expected to overcome. However, the gains in high bee densities come at the cost of rapid depletion of floral resources (Strickler and Freitas, 1999), and reductions in brood survivorship, number of cells produced per bee, percentage of cells producing viable prepupae by summer's end and percentage of cells producing adults (Pitts-Singer and Bosch, 2010). One consequence of this state of affairs is that US production of live bee progeny is not self-sustaining (Pitts-Singer and James, 2008), and US growers consistently purchase bees from Canada to maintain current stocking levels. For several decades now there has been a tension at work in the bee–alfalfa propagation system in which the grower must compromise the speed with which the crop is pollinated with the desired increase in bee numbers (Stephen, 1981).

9.1.4. Rearing and managing alfalfa leafcutting bees

ALCBs accept artificial nesting tunnels made from a variety of materials. Taking advantage of this behaviour, growers and researchers have developed methods for mass-rearing of ALCBs. Management is based upon protecting cocoons during their dormancy, activating dormant bees in time to pollinate the crop, and providing nesting holes and shelters in seed fields while bees forage, pollinate and provision their nests. The *loose-cell* and *solid board/ phaseout* rearing systems are the most common methods used in North America.

9.1.4.1. Cold storage and incubation

Cells of immature ALCBs are kept in cold storage for most of the year and then incubated to synchronize bee emergence with crop bloom.

Nesting boards are removed from fields in mid-August to September and held at ambient temperatures for 2–3 weeks. This gives the larvae time to moult into prepupae and spin cocoons. Cocoons are then kept at 4–5°C and 40–60% relative humidity, either in their boards (solid board/phaseout system),

or punched or stripped out of the nesting boards and stored as loose cells (loose-cell system), until the following spring.

In spring, nesting boards or loose cells are switched from cold storage to incubation at 30°C and 50–60% relative humidity; this process begins about 21 days before the anticipated beginning of crop bloom. Wasp parasitoids emerge at this point, preceding bee emergence, and this is the grower's best chance to control them for season-long benefit (see section 9.1.4.5). If the weather turns cool and crop bloom is delayed, bee emergence can be delayed by lowering the incubation temperature to 15–20°C during days 15–19 of incubation (Rank and Goerzen, 1982). The goal is to synchronize bee emergence with crop bloom at the 25–50% flowering stage (Baird and Bitner, 1991). For the loose-cell system, bees are moved into field shelters when ~75% of females have emerged. With the solid board/phaseout system, growers generally move boards to shelters upon emergence of males, letting the rest of the males and females emerge in the field. In any case, moving bees to field shelters should only be done on warm, still days. If temperatures are cold the bees will be sluggish and birds may eat them, and if conditions are windy the bees will have trouble orientating to their nests.

9.1.4.2. Nesting materials and shelters

Bees must be provided nesting materials with some kind of artificial tunnels. Nesting materials and the bees are, in turn, housed in a field shelter to protect bees from weather and provide them an orientation landmark while they provision their nests. After the pollination period, cocoons are retrieved and stored for the next season.

Here are the most common nesting materials:

- *Solid board* – a solid wooden board, ~120×15×7 cm with ~2000 drilled holes per board, each hole 5–7 mm diameter × 95–150 mm deep (Fig. 9.3). These are used in the solid wood/phaseout system (section 9.1.4.4).
- *Removable-back solid board* – similar to above except the back can be removed to punch out cells. These are used in the loose-cell system (section 9.1.4.3).
- *Laminated grooved board* (Fig. 9.4) – a pair of wood or polystyrene boards with opposing grooves that form tunnels when strapped together. These nests can be taken apart to remove cells and

Fig. 9.3. Solid wood nesting boards in field shelter. Photo courtesy of James Cane.

Fig. 9.4. Wood laminate nesting board. Photo courtesy of Theresa Pitts-Singer.

sanitize the boards. These boards are suitable for research or for monitoring, but it is hard to bundle them tightly enough to exclude the tiniest parasites, and their use is limited in commercial production. They are appropriate for the loose-cell system.

● *Paper nest board* – like laminated nest boards but designed to be used only once.

Nesting materials and bees are housed in proper field shelters to ensure good bee activity and propagation. Most shelters have three sides, a roof and a floor (Figs 9.3 and 9.5), but designs are variable

and include modified tractor trailers and school buses. The roof should have a 30–46 cm overhang to protect nest boards from direct sunlight; some growers use an awning to further protect nest boards. There should be a 10–15 cm gap between the top of the sides and the roof to let hot air escape from inside the shelter. Shelters should be painted yellow, blue or green and marked with various geometric symbols to help bees find them (Richards, 1996). Bees orientate best to large objects in a field and work most efficiently when they are housed in large shelters. It is convenient to build shelters on trailers so that they can be moved to other late-blooming fields or away from insecticide applications. If there is a problem with birds preying on bees, the fronts of shelters can be covered with a 5 cm mesh wire screen. A smaller mesh size may injure the bees as they fly through the screen.

Shelters should be large enough to accommodate at least 60,000–80,000 nest holes (Stephen, 1981). Shelters with 30,000–90,000 holes can provide enough female bees for 0.8–2.4 ha, but their small size increases the tendency for bees to drift, especially if there are larger shelters nearby. Large shelters with 150,000–750,000 nesting holes can provide for 14–20 ha, but they are costly to build and may increase problems with bee parasites and diseases.

Fig. 9.5. Field shelter for alfalfa leafcutting bees. Photo courtesy of James Cane.

Nest boards are placed inside the shelter, arranged back to back in rows (Fig. 9.5). There should be at least 60 cm between facing nest boards; ALCBs must be able to fly freely within the shelter. The faces of nest boards are painted with various shapes and symbols to help bees orientate to their own nest holes (Fig. 9.6).

Shelters in most regions should face south or south-east so that the nests are shaded after 10 a.m. Some growers attach an awning to the front of the shelter so it can be faced south or south-east to increase interior ventilation by prevailing winds. Bees fly only after sunrise and when ambient temperatures are at least 21°C (Lerer *et al.*, 1982). Owing to the comparatively short foraging range of ALCBs (Strickler, 1996, 1997), it is a good idea to evenly distribute shelters within the field as well as along the edge.

9.1.4.3. Loose-cell rearing system

Cells are removed from nest boards with this system, being stored and incubated loosely in trays. Bees and emerging cells are released in field shelters in spring and provided with clean, empty nest boards for nesting. This method allows for better control of bee parasites and nest destroyers, reduces space requirements during cold storage, and reduces spread of disease. It is, however, a labour-intensive method.

The loose-cell method requires a removable-back solid nest board or a grooved laminate board (section 9.1.4.2). The nest boards are opened and the cells punched or stripped out after the larvae have moulted to prepupae at the end of the nesting season. Cells are then sifted through a screened tumbler in order to remove loose leaf pieces, chalkbrood-infected cells and many of the insect enemies. Cells are put in large, covered containers and placed in cold storage.

In spring the loose cells are poured into shallow trays with screen lids at 30°C (section 9.1.4.1). Each tray is 61×61 or 61×91 cm and holds about 7.6 l of cells. Incubation rooms are fitted with racks to hold large numbers of trays. There should be about 3.8 cm between trays to allow for air circulation. It is important to maintain temperatures carefully because loose cells are more vulnerable to temperature extremes than are cells insulated in

Fig. 9.6. Nest boards with letters and symbols to encourage accurate bee navigation to nest holes. Photo courtesy of James Cane.

nest boards. By days 21–24 the first males, and maybe some females, are starting to emerge and the trays are ready to go to the field shelters.

Before trays and empty nest boards are put into the field shelters, the empty nest boards are sanitized by kilning them at 127–149°C for 24 h (Stephen, 1982). It is also a good idea to spray field shelters with a 3–6% calcium hypochlorite (bleach) solution to help control chalkbrood disease. Trays are placed inside shelters in the shade, and the bees are released by removing the top screen. Bees will mate and females begin nesting in the nest boards, and trays will be empty of live bees within 1 week or so. If the nesting rate is high the grower may need to add more nest boards to the shelter. It may be necessary to move the shelter at night to another blooming field if the field finishes blooming before the last bee cells are provisioned.

9.1.4.4. Solid wood/phaseout rearing system

Diapausing larvae are kept in their solid nest boards during cold storage, and nest boards are reused repeatedly (Fig. 9.3). The disadvantage is that reused nesting material accumulates disease spores and insect enemies over time. Hence, old nest boards should be *phased out* every other year; this happens when bees are returned to field shelters at the start of crop bloom.

To phase out old boards, the boards are first switched from cold storage to incubation 21 days before bloom as usual. When the boards are taken to the field, they are placed in *phaseout boxes* next to the field shelters. A phaseout box can be any kind of large enclosure (a tractor trailer works well) with numerous slits 5×15 cm cut into the walls. The field shelters are fitted out with newly sanitized empty nest boards. As bees emerge from their nest boards in the phaseout box, they are attracted to light, fly out through the slits, and begin nesting in the clean boards in the nearby shelter. The old nest boards are then reamed out with a mechanical corkscrew device to remove nest debris. Boards are then sanitized either by kilning them (wood boards only) at 127–149°C for 24 h (Stephen, 1982), by dipping them in bleach (Mayer *et al.*, 1988), or, in Canada, by fumigating them with paraformaldehyde; in the USA neither paraformaldehyde nor methyl bromide are registered for this use (James, 2005).

Boards full of bees but not yet scheduled for phaseout are placed directly in the field shelters from incubation. The grower should provide 1–1.5 empty sanitized boards in the shelter for every full board to allow for population expansion.

9.1.4.5. Alfalfa leafcutting bee enemies

Up to 20% of ALCB cells produced in the USA are parasitized by wasps (Pitts-Singer and James, 2008), chiefly the European *Pteromalus venustus* and the native *Tetrastichus megachilidis* (Pitts-Singer and Cane, 2011). The end of season tumbling process does not eliminate these parasites as they are protected inside the cocoons. Rather, they are controlled in spring with dichlorovos insecticidal strips placed in the incubation chambers or by ultraviolet lights placed over pans of soapy water that attract and drown the wasps (Pitts-Singer and Cane, 2011). Many nest destroyers are removed in loose-cell systems during the autumn tumbling procedure. With solid board/phaseout systems, the phaseout box exit slits can be fitted with special excluders that let the bees escape but trap certain nest destroyers such as chequered beetles.

It is important to stress the importance of tumbling cocoons (section 9.1.4.3) and sanitizing nesting materials (section 9.1.4.4) to restrict chalkbrood disease, the most significant brood pathogen of ALCBs. With either rearing method, it is good practice to spray field shelters with a 3–6% bleach solution before adding new boards and bees. The chalkbrood pathogen *Ascosphaera aggregata* infects and kills larvae and persists as infective spores in dead cocoons. The primary mode of spread is in spring when emerging bees chew through cells of dead siblings and pick up spores, which they in turn spread to other tunnels as they search for nest sites (Rank *et al.*, 1990). There is another surge of contagion in those regions where a second generation of bees emerges in summer and again chews through the cells of infected siblings (Vandenberg *et al.*, 1980). The loose-cell system removes the need for emerging adults to chew through infected siblings, but there is still enough spore release during routine handling to infect new cocoons (James and Pitts-Singer, 2005). Treating cocoons in a loose-cell system with the fungicide iprodione did not significantly reduce field incidence of disease (James, 2011). In short, practical chalkbrood control is a nagging problem in commercial ALCB culture.

9.2. Alkali Bees[i]

The alkali bee (*Nomia melanderi*, family Halictidae) is a native of western North America and the world's only intensively managed soil-nesting bee. It is a floral generalist but has distinguished itself as a pollinator of commercial alfalfa (lucerne) seed and to a lesser extent onion seed.

9.2.1. Biology

The alkali bee is solitary and nests gregariously in places where the soil is sub-irrigated over a hard-pan layer with relatively bare alkali spots and salty surfaces (Fig. 9.7). Due to its narrow soil-nesting requirements, its distribution is patchy, but nest sites can nevertheless accumulate into huge metapopulations. For 8 years, Cane (2008) monitored 24 nest sites in a 240 km² watershed in the Touchet Valley in south-east Washington, during which time the bee numbers increased ninefold to 17 million females, the largest reported metapopulation of non-social bees in the world. This population was shown by later genomic analyses to be largely panmictic (freely interbreeding), but to also show signs of low genetic diversity and recent genetic bottleneck events ranging in time from glacial floods to recent pesticide kills (Kapheim *et al.*, 2019).

Bee beds can be long lived, in fact exceeding records for any other soil-nesting bee in the literature. Cane (2008) documents specific bee beds continually active for 33, 34 and 50 years and notes cases in which beds inundated by flood silting were rapidly restored following deliberate bee reintroductions (see 9.2.10). The ecological robustness of this native bee is one of its most important assets for growers who use it in commercial pollination. Populations exported to California did not fare so well (Wichelns *et al.*, 1992), in part because the success of *N. melanderi* in its native habitats owes no small measure to a culture of bee awareness and good stewardship by the farmers there.

The original description of the species was based on a specimen collected at Yakima, Washington. Adults are about two-thirds the size of a worker honeybee and black with metallic-coloured bluish, greenish or yellowish bands circling the abdomen. Females (Fig. 9.8) have stingers but rarely use them. Males have white-coloured faces and long pointed antennae.

Alkali bee adults emerge from the soil in late spring or early summer, depending on temperature and moisture of the soil. Emergence is delayed if temperatures are cool or the ground wet. The active season varies from late May to early September. As typical for most solitary bees, males emerge first by up to 1 week or more. Newly emerged females are receptive, and mating takes place for about 30 s on the ground after which females are no longer receptive. Males continue harassing females in attempts to copulate, but single mating events for females

---

Fig. 9.7. Habitat and alkaline soil types typical of *Nomia melanderi* nesting sites. Note the white salty surface of the bee bed. Photo courtesy of James Cane.

Fig. 9.8. A female alkali bee, *Nomia melanderi*.

seem to be the norm (Mayer and Miliczky, 1998). Females begin nesting soon after they mate. They prefer to use pre-existing holes in the ground but readily excavate new ones if necessary. Each female makes and provisions a single nest, constructing 15–20 cells over a course of about 30 days (Bohart, 1955). Females provide rudimentary brood care. They remove dirt from larval provisions (Batra, 1970) and express a form of defence against pathogens, opening cells infested with fungi and packing them with soil to restrict fungal sporulation (Batra and Bohart, 1969).

The nest consists of a vertical shaft with a lateral tunnel that has oval cells branching from it (see Fig. 2.5, this volume, for general design). Cells may be as deep as 30.5 cm below the surface, but most are 5.1–20.3 cm. They are deeper during dry seasons and shallower during wet seasons. Females line cells with a moisture-resistant glandular secretion. Soil removed from the nest is deposited around the entrance hole to form a *tumulus* (Fig. 9.9).

Females eat nectar and pollen, and begin collecting pollen to provision a nest about 2–3 days after emergence. Daily pollen feeding by an adult female is maintained during her first 2 weeks of intense nesting, but the sum of pollen consumed for her own needs amounts to no more than one-third of that for a single larval provision (Cane *et al.*, 2017). Pollen is

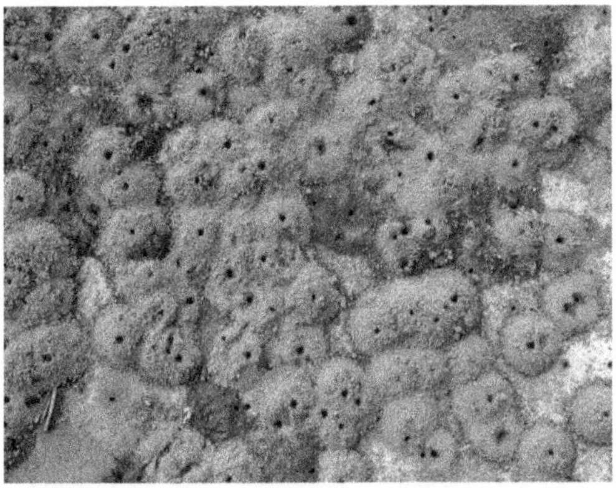

Fig. 9.9. Surface tumuli of subterranean *Nomia melanderi* nests showing salty surface of typical bedding sites. See also Fig. 2.5 (this volume). Photo courtesy of James Cane.

placed at the bottom of each cell in a round flattened ball, about 0.6 cm in diameter and 0.3 cm thick (see Fig. 2.4B, this volume). Females 1–4 weeks old collect enough pollen each day to provision one cell. At night the egg is laid, the cell is plugged, and a new cell prepared for the next day's pollen collection. Eggs are found in cells in Washington from about 17 June to 26 July. Eggs hatch in 2–3 days and larvae consume the whole pollen ball in 7–10 days, growing progressively larger.

The species is *univoltine*, producing only one generation per year. There are five instars, and feeding larvae are found in cells in Washington from about 23 June to 14 August. The fifth instar is divided into the *predefaecating stage* and the *prepupal stage*. During the predefaecating stage of 4–6 days, the body appears bloated and brown faeces are visible inside the larva. Larvae defecate by smearing faeces on the sides of the cell wall over 2–3 days. The prepupa is opaque white, has a sharp angle between the head and the thorax, and has prominent humps on the back. This is the overwintering diapause stage of the alkali bee which lasts 10–11 months. Prepupae are first found in Washington cells from about 3–9 July.

Increasing soil temperature in spring breaks diapause, and the prepupa changes to the pupa around the end of May. This stage lasts 15–20 days, with the pupa gradually darkening into adult coloration. The pupa changes to the adult which remains in the cell for a short time before digging to the surface. This generally occurs during June in Washington. Individual females live 4–6 weeks. Females will be present and active in a given area for about 60 days under good weather conditions.

9.2.2. Alkali bees as pollinators

The native alkali bee is a major pollinator of the introduced alfalfa (lucerne) plant and enjoyed its heyday in the 1960s and 1970s before ALCBs became more widely used as pollinators (Bohart, 1972; Torchio, 1987).

An alkali bee female trips an average of 92% of the alfalfa (lucerne) flowers she visits as she gathers pollen for her nest; 48% of tripped flowers set pods and each pod averages 2.7 seeds (Cane, 2002). The significance of male alkali bees is not to be overlooked. Cane (2002) found that a male had a 41% chance of tripping a flower, significantly better than a single honeybee visit at 3%. Curiously, the agency of bees at moving pollen remains unclear. Cane (2002) cites his own unpublished observations that 15 flowers visited once by pollen-foraging *N. melanderi* retained as much pollen as 15 flowers that had been manually tripped. If simply exposing pollen is the outcome of tripping, then nectar foraging bees may be as good as pollen foragers at effecting cross-pollination.

Under good conditions, each female will trip about 25,000 alfalfa (lucerne) flowers during her lifetime which translates to ~73 g of clean seed. A female will provision 10–20 nest cells in her lifetime (Hobbs, 1956; Johansen *et al.*, 1978). Bees tend to limit foraging to within 1.6 km of the nest, although distances up to 11 km have been noted.

Alkali bees seek out untripped flowers and flowers deep inside dense lower foliage (Batra, 1976) and fly in cool, windy weather. Alkali bees prefer alfalfa (lucerne) for its pollen whereas honeybees prefer it for its nectar (Torchio, 1966). Alkali bees work alfalfa equally well in fields that are water-stressed or well-watered, but honeybees have trouble tripping flowers in well-watered fields. Thus, pollination with honeybees requires more careful water management to provide optimum pollination conditions. This is an important advantage to alkali bees in areas like the San Joaquin Valley of California where water stress is difficult to manage (Wichelns *et al.*, 1992).

The value of alkali bees was highlighted in a case history from a central California water district (Wichelns *et al.*, 1992). During the 1960s, several alfalfa (lucerne) growers began installing artificial alkali bee beds and importing bees from Washington and Oregon. During this time, average seed yields jumped from 627 kg/ha in 1960 to 1064 kg/ha in 1971. In 1970/71, however, alkali bee populations declined sharply; possible causes were insecticide applications in the area, insecticide residues in the water used to dampen bee beds, and bee parasites. Alfalfa yield and acreage declined rapidly thereafter, and alfalfa acreage has never recovered to pre-1971 levels. Thus, we see the rise and fall of a local alfalfa industry coincidental with the presence of alkali bees.

Today, alkali bees are largely supplanted as commercial pollinators by the exotic ALCB. However, for two or more decades theirs was a remarkable case history of 'a new pollinator brought up from the ranks of native bees' and 'domesticated' for commercial pollination purposes (Buchmann and Nabhan, 2012, p.188). As the world's most intensively managed soil-nesting bee and a native to North America, alkali bees are both a valuable natural resource for the western USA and a model for managing native soil-nesters elsewhere.

9.2.3. Recommended bee densities

Alfalfa (lucerne) seed growers purchase ALCBs to pollinate the crop (section 9.1), but growers in alkali bee areas also adjust their ALCB purchases according to projections of local alkali bee populations. Bee managers have relied on labour-intensive soil coring methods to direct-count pre-pupae and to project bee numbers for the following season. Vinchesi and Walsh (2014), however, developed a much simpler technique based on counting number of emergence holes in 0.5×0.5 m quadrats on the surface of bee beds during peak *N. melanderi* activity in July. Their results closely correlate with results of the coring method and can predict bee populations for a bee bed. The quadrat method has the added advantage of not requiring destructive sampling. A 0.5×0.5 m sampling quadrat is made from 1.27-cm diameter polyvinyl chloride pipe. The quadrat is randomly tossed on to the bee bed during active flight in July and the number of exit holes per quadrat counted. This is repeated 24 times per bed to obtain a robust average quadrat count (*mq*) per 0.5 m², which is multiplied by ⅔ to adjust for the vacancy rate observed by Cane (2008). A modification of their Equation 1 (Vinchesi and Walsh, 2014) is shown here for deriving abundance of *N. melanderi* (*Ab*) at bee bed *x*:

$$Ab = \left(\left[\frac{mq}{0.5} \right] \times [2/3] \right) \times A \qquad (9.1)$$

where A = area of bee bed x in m². According to Johansen and Eves (1971), good natural nest sites should contain 2.5 million or more nests per hectare with the most heavily nested portions approaching 8 million nests per hectare. Maximum abundance in artificial beds can approach 13.6 million nests per hectare. The number of foraging female bees required in an alfalfa (lucerne) field is not fully known, but it should probably exceed 7410 per hectare.

9.2.4. Qualities of good nesting sites

There are three important factors determining the quality of alkali bee nesting sites, whether natural or managed: (i) soil moisture; (ii) soil composition and texture; and (iii) vegetation.

9.2.4.1. Soil moisture

Soil must be moist down to at least 31 cm. Soil moisture in good sites varies from 8–32% depending on soil type. Soil moisture can be measured using a tensiometer. A reading of 15–25 centibars indicates adequate moisture regardless of soil texture.

Dry nesting sites have been a limiting factor in alkali bee production. Good natural moisture conditions are associated with shallow layers of calcareous hardpans lying a few inches to several feet below the surface. Seepage water may sub-irrigate nearby nesting sites when a shelf of this impervious material lies along a river, canal or pond. Where this occurs, populations of alkali bees may build up naturally with little management. Most nesting sites are man-made and require an artificial water supply provided with shallow ditches dug across or around beds, or some kind of subsurface distribution system (see section 9.2.5).

9.2.4.2. Soil composition and texture

The goal of soil texture management is a moist and moderately compact upper soil horizon which persists throughout the active bee season. The best alkali bee beds have soil classed as silt loams with 2–6% fine silt and 42–68% coarse silt. They contain 13–24% clay and 10–40% sand.

9.2.4.3. Vegetation

The surface should be essentially bare with sparse vegetation. Plants use up soil moisture and bees prefer to nest in bare ground. Nevertheless, a little vegetation can help protect bees from summer rains and reduce wind erosion. See also section 9.2.9.

9.2.5. Building or enhancing bee beds

Until the late 1950s, most nesting sites were of two basic types – *natural/semi-natural* or totally *artificial*. Beginning around 1958, growers in Washington developed a third type – the *semi-artificial* bed.

9.2.5.1. Natural/semi-natural (open-ditched) beds

Many areas in the Great Basin of the north-western USA are naturally ideal for alkali bee nesting. Such areas usually occur in low lying regions of larger valleys where subsurface drainage is poor (on alkali spots). Some of these areas support small populations of non-managed alkali bees. Alfalfa (lucerne)-growing lands that border on rivers or include large alkali spots may be at a special advantage with regard to pollination.

Auxiliary water supplies may be needed at natural nest sites if there is typically a marked drop in soil moisture during the bee nesting period. One can maintain proper moisture conditions in sloping sites by digging basins 46–91 cm deep along the upper border of the bed. Water from these reservoirs seeps downhill under the site. Hardpan or caliche layers at 31–46 cm depth can help direct the lateral movement of water below the surface. It is important not to penetrate these layers when excavating seepage basins.

For beds on level ground, one can dig a series of 31–46 cm ditches around the perimeter of the bed and at 4.6-m intervals through the middle. The objective is to maintain water in these ditches starting about 1 month before bee emergence and continuing through most of the nesting period. This is especially important during dry seasons. Flood irrigation water is available throughout the summer at many locations in south-western Idaho and eastern Oregon. In Washington, several large and highly productive natural sites in the Touchet-Lowden-Umapine region have near-perfect moisture conditions. Surface moisture at these sites remains at optimum levels throughout the nesting

period, and only token auxiliary water supplies are necessary. Ideal situations like this are rare.

9.2.5.2. Semi-artificial (pipeline) beds

Semi-artificial beds have some important advantages:

- they can be built anywhere there is suitable soil and adequate water supplies;
- compared to natural beds, they give a grower greater control over a pollination programme; and
- they are less expensive to install than artificial beds.

The first step is to determine the size of bed required for the crop acreage to be pollinated (see section 9.2.3). The bed must be located as near as possible to the alfalfa (lucerne).

A large and continuous supply of water is necessary for semi-artificial sites. One must plan for 1.9–7.5 million l/ha to moisten the upper layers of a bed before nesting season. Most growers start adding water in April or May, depending on beginning moisture levels. After nesting begins, smaller amounts of water must be kept flowing in the lines throughout all but the last week of the nesting period.

Trenching can begin when the ground is dry enough to accommodate lightweight equipment. Most growers in the Touchet-Lowden-Umapine area dig trenches 61–76 cm deep and 15–20 cm wide. Parallel trenches should be dug across the bed at 2.4–3.1-m intervals. This degree of spacing permits water to distribute uniformly.

A number of piping materials can be used. Rigid PVC pipe, 6.4–7.6 cm in diameter, is easy to work with and can be easily cut and joined to other sections with cementing compound. Black flexible plastic piping, 2.5–3.8 cm in diameter, is less expensive and can be purchased in large rolls and easily transported to a site. With either piping material, it is necessary to drill one or two 1–1.3 cm holes in the pipes at 15–61-cm intervals along their full lengths before laying them in the trenches. This allows for an even flow of water into the surrounding bed. Some growers use 7.6–10.2 cm diameter corrugated polyethylene drainage tubing which comes with 24 perforation slots per foot. For a little more money per foot, it is possible to get this material with a fine mesh wrap which reduces silting.

A 25 cm layer of clean round gravel is laid inside the trench to facilitate movement of water. A thin layer of straw or coarse sand placed over the gravel can help prevent downward movement of fine soil. Rigid PVC piping can be laid flat in the trench on top of the gravel layer. Elbow sections with short vertical pieces that extend to the bed surface are attached at both ends. One vertical end-piece is the downspout; the other is a breather tube. If one is using flexible pipe material, it is laid in the trench so that both ends angle gradually to the bed surface. Pipes should not exceed 30.5 m in length because longer lengths impair good water distribution. Once pipes are in place, soil can be returned to the trenches and watered down to settle the matrix into the trenches. It is important not to use too much water or fine soil may wash into the gravel area.

A water main must be brought to the new bed from a well or other source. Some system is needed to deliver water equally to all downspouts. Most growers use a mainline supply trunk lying along one edge of the bed with a separate spigot and hose serving each downspout. Watering must begin in May or earlier depending on the dryness of the soil. Large quantities of water will be needed where soil is dry (above 25 centibars).

9.2.5.3. Artificial (plastic-lined) beds

Artificial bee beds are an effective way to concentrate populations of alkali bees at desired locations. However, there are some important limitations:

- Construction costs per unit area are higher than for semi-artificial beds.
- Due to high costs, artificial beds are usually smaller and generate smaller numbers of bees.
- Moisture management is more difficult in artificial beds. They can quickly become too wet or too dry.
- Artificial beds require renovation more often and tend to go through 'boom and bust' cycles.

First, an artificial nest site must be prepared by excavating soil to a depth of 46–91 cm with a backhoe. Artificial beds are commonly sized 9×18 m which happens to correspond to the size limit of the commercially available polyethylene plastic material used to line the floor of beds. Some growers overlap sheets of plastic to increase the size of their artificial beds.

The floor of the pit must be carefully levelled once excavation is complete. This helps ensure uniform distribution of water. The floor and sides of the pit are lined with a sheet of 0.15–0.2 mm polyethylene plastic. It is very important not to tear the plastic liner from this point on. It must provide an impervious underground reservoir. Clean round gravel, 2–2.5 cm in diameter, is spread over the plastic to a depth of 20–30 cm. On top of that goes 5–8 cm of coarse sand to help protect the gravel from clogging with fine soil.

Concrete or clay downspouts, each 20–25 cm in diameter, are placed vertically in the bed for adding water. These should extend from the gravel layer on the floor to the surface. Downspouts are set upright on mounds of gravel raised 7.6–10.2 cm above the rest of the gravel layer. The recommended density is one downspout for every 37–56 m^2 of bed surface.

Soil of proper texture is then placed into the prepared pit (see section 9.2.4). The amount of water that must be added through the standing downspouts depends on the condition of the site in late spring. When a bed is new and the backfill soil is relatively dry, one can estimate the required amount of water on the basis of soil volume in the bed. A bed with optimum amounts of sand and clay should have 20% moisture at the brood cell level (5.1–20.3 cm) in late spring. It is important to test the moisture level of backfill soil and, if necessary, make cautious amendments. Adding too much water and saturating the nesting ground can cause serious problems, especially if warm weather does not dry the area sufficiently. High moisture levels can delay emergence of bees, encourage growth of pathogenic fungi and cause female bees to fly away to better locations. In established beds, one watering through the standing downspouts in late May will usually last all season.

9.2.6. Surface moisture

Any of the three types of nesting sites – natural/semi-natural, semi-artificial and artificial – can be sprinkled with 1.3–2.5 cm of water before bees emerge, to provide an attractive, dark surface at the start of the nesting season. This is advisable only if the surface is excessively dry and if sprinkler action will not harm the soil structure.

9.2.7. Late-season moisture

Additional moisture is not needed after bees have finished nesting. If there is subsequent high precipitation during winter and early spring the excessive moisture levels can increase the rate of prepupal deaths. It is preferable to let sites dry out during late summer to reduce microbial growth and spoilage problems.

9.2.8. Surface salting

Evaporation causes alkali spots to develop on the soil surface of natural beds due to the deposition of salts. The salt is beneficial because it seals in moisture and reduces weed growth. Surface applications of supplemental sodium chloride (common salt) are required at about 4.8 kg/m^2 in the soils of southeastern Washington. New sites may require up to twice that rate. An annual application of 0.6–2.4 kg/m^2 ($\frac{1}{8}$–$\frac{1}{2}$ lb/ft^2) is usually sufficient for maintaining established beds. A light, fluffy soil layer on the surface indicates high calcium content and a need for sodium salts. Soils naturally high in sodium become too hard and crusty if salt is added to them.

9.2.9. Vegetation management

Alkali bee nest sites should principally be kept bare with, at most, only a light cover of vegetation. Heavy plant growth can interfere with nesting activity and use up soil moisture. Roots can penetrate nest cells, killing prepupae. However, a little plant growth, preferably in strips, is desirable because it helps protect bees from wind and rain. Chemical herbicides are the most effective way to control weeds. Heavy applications of long-lasting residual soil herbicides do not seem to harm any stage of the alkali bee.

9.2.10. Attracting and establishing bees

The final and critical step in developing a new bed is establishing a population of bees. This may be easy if the new site is close to existing beds which support large and growing bee populations. Surplus bees from these established sites will move to the freshly prepared surfaces which are relatively clear of vegetation.

If a new bed is isolated from existing populations, bees must be introduced to the area either as adults or as immatures. Several methods have been developed for transplanting bees from one location to another, sometimes over great distances.

The most successful and widely used method is to transplant cores or blocks of soil containing

prepupae. This must be done in spring (April) before the immatures transform into pupae or adults. Cores are placed on pallets and loaded on to a truck for transport. They should be covered with moist canvas or burlap if they are being transported long distances. In this manner, several thousand cores can be moved by one semi-truck. When the cores arrive at a new site, they should be buried in 30 cm trenches and puddled in with sparing amounts of water. Soil at the new site should be properly prepared before the transplants are installed. It should have about the same moisture content as the core. Cores should be installed in straight lines at least 10.2–15.2 cm apart to ensure good soil moisture contact. They should never be stored for any length of time before burying them; this increases the risk of desiccation and decreases the viability of prepupae.

9.3. Orchard Mason Bees

Bees in the genus *Osmia* (family Megachilidae) have proven themselves effective pollinators of apples and other orchard fruits.

9.3.1. Biology

These solitary bees nest in hollow reeds or pre-existing holes in wood, but they can be found in abandoned nests of mud dauber wasps or ground-nesting bees. Some species build free-form mud nests against rocks or other solid surfaces (Cane *et al.*, 2007). They nest in large aggregations if nest holes are abundant. In a pattern we have encountered before, *Osmia* spp. bees partition their linear cells and seal their nests with substrate collected from the environment, in this case mud, chewed leaf material or a mixture of both – hence, they are called *orchard mason bees*.

As far as their biology and pollination management are concerned, the four most well-known species are *Osmia cornifrons*, native to eastern Asia (Batra, 1982, 1989); *O. lignaria*, native to North America (Torchio, 1976); *O. cornuta*, native to Europe (Bosch and Vicens, 2006); and *O. bicornis* (formerly *O. rufa*), also native to Europe (Gruber *et al.*, 2011). These four commercially managed species belong to the subgenus (*Osmia*) which contains 25 species in total and belongs to a clade conspicuous for its polylectic foraging habits (see section 1.5, this volume). In fact, polylecty is the ancestral state of the genus with oligolecty having evolved only twice independently in two diverging lineages (Haider *et al.*, 2014). It is worth considering, however, whether the consistent preference of *Osmia* for plants in the family Rosaceae, including many important orchard fruits, better qualifies them as mesolectic foragers. In any case, the *Osmia* spp. are an interesting group for studying the evolution of foraging specialization, and for our purposes the relative polylecty of the four commercial species adds to their value as crop pollinators.

The horned-face bee (*O. cornifrons*) was introduced from Japan into Utah in the 1960s, and from Utah to Maryland by 1978 (Batra, 1989); it has since become established in many areas of the eastern USA and Canada, but its use as a pollinator in North America remains poorly developed (McKinney and Park, 2012). In Japan it is the primary pollinator of commercial apples (Sekita, 2001). Even though *O. cornifrons* is an exotic species in North America, its preference is for crop families in the Rosaceae, themselves non-native plants to the continent (Vaudo *et al.*, 2020).

The most important orchard mason bee in North America is the native blue orchard bee *Osmia lignaria*. The eastern subspecies (*O. lignaria lignaria*) occurs from the eastern slopes of the Rocky Mountains to the Atlantic. The western subspecies (*O. lignaria propinqua*) occurs from the western slopes of the Rockies to the Pacific Ocean. Females of the blue orchard bee have a pair of horn-like projections extending from the lower face. The blue orchard bee is shiny blue/black and about two-thirds the size of a honeybee (Fig. 9.10). The male is about a third smaller than the female and has a white patch of hair on the face and long curved antennae. Females have no white on the face and

Fig. 9.10. A female North American blue orchard mason bee, *Osmia lignaria*. Photo courtesy of Jason Gibbs.

their antennae are about half as long as those of males. *O. lignaria* has been developed as a pollinator of apples and plums (Torchio, 1976), almond (Torchio, 1982) and cherry (Bosch and Kemp, 1999; Bosch *et al.*, 2006).

The orange orchard bee (*O. cornuta*) was introduced from Spain to California almond orchards in the 1980s (Torchio, 1987). It remains a significant pollinator of almond in Spain (Bosch, 1994a) where it has also compared favourably with honeybees as a pollinator of apples (Vicens and Bosch, 2000b). The female is slightly larger than the female blue orchard bee. Its most distinguishing characteristic is an abdomen coated with beautiful, bright orange hair. It also has a pair of horns on the lower face.

Among our four commercial species, the European *O. bicornis* (formerly *O. rufa*) is the most catholic in its polylecty, with foraging records on 19 different plant families (Haider *et al.*, 2014) yet a confirmed affinity for commercially important Rosaceae (Schindler and Peters, 2011). It has been developed as a pollinator of apples (Gruber *et al.*, 2011), sweet cherries (Ryder *et al.*, 2020) and strawberries (Klatt *et al.*, 2014).

Other species, including *O. excavata*, *O. pedicornis*, *O. taurus* and *O. tersula* have shown promise in tree fruits (Sedivy and Dorn, 2014), and the south-western US *O. ribifloris biedermannii* is a potential pollinator of highbush blueberry (Torchio, 1990). Wild populations of *O. atriventris*, *O. inspergens* and *O. tersula* are known visitors of lowbush blueberry in Québec, but they are probably minor pollinators (Morrissette *et al.*, 1985). In Nova Scotia, *O. inermis* and *O. proxima* provision their cells with lowbush blueberry pollen (Finnamore and Neary, 1978).

The general life histories of these species are similar (Bosch and Kemp, 2002; McKinney and Park, 2012; Sedivy and Dorn, 2014). Male and female orchard mason bees are among the earliest bees to emerge in spring, typically on days when temperatures exceed 10°C. Males emerge earlier than females and patrol nesting sites looking for females. When a new female emerges, a waiting male pounces on her to mate. A female may mate with several males.

The female finds a suitable nest tunnel and begins making cells in the back-to-back pattern typical of many solitary bees. Usually only one female nests in a given hole. Females do not forage for cell provisions unless temperatures exceed 12.8°C. The female collects nectar and pollen and makes a pollen mass in each cell. *Osmia* spp. carry pollen with an abdominal scopa. It takes an average of 19.8 foraging trips and ~222 foraging min to collect the requisite pollen; 11.5 foraging trips and ~33 min to collect mud for partitions; and ~51 min in-nest labour to provision one cell (McKinney and Park, 2012). The female lays an egg 3 mm long and sausage-shaped, with one end embedded in the pollen mass. After laying the egg, the female partitions the cell from others with a thin wall of mud or chewed leaf material (Fig. 9.11). Female eggs are laid in cells toward the bottom of the tunnel and males are produced near the entrance; the average sex ratio is 1.7 males per female, but females receive larger pollen and nectar stores as immature forms and are thus larger (Bosch and Kemp, 2002). After she fills the nest tunnel with cells, the female covers the entrance with a thick mud cap. All stages of development occur between 15–30°C. Eggs hatch at about day 7. The larva eats pollen for about 30 days and then defecates. It rests for several days then starts to spin a pinkish-white silk cocoon around itself, weaving faecal pellets into the cocoon's outer layer. In a few days the cocoon turns a dark brown colour, and after about 30 more days the larva pupates. After 2 weeks the pupa moults into an adult. It is these new adults that overwinter in dormancy. Temperatures in winter must be lower than 4.4°C or adults will not be able to break dormancy and emerge in spring. Adult females provision cells and pollinate for 4–6 weeks then die. There is only one generation per year.

Under conditions of unlimited food and ideal nesting sites, such as those found in well managed greenhouses, one *Osmia* spp. female may produce 30 brood cells in her lifetime with up to a third of those becoming females. Conditions are much less predictable in ambient orchard conditions, however, and females there are more likely to produce 2.5–6 cells in her lifetime (Bosch and Kemp, 2002).

9.3.2. Orchard mason bees as pollinators

Osmia spp. have an excellent record as pollinators of tree fruits, especially in the family Rosaceae (Bosch and Kemp, 2002; Sedivy and Dorn, 2014). The commercial pollinating species have extraordinarily high rates of legitimate flower visitation; high fruit-set rates as measured by single-bee flower visits; high foraging fidelity for Rosaceae pollen; short flying ranges which focuses them on the crop; robust flight even in marginal weather conditions; and high rates of crossing between rows and trees which is important for effecting pollination between main varieties and pollenizers. Females only remove about 90% of the pollen from their bodies when they return to a cell, leaving ~10% available on their bodies to effect further cross-pollination (Matsumoto *et al.*, 2009). Males, which occur at higher numbers than females, do not participate in nest provisioning and forage only for nectar, yet they still contribute significantly to moving pollen (Bosch, 1994b).

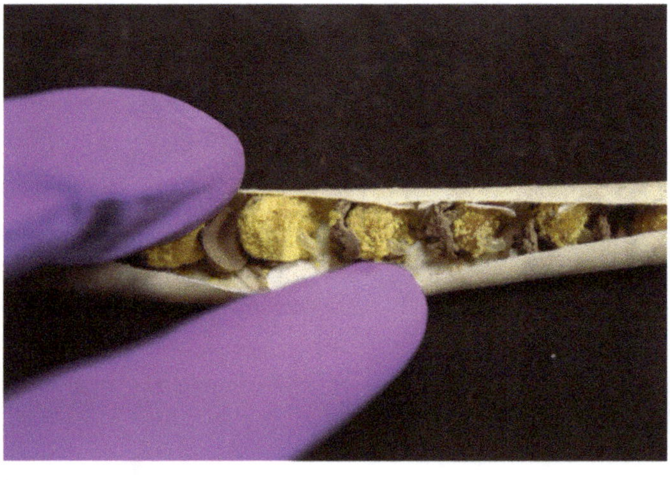

Fig. 9.11. *Osmia lignaria* nest showing the succession of mud-partitioned brood cells enclosed in paper lining. Photo courtesy of Theresa Pitts-Singer.

These superlatives are partially offset by a recurring problem in *Osmia* spp. culture: low bee reproduction on the crop and unpredictable results at establishing local populations. One problem is the high number of females that disperse away from the orchard when they are released in spring. This has been documented with *O. lignaria propinqua* in California almond, where up to 50% of females disperse from the release site before nesting (Torchio, 1982); with *O. cornuta* in California almond (Torchio, 1987); and in North Carolina orchards of Delicious apples with *O. lignaria lignaria*, *O. lignaria propinqua* and *O. cornifrons* (Kuhn and Ambrose, 1984). The problem is made worse when bees are released as loose cocoons (Bosch, 1994c) or when bees are released in habitats with poor floral resources (Bosch, 1995). Another issue is low emergence rates of cocoons introduced to orchards (Monzón *et al.*, 2004), a problem that may be associated with starting wintering too late in autumn (Bosch, 1994a, 1995).

Good pollination management with *Osmia* spp., therefore, goes hand in hand with 'seeding' the habitat with bees and making conditions favourable for their local nesting and population increase. These aspects of their pollination management are covered in the following section.

9.3.3. Rearing and managing orchard mason bees

As discussed in section 9.3.2, one of the persistent problems in *Osmia* spp. pollination management is slow or even negative growth of orchard bee populations. Unsustained local populations at a commercial scale is a sum of adult exodus, poor temperature and rearing conditions, and high post-emergence mortality.

These losses can be simply built into one's management scheme as in Japan, where stocking rates are made anticipating no more than 50% female establishment (Maeta, 1990). However, establishment rates as high as 75% are possible when optimal rearing and storage conditions are met and bees are released directly from their natal nests or from cocoons inserted into nesting tunnels (Bosch, 1994a,c; Bosch and Kemp, 2001).

This leads us to effective target bee densities for crops. It is no exaggeration to state that *Osmia* spp. are unexcelled among crop pollinators when it comes to efficacious flower visitation and pollination efficiency. Effective orchard yields are recorded at bee densities as low as one female bee per 5.5 trees (Table 9.1). As bees are central place foragers (section 3.4.1, this volume), however, and foraging ranges for solitary bees are generally small (Gathmann and Tscharntke, 2002), this table underscores the importance of localized efforts at establishing nesting populations of *Osmia* spp. at the orchard level. Although maximum homing distances are recorded at 1200 m for *O. lignaria* (Guedot *et al.*, 2009) and 1800 m for *O. cornuta* (Vicens and Bosch, 2000a), these measures are derived by marking bees at the nest, moving them prescribed distances, releasing them, and noting bee return rate at the nest. More realistic foraging ranges with an orchard in peak bloom are 100–200 m from the nest (Vicens and Bosch, 2000a). Near-orchard nesting and population focus are the goals.

Osmia spp. bees are gregarious nesters and prefer to nest in near company to conspecifics. Orchard mason bees accept a variety of manufactured nesting materials, the most common being containerized cardboard tubes (Fig. 9.12), solid blocks of wood or polystyrene with drilled holes, or grooved

Table 9.1. Some published effective densities for female *Osmia* spp. bees in crops. From Torchio, 1985, 1990, 1991; Bosch, 1994b; Vicens and Bosch, 2000b; Bosch and Kemp, 2002; Maccagnani *et al.*, 2003; Ladurner *et al.*, 2004; Bosch *et al.*, 2006; West and McCutcheon, 2009; Sampson *et al.*, 2013.

Osmia species	Almond	Apple	Cherry	Pear	Blueberry
O. cornifrons		550 ♀ per hectare			8.3 ♀ per mature bush
O. lignaria	740 ♀ per hectare	618 ♀ per hectare 625 ♀ per hectare	1290–1857 ♀ per hectare		
O. cornuta	3 ♀ per tree 750 ♀ per hectare	530 ♀ per hectare 1 ♀ per tree 1 ♀ per 3.3–5.5 trees		2.5 ♀ per tree	
O. ribifloris					741 ♀ per hectare 2000 ♀ per hectare

Fig. 9.12. *Osmia lignaria* shelter in a field flight cage, with carboard tubes and paper liners. Photo courtesy of Christine Cairns Fortuin.

laminates strapped together to form intact tunnels. Regardless of the substrate, it is advisable to line the tunnels with purpose-made single-use paper linings which improve sanitation and reduce build-up of pathogens. Companies now offer a variety of nesting materials and associated supplies online.

Tunnel dimensions have been a focus of intense research interest. There is a range of tunnel diameters and lengths that have consequences to the biology and reproductive success of the bees. Tunnels too narrow risk a surplus production of males, and tunnels too wide risk rejection by nesting females. With *O. cornuta*, production of female cocoons was highest at tunnel diameters of 9 mm, but tunnel acceptance was highest at 8 mm (Bosch, 1994a). Tunnel length also has consequences on sex ratio. Female investment in daughters is constrained when tunnels are too short to adequately protect offspring from parasites; in such cases females are more apt to produce sons who are fed less copiously and thus represent lower investment costs (Seidelmann, 2006; Seidelmann *et al.*, 2010). In *O. bicornis*, a tunnel length of 15 cm or more provides significant improvement in parasite resistance and, consequently, higher investment in female cocoons (Gruber *et al.*, 2011). Overall, tunnels no shorter than 15 cm, closed at one end, and with 8 mm diameters are a good compromise between female acceptance and optimum production of female progeny.

Solid wood nests seem to be preferred by nesting females, as shown in studies with *O. cornuta*

(Bosch, 1995) and *O. bicornis* (Wilkaniec and Giejdasz, 2003), followed by natural reed and grooved laminates. The laminates are noteworthy for their ease of removing cells for overwintering and parasite control. Females also seem to prefer their natal nest over clean cavities provided by the manager, a response it seems to chemical cues in female cocoons (Pitts-Singer, 2007), but reuse of natal materials risks the accumulation of parasites and infectious diseases.

Nest materials must be housed in some kind of weatherproof shelter. It is possible to make an adequate shelter by stretching a tarpaulin over a frame or by using a lean-to, clean empty drums, garbage cans or overhanging eaves of buildings. Shelters must shield bee nests from rain and direct afternoon sun. They should be secured so they do not sway in the wind. The shelter's opening and nest entrances should face east, south or south-east so that the morning sun warms bee nests and stimulates early flight. Shelters should be ventilated to prevent excess heat build-up. They should be painted a light colour, preferably blue or yellow, but not in a bright metallic finish because the shine may repel bees. It is a good idea to cover shelter entrances with bird netting or with 3.8–5.1 cm poultry mesh in order to repel birds, racoons or other animals that may attack the nests. Bird netting is the preferred material because chicken wire can damage bees' wings.

Owing to the short foraging distances of orchard bees, it is important to distribute nesting shelters regularly throughout the orchard. Vaudo *et al.* (2020) point out that the mesolecty of *O. corni-frons* and its preference for Rosaceae crops suggest that growers should concentrate *Osmia* spp. shelters in orchard interiors, into which wild bees from surrounding habitat rarely penetrate. This of course is a concession to cost and presumes a robust population of naturally occurring wild bees available to service the edges of orchards.

Osmia spp. bees need a steady source of mud near their nests for making cell partitions. If not naturally available, mud must be artificially provided and kept continuously damp. Bee supply houses sell mud packets designed to provide a gradient of soil moisture levels to accommodate most species' needs (Fig. 9.13). These products are especially useful in greenhouses where sources of mud are not normally available.

If the objective is to encourage and augment existing natural populations, one simply needs to put out shelters full of clean empty nesting material in early spring, as soon as bees become active. Chances are good that bees will nest in the shelters. If one is releasing dormant bees overwintered in nest tubes, bees and shelters should be released a few days before bloom to allow time for males to emerge and females to complete development. Bees should not be released as adults because the rate of

exodus from the orchard is extremely high (Sedivy and Dorn, 2014). A large number of small shelters evenly distributed across the orchard not only minimizes adult exodus from the orchard but improves pollination outcomes (Sedivy and Dorn, 2014). Adult exodus is further minimized when females are released in their natal nesting materials or, alternatively, if they are first put inside paper liners as loose cells and inserted in tunnels of nesting material. 'Seeding' female cocoons in nesting materials apparently simulates natal emergence and increases local nesting (Monzón *et al.*, 2004). Male cocoons can be released *en masse* inside shelters protected from rain. If the weather turns cold and bloom is delayed, bee emergence can be delayed by putting the nest material full of bees in a refrigerator at 3.9–4.4°C. Bees must be removed from refrigeration 3–7 days before expected crop bloom. Adult males begin emerging shortly thereafter, and females begin emerging 2–3 days later. It is best not to move shelters once females start nesting.

Nest materials full of developing bees should be removed from the orchard after nesting season and stored, protected from parasites and nest enemies, in a cool, unheated, dark place. At least one manufacturer makes whole-shelter nets for enclosing shelters in place after foraging season to protect the developing brood from parasites. Small-scale

Fig. 9.13. A source of clean moisturized mud is necessary for nesting *Osmia* spp. This product is made to provide a gradient of mud moisture levels to accommodate a variety of nesting needs in greenhouses.

operators may wish to employ a loose-cell overwintering regimen which offers the best chance to control parasites and pathogens. Cocoons are removed from paper liners in late summer and sexed and separated: female cocoons are larger and males significantly smaller. This is the opportunity to cull cells visibly parasitized by drosophilid flies, *Chaetodactylus* mites and chalkbrood fungi (Sedivy and Dorn, 2014). Loose cocoons can be protected against chalcid wasp parasites by covering cocoons in sawdust and attracting and killing wasps with a black light trap. The black light trap will also kill other parasites that emerge during storage.

Dormant bees must experience some freezing temperatures before they can break dormancy in spring, and ambient winter temperatures will work adequately except in areas that experience cold spells below −15°C. During these cold snaps, it is advisable to store bees in slightly warmer conditions at −12.2–4.4°C. Some growers leave dormant bees outdoors in their shelters all winter. If this is practised, it is important to screen shelter entrances to exclude nest enemies, including mice and other vertebrates.

Bees will reuse old nest tunnels, but nest materials should be replaced every 2–3 years to avoid a build-up of diseases and enemies. One way to phase out old nest material is to place it, full of bees, inside a black plastic bag. When bees are ready to be released in spring, the bag of old, occupied nest material is placed in the shade toward the rear of the shelter (this method works best with spacious shelters such as old drums). The mouth of the bag is reduced just enough to let air in and bees escape, and it is turned to face towards the shelter entrance. New, clean, empty nest material is placed in front of the old material and closer to the shelter entrance. When bees return to the shelter they tend to occupy the new nest holes, ignoring the old ones at the rear. The old nest material is then discarded or, in the case of drilled solid wood blocks, sanitized in an oven.

Note

[i] I acknowledge the contributions of my co-author to the first edition of this volume, Dan Mayer, for sections 9.1 and 9.2.

10 Wild Bees

It was mentioned in section 2.4 (this volume) that when we are talking about managed bees we are in the realm of husbandry (Chapters 7, 8, 9), and when we are talking about wild bees we are in the realm of land management and conservation. Applied bee conservation is covered in Chapter 6; wild bee biology is covered in Chapter 2 and sections 8.2, 9.1.1, 9.2.1 and 9.3.1. In section 3.2 wild bees are used to illustrate principles of bee pollinator efficiency, leaving us in this chapter to concentrate on the evidence for the pre-eminence of wild bees as agriculture's premier pollinators.

10.1. Wild Bees as Pollinators

Wild bees are largely solitary with comparatively few species expressing simple or complex eusociality (see section 2.3 and Fig. 2.14, this volume). Herein lies the biology grounding the profligate efficiency of wild bees as pollinators; the best word, I think, is 'tempo'. Lacking the resources of partners in labour and the food reserves of a colony, a solitary forager quite simply has more at stake with each action she takes. The death or survival of her progeny hangs on her success at excavating a nest and provisioning one or more cells with enough food to feed the larvae, and enough nesting substrate to insulate and protect them from parasites and temperature extremes. We are talking about direct fitness, not indirect fitness of the kind that animates the sterile forager of a complex eusocial colony. As such, speed and efficiency are at a premium. A solitary bee who completes ten brood cells has a higher fitness return than one who completes two. As the foraging season of a solitary bee is brief, there is selfish adaptive advantage to helping those plants reproduce whose bloom interval coincides with the bee's active period. The sum of these evolutionary imperatives is a pollinator who works efficiently and fast.

These are the biological foundations for the derived flower visiting traits described in section 3.2 (this volume): variable tongue length, hairiness and the ability to sonicate flowers. Ecological dynamics weigh in as well, however. It is thus at many levels, in evolutionary time and ecological time, that wild bees deliver their keystone actions as pollinators. Table 10.1 summarizes some of these effects for crop plants.

10.2. Drivers of Wild Bee Abundance and Pollination Performance at Crops

In the case of 'wild' bees, by extension we look to principles of ecology rather than husbandry for boosting their numbers and performance in crops. This subject is covered more practically in Chapter 6 (this volume), but Table 10.1 points out some principles worth developing here.

One driver that consistently affects stability of pollinated systems is the distance of high-quality natural nesting habitat from the focal crop. This is pinned to the fact that bees are central place foragers (section 3.4.1, this volume) and are constrained to make repeated flights to a fixed point in space – the nest. Ecological cost and benefit optima dictate that bees will minimize costly long-distance foraging bouts. Solitary bees, lacking the energy reserves of a social colony, are vulnerable to landscape alterations that lengthen the average foraging trip. We see this in intensified agriculture in the form of fragmented landscapes that increase average distance from suitable nesting patches to suitable foraging patches. Large individuals among the solitary bees can make extraordinarily long journeys, but most foraging bouts are no longer than 300 m from the nest (Zurbuchen et al., 2010).

Distance to natural nesting sites, in turn, strongly affects two more measures of the system's pollination stability – flower visitor *abundance* (number of individuals) and flower visitor *species richness*

© Keith S. Delaplane 2021. *Crop Pollination by Bees,* 2nd Edition, Volume 1 (K.S. Delaplane)
DOI: 10.1079/9781786393494.0010

Table 10.1. Some pollination metrics for wild bee pollinators on crops.

Wild bee taxa	Crop(s)	Location	Pollination results	Reference
3 families 8 genera 18 species	almond	California, USA	Wild bees caused *Apis* spp. to switch rows more frequently, increase cross-pollination, and increase yield	Brittain *et al.* (2013)
78 species	apple	Wisconsin, USA	Wild bee species richness and abundance unaffected by presence of *Apis* spp. Fruit-set not different between farms that used *Apis* spp. and farms that did not Wild bee species richness, but not abundance, increased fruit-set	Mallinger and Gratton (2015)
Andrena carlini *A. carolina* *Bombus impatiens* Halictidae *Xylocopa virginica*	blueberry	Michigan, USA	Small fields received 82% of pollination from wild bees Large fields received 11% of pollination from wild bees	Isaacs and Kirk (2010)
native bees	cantaloupe	California, USA	11 native bee species make 8% of total bee visits Fruit-set directly increased by wild bee visitation and indirectly by increase in high-diversity bee habitat	Kremen *et al.* (2002)
92% visits *Andrena* spp. 2% visits *Bombus* spp.	cherry, sweet	Hesse, Germany	Fruit-set unaffected by increase in honeybee visitation	Holzschuh *et al.* (2012)
29 pollinator species including *Apis* spp. and 7 hoverfly spp.	cherry, sweet	Flanders, Belgium	Fruit-set positively affected by both species richness and abundance Fruit-set unaffected by *Apis* spp. abundance Bee species richness and abundance positively associated with quality bee habitat ≤100 m	Eeraerts *et al.* (2019)
native bees	aubergine	California, USA	5 native bee species make 74% of total bee visits	Kremen *et al.* (2002)
wild bees (*Apis* spp. <2% of captures)	oilseed rape	Alberta, Canada	Bee abundance followed order: organic > conventional > GM fields Pollination deficit followed order: GM > conventional > organic fields	Morandin and Winston (2005)
wild bees	oilseed rape	Alberta, Canada	Seed-set positively associated with bee abundance Bee abundance highest in fields with uncultivated land ≤750 m of field edge Seed-set increased with bee abundance Yield and profit predicted to maximize when 30% of land uncultivated ≤750 m of field edge	Morandin and Winston (2006)
native bees	strawberry	California, USA	16 native bee species make 96% of total bee visits	Kremen *et al.* (2002)

Pollinators	Crop	Location	Findings	Reference
wild bees	strawberry	New York, USA	Wild bees 93% of collected individuals; Landscape simplification reduced wild bee abundance and species richness; Strawberry yield increased by greater pollinator abundance, not species richness	Connelly et al. (2015)
native bees; 3 families; 21 genera; 33 species	sunflower; sunflower	California, USA; California, USA	25 native bee species make 37% of total bee visits; Wild bees increased Apis spp. pollination efficiency 5 times; Wild bee efficiency <1–19 seeds per visit; Apis spp. efficiency increased from 3 seeds per visit in absence of wild bees to 15 seeds with wild bees	Kremen et al. (2002); Greenleaf and Kremen (2006b)
native bees; 60% visits *Anthophora urbana*, 34% *Bombus vosnesenskii*, 4% *Lasioglossum* spp., 2% *Halictus* spp., <1% *Apis* spp.	tomato; tomato	California, USA; California, USA	6 native bee species make 98% of total bee visits; Wild bees dominant visitors and effected pollination; *Bombus vosnesenskii* visitation rates correlated to near natural habitat; *Anthophora urbana* visitation rates not correlated to natural habitat	Kremen et al. (2002); Greenleaf and Kremen (2006a)
5 families; 17 genera; 53 species	tomato, bell pepper, cantaloupe, watermelon	New Jersey, Pennsylvania, USA	62% flower visits by wild bees; 38% by honeybees; Tomato consistently visited more frequently by solitary bees; Wild bee abundance positively associated with weedy flowers in field	Winfree et al. (2008)
native bees; 5 families; 14 genera; 45 species	watermelon; watermelon	California, USA; New Jersey, Pennsylvania, USA	30 native bee species make 27% of total bee visits; 45 native species visited watermelon; Native bee visitation positively correlated to pollen deposition; *Apis* spp. flower visitation not correlated to pollen deposition; Native bees responsible for 62% pollen grains deposited on female flowers	Kremen et al. (2002); Winfree et al. (2007)
biotic pollinators	16 cropping systems	global, 5 continents	Pollinator visitation drops to 50% of maximum at 600 m from natural habitat; Species richness drops to 50% of maximum at 1500 m from natural habitat; Visitation rate drops more steeply in tropics	Ricketts et al. (2008)
wild bees	41 cropping systems	global, 600 sites	Mixed effects on crop yields; Positive associations between visitation universal; Positive associations between fruit-set and wild insect visitation; Positive associations between fruit-set and *Apis* spp. visitation in 14% of systems studied	Garibaldi et al. (2013)

or *diversity* (number of species). The most proximate, direct measures of the system's stability are stigmatic pollen deposition rates, fruit-set and yield.

We can predict, supported by evidence, that increasing bee species richness and abundance are associated with nearness of undisturbed nesting sites to the focal crop. The studies in Table 10.1, however, show that these effects on pollination can be local and contextual. Nearness of natural habit served to increase abundance and pollinating performance for *Bombus* spp. but not for the solitary bee *Anthophora urbana* in California tomato (Greenleaf and Kremen, 2006a). Nearness of nesting habitat did not impact wild bee visitation to vegetable crops in New Jersey and Pennsylvania (Winfree *et al.*, 2008). In contrast, nearness of habitat increased wild bee visitation and pollination performance in two independent studies of sweet cherry in Germany (Holzschuh *et al.*, 2012) and Belgium (Eeraerts *et al.*, 2019) and oilseed rape in Canada (Morandin and Winston, 2006).

Similarly, the comparative benefits of bee abundance versus species richness are contextual. In the studies cited above we see benefits from wild bee richness in apple in Wisconsin (Mallinger and Gratton, 2015), wild bee abundance in strawberry in New York (Connelly *et al.*, 2015), and both wild bee richness and abundance in sweet cherries in Belgium (Eeraerts *et al.*, 2019).

Large local effects like these may simply reflect differences in the most limiting factors among sampled habitats. Imagine the effect of a parcel of natural bee habitat at three farms: one farm is nestled in a region of small-scale agriculture interwoven with semi-natural land parcels; the second is an island in an ocean of monoculture; while the third is intermediate with mixed monoculture and natural patches. The effect size of the habitat parcel is predicted to be greatest at the intermediate farm (Tscharntke *et al.*, 2005). In the first case the conditions for bees are already good and a habitat restoration will make little measurable impact. Conditions at the monoculture may be so bee-impoverished that few pioneers are available to exploit the habitat, whereas bee populations in the intermediate scenario are viable enough to benefit from augmentation. Similarly, the effect size of bee abundance versus diversity may be highly crop specific. Bee abundance may be of relatively small consequence for a crop like tomato that requires a sonicating flower visitor; far better in this case to have a diversity of flower visitors – some that sonicate the flower and liberate pollen and others that help move the pollen to other stigmas.

In an attempt to generalize the relationship between distance from nest habitat and pollinator performance, Ricketts *et al.* (2008) performed a meta-analysis of 23 published studies of 16 crops on five continents. The investigators found strong exponential declines in pollinator visitation rate (abundance) and species richness with increasing distance from natural nest habitat, but the relationships were not parallel. Species richness reached 50% of the maximum at 1500 m from the natural habitat, whereas visitation rates fell more rapidly, reaching 50% of the maximum at 600 m. In tropical latitudes, visitation rates dropped even more steeply. Ricketts *et al.* (2008) suggest that this visitation drop may be overinfluenced by social pollinators in the tropics and could be an artefact of nest substrate requirements which, if verified, could represent a special vulnerability for crops dependent on meliponines. Stingless bees tend to nest in hollow trees, thus favouring forests (Wille and Michener, 1973; Eltz *et al.*, 2003), while solitary bees are more likely to nest in soil or hollow reeds (section 2.2, this volume), sites more consistent with disturbed intensive agriculture. Another explanation may be that wild social bees with long-lived colonies require the phenological diversity of a succession of blooming plants like those found in forests, while solitary bees are able to forage, and nest near, crops that synchronize bloom with the bees' annual cycle.

Curiously, Ricketts *et al.* (2008) found inconsistent effects of distance from nest habitat on fruit- and seed-set. The authors speculate that introduced honeybees (which were excluded from their visitation analyses) may provide satisfactory pollination across agricultural landscapes, or else that wild bees are sufficient to pollinate distant fields even when the bees occur at low abundance and richness.

A subsequent global effort at generalizing the relationships among pollinator abundance, richness and pollination performance was performed by Garibaldi *et al.* (2013) who collected original data from 41 cropping systems at 600 sites on all continents except Antarctica. The importance of wild pollinator (bees + other insects) abundance was direct and clear: fruit-set increased significantly with wild insect visitation in 100% of the study systems, whereas with honeybees the same was true for only 14% of systems. Moreover, when impacts of stigmatic pollen deposition were tracked

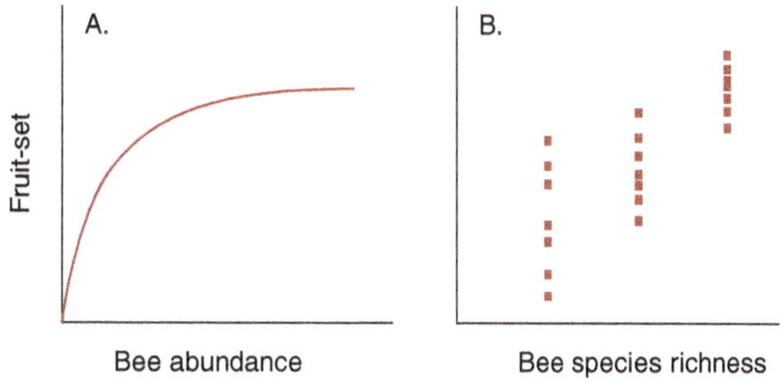

Fig. 10.1. Expected effects of increasing bee abundance (A) and species richness (B) on fruit-set. Increasing abundance is expected to increase fruit-set at a decelerating rate until additional visits effect no additional pollination. Increasing species richness is expected to increase the mean and reduce the variance about that mean.

all the way to fruit-set, pollen deposited by wild pollinators had a higher likelihood of actually setting fruit and effecting cross-pollination. Not only did pollinator visitation (abundance) increase with increasing pollinator species richness, but fruit-set also increased with increasing pollinator species richness. Finally, even though both honeybees and wild bees effected pollen deposition and fruit-set, this study found no evidence of a positive synergy. Pollination by each group was accomplished independently, but the overwhelming superior performance of wild pollinators led the investigators to conclude that honeybee pollination is properly considered a supplement to wild bee pollination, not a replacement.

From the discussion so far, readers can see that effects between near nest sites, pollinator abundance, pollinator richness and pollinator performance do not necessarily scale evenly from a local situation to a global view. There is nothing more 'true' in the results of a well-executed global study than in the results of a well-executed local study. The global analyses simply infer what is most often true in most cases, and that is: wild bees are the mainstay of agricultural pollination; proximate high-quality natural habitat is expected to increase pollinator abundance and species richness at the crop which, in turn, promote fruit-set. It is not remarkable when local vagaries supply exceptions and variations to these baseline expectations. We know, for instance, that widely divergent quality of food patches can upend predictions based on distance alone (Olsson *et al.*, 2015).

Across scales of geography, if there are generalizations to be made about bee abundance and species richness, they probably settle on the manners in which pollination is accomplished. Increasing

bee abundance is associated with a plateau effect, such that fruit-set increases geometrically with increasing visits until it slows and levels out (Morandin and Winston, 2006; Morris *et al.*, 2010). Conversely, increasing bee species richness tends to increase mean fruit-set and to decrease variance about that mean (Fig. 10.1; Winfree and Kremen, 2009), a pattern believed to reflect ecological *complementarity* and *facilitation*.

Complementarity is seen in pollination when members of a diverse guild of flower visitors partition themselves spatially or behaviourally so as to access the floral resources differently. Species of visitors may vary in preferred flower height off the ground, time of day, or speed at which they visit flowers (Hoehn *et al.*, 2008). An example of complementarity improving crop pollination performance is seen in Québec where medium-sized honeybees concentrate on the apex of the strawberry receptacle while smaller native bees concentrate on its base. The result is more even distribution of pollen and more shapely marketable fruit under conditions of increased species richness (Chagnon *et al.*, 1993).

Facilitation is seen when the presence of one species positively alters the behaviour of another, thus improving net pollen delivery. In crop contexts this is observed when wild bees stimulate honeybees to increase the rate at which they move between different sunflower heads (Greenleaf and Kremen, 2006b; Carvalheiro *et al.*, 2011) or to cross rows in almond orchards (Brittain *et al.*, 2013). Facilitation is seen in blueberry crops when sonicating bumble bees free up pollen from poricidal anthers, making it available to the more numerous honeybees who then aid its distribution in the orchard (Drummond, 2016).

One more idea needs to be highlighted here, and that is the stabilizing effects on a pollinated system procured by permanent, near, high-quality bee habitats and, by extension, the increase in bee species richness and abundance supported by that permanence. If left undisturbed, the occupancy of nesting sites of soil-nesting solitary bees can be measured in decades, as shown for *Nomia melanderi* (Cane, 2008) and *Peponapis pruinosa* (Mathewson, 1968).

Working with 22 watermelon fields in California, Kremen *et al.* (2004) were able to duplicate the kinds of results already presented here – namely, that pollination services from native bees are positively associated with the proximity of quality natural habitat. This study proceeded to show, however, that pollination services from wild bees were associated *with no other variable*, including farm type (organic versus conventional), insecticide use, field size and honeybee abundance. The stability of this effect across time and space permitted the investigators to make a model for predicting the area of natural habitat a farmer needs for procuring sufficient pollen deposition to set a marketable fruit. In this system, a farm needs to be situated in an area with ≥40% natural habitat within a 2.4 km radius, or ≥30% within a 1.2 km radius, to achieve total reliance on wild pollinators. Although the contributions of wild bees can persist with shrinking proportions of natural habitat – even up to 0% (Kremen *et al.*, 2004), there is a confirmed deterioration in the stability of the pollinated system as average distance from natural habitat increases (Garibaldi *et al.*, 2011). Clearly, the magnitude of benefit of near natural habitat combined with the natural permanence of bee nesting sites underscore the importance of identifying and conserving such landscape features near the farm.

The work of Morandin and Winston (2006) suggests that farmers can push this principle even further – from passive habitat conservation to active substitutionary habitat installation. These investigators, working with oilseed rape in Alberta, demonstrated strong correlations between bee abundance, seed-set and area of uncultivated land within 750 m of field edges. What made this work remarkable, however, is the authors' demonstration that the yield-enhancing benefits of natural bee harbourage extend so far as to justify a grower converting *cultivated* area to bee sanctuary. The authors applied their results to an hypothetical 4 km^2 agricultural landscape containing five 800×800 m fields, with 64 ha uncultivated land within 750 m of field edges. They showed that the grower could double the uncultivated area to 128 ha within 750 m of field edges, cede it to bee habitat, eliminating one 800×800 m field from cultivation, and still increase profit by 10% from the same 4 km^2 landscape.

11 Stingless Bees, Tribe Meliponini

The stingless bees, collectively the Meliponini, are one of four tribes in the family Apidae making up the so-called 'corbiculate' bees, united by their common possession of a pollen-carrying corbiculum on the hind legs (Figs 2.1 and 2.2, this volume). The corbiculate tribes have long been of interest to students of social evolution because in these four tribes we find the full range of sociality from solitary to simple eusociality, to complex eusociality. Their molecular phylogeny does not resolve along expected lines of social development, however, awkwardly lumping together as sister groups the highly eusocial Meliponini with the simple eusocial Bombini, and the highly eusocial Apini with the solitary Euglossini (Kawakita *et al.*, 2008). Among other inconveniences, this phylogeny demands that complex eusociality has evolved independently on two occasions (Fig. 2.1). The only character in this arrangement that is superficially satisfying is a kind of similarity between the nests of the Meliponini and the Bombini, in each of which food materials are stored in 'pots' instead of 'cells'.

For students of pollination, the possession of a corbiculum in bees signals that meliponine pollination is chiefly a matter of vectoring pollen to stigmas from 'safe sites' on the bee's body where the bee cannot groom itself clean of pollen and pack it into its corbiculae for transport to the nest. Once dampened by nectar and packed away, the viability of pollen for setting fruit drops significantly (see section 3.2, this volume). However, there is no evidence, in stingless bees or other corbiculates, that possession of a corbiculum is a serious impediment to pollination (see Fig. 3.1).

The meliponines are the least studied and, for purposes of crop pollination, underutilized bees in the world. Whereas the famous honeybees consist of only 11 species all in one genus, the meliponines consist of over 500 species, many rare, others regionally ubiquitous, divided among 32 genera (Michener, 2013). The group is exclusively pantropical in distribution, but there are species adapted to temperate-like habitats at high elevations (Roubik *et al.*, 1997).

Stingless bees are deeply entrenched in the values, food, craft, religion, medicine and culture of indigenous humans in the bees' natural ranges (Quezada-Euán *et al.*, 2018). Maya codices and colonial era Spanish documents attest to the practice of meliponiculture in pre-Columbian Meso-America. Honey and wax were harvested from wild colonies and used for tribute and for religious rituals (Ransome, 1937). In modern times meliponiculture is practised by Maya in southern Mexico and Guatemala. The Kayapo of the Amazon basin sustainably harvest honey and wax from wild colonies and practise forms of semi-domesticated bee management (Cortopassi-Laurino *et al.*, 2006). The Abayanda of Uganda have a sophisticated indigenous taxonomy for their local stingless bees, a transmitted knowledge of their ecology, and keen awareness of stingless bee pollination and its value to the forest ecosystem (Byarugaba, 2004). Today, meliponiculture is most widely practised in the neotropics, and pockets of innovation are emerging in Australia, Brazil (Cortopassi-Laurino *et al.*, 2006) and Thailand (Chuttong *et al.*, 2014).

Despite these deposits of indigenous knowledge around the world, meliponiculture is, with some exceptions, in a fragile state (see section 11.3). If this trend is to be reversed for the conservation of these bees and their greater incorporation into tropical agriculture, the home-grown traditions and knowledge of indigenous peoples must be sustainably and justly accessed.

11.1. Stingless Bee Biology

Except for being unified by their common eusociality and tropical biogeography, the Meliponini are otherwise astonishingly diverse. Their body size

DOI: 10.1079/9781786393494.0011

ranges from 2–14 mm; their colony populations from a few dozen individuals to tens of thousands; their body shape slender to robust; and their nests, while following a general *bauplan* described below, vary in complexity and location. Some nests are constructed in tree limbs and exposed, whereas most are built in cavities as varied as hollow trees, subterranean cavities, or abandoned nests of ants, termites, even birds. Although they are forest dwellers, some stingless bee species thrive in urban landscapes.

Nests are made of *cerumen* – beeswax mixed with varying proportions of plant resins (also called *propolis*), mud, plant fibres or even animal manure. The resin fraction is increased in those parts of the nest serving as external protective sheets or interior structural supports. The wax fraction is increased, making a softer more pliable amalgam, in those parts forming brood cells and storage pots.

Meliponine nests are distinctive for their strong compartmentalization. The design in Fig. 11.1 shows a tight integration between brood and food storage areas, but the two can equally be discontinuous as shown in Fig. 11.2. In the generalized plan (Fig. 11.1), brood cells of soft cerumen are in the centre, often surrounded by a soft sheath of the same material called the *involucrum* (Fig. 11.2). The food storage pots bear a striking resemblance to the same features in nests of *Bombus* spp., but unlike those of the bumble bees, these pots are purpose-built and are not recycled brood cells. Thus, uncoupled from the reproductive process, the food pots of meliponines are disproportionately large relative to the body size of the bees. Enveloping the brood and food storage areas is a tough resin-rich structure called the *batumen*, sometimes laminated with multiple layers and featuring interspersing cavities (Fig. 11.1). Entrances have turrets or other elaborated structures made of wax or batumen, functioning to resist invaders and inundation. These turrets are often ringed with defensive guards (Fig. 11.3).

For persons familiar with the life history of the honeybee *Apis mellifera*, there are both parallels and striking differences with stingless bees. Each is perennially social, and each possesses a queen, workers and drones. However, the overarching survival strategy and mechanisms of colony reproduction are markedly different. The development times for immature forms are over twice as long for stingless bees, the longest recorded being ~53 days for a young queen, 55 days for a worker and 57 days for a male, as shown for *Melipona colimana*

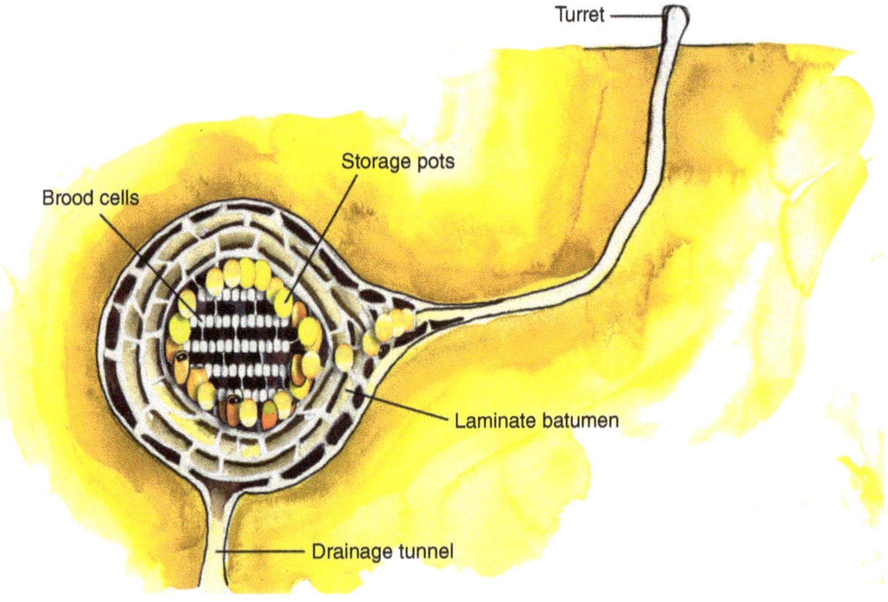

Fig. 11.1. A subterranean nest of *Trigona (Trigona) recursa* in Brazil illustrating the nest *bauplan* of Meliponini.

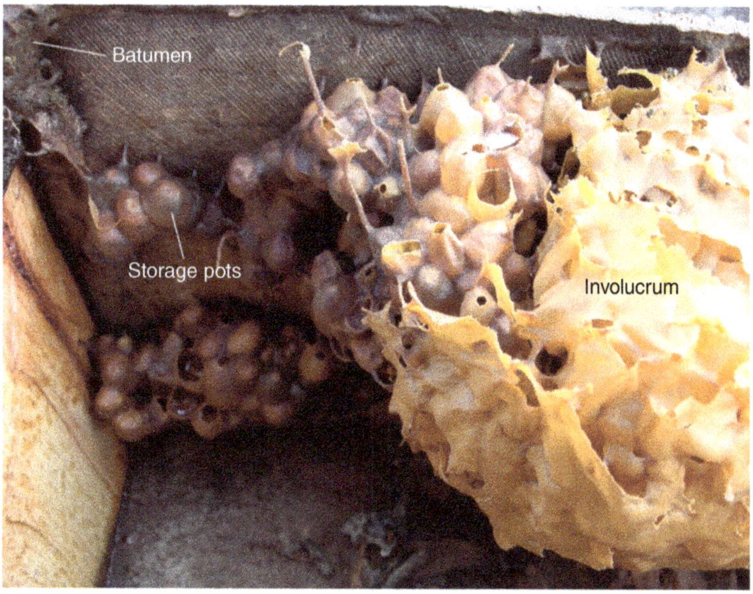

Fig. 11.2. A hived nest of *Tetragonisca angustula* in Honduras, commonly known as *angelita* (little angel) and the second most widely domesticated honey-making bee in the Americas. The compartmentalized feature of the meliponine nest is evident. The soft wax involucrum envelopes the brood area, and the tough batumen is plastered against the man-made hive walls. Identification: David Roubik, José Octavio Macías-Macías.

Fig. 11.3. Nest entrance turret and guards of *Nannotrigona perilampoides*, a species common in Mexico and noteworthy as the furthest north naturally occurring stingless bee in the Americas (Ayala, 1999). Photo courtesy of José Octavio Macías-Macías.

in Mexico (Macías-Macías and Quezada-Euán, 2015). These long development times may be an effect of the comparatively low digestibility of meliponine larval food compared to that of *Apis* spp. (Hartfelder and Engels, 1989). The interval between colony reproductive events is also much longer in the stingless bees – as much as 20–25 years (Slaa, 2006) – an adaptation to lower availability of nesting sites and higher rates of competition from other colonies (Roubik *et al.*, 2018). As a result, the acquisition of food is oriented toward surviving dearth periods rather than driving an annual colony-level reproductive fission as in *Apis* spp. Meliponine queens are not significantly larger than workers, but they are morphologically distinct, with mature queens expressing *physogastry* – a grossly distended abdomen with the corollary effect of precluding flight. As a consequence, colony founding in the meliponines is exclusively with a young, unmated, pre-physogastric queen, the exact opposite of *Apis* spp. whose primary swarm is headed by the old queen.

Meliponine colony reproduction begins with scout workers searching for a new nest site. Once a suitable cavity is found, workers prepare the site, sealing its cracks with batumen and building its entrance with the architectural features peculiar to the species. In another stark departure from *Apis* spp., the workers take nest materials directly from the parent nest, move it to the new site, and use it to build the batumen, involucrum, storage pots and initial brood cells. In later stages, even honey and pollen are forfeited by the parent colony to provision the storage pots of the new nest. Males, including members of other nests in the area, begin flying or resting about the new nest entrance. Eventually, a young queen emerges from the parent nest, flies to the new nest with or without a cohort of accompanying workers, takes a mating flight, and assumes egg production in the new nest. Initial mating flights appear to result in monogamous single matings, but there are records in some species of physogastric queens mating throughout life.

When it comes to evolved complex sociality, meliponines are the peers to *Apis* spp. by every measure, with some striking divergences. Meliponines are mass-provisioners, stocking each brood cell with a food mass and sealing it immediately after the queen deposits an egg, whereas *Apis* spp. are progressive provisioners, leaving brood cells open and feeding larvae continuously as they grow (Faustino *et al.*, 2002). Meliponine queens and workers are reared side by side in visually indistinguishable cells (Ratnieks and Helanterä, 2009), whereas cells of the two are very different in *Apis* spp. Furthermore, the latter species communicates resource quality, direction and distance with symbolic dance language, whereas meliponines recruit nestmates to resources with pheromone trails (see section 3.4.2, this volume). However, meliponine workers share with *Apis* spp. workers a complex and effective division of labour, including task profiles based on age, specialism and 'elitism' – the presence of overachievers who excel at numerous tasks (Hammel *et al.*, 2016; Mateus *et al.*, 2019).

11.2. Stingless Bees as Pollinators

The use of meliponines as deliberate inputs in industrial agriculture in the tropics is in its earliest stages of development. Their diversity of ecological adaptations and foraging behaviours means that some species are good candidates for crop pollination, others less so, and still others outright antagonistic to crop production. As an example of the latter, foragers of *Trigona amalthea* and *T. truculenta* in Peru strip fibres from the physic nut, a drought-resistant shrub used in traditional medicine, the seeds of which can be used to produce biodiesel (Rasmussen *et al.*, 2009). Literature has been reviewed recording economic damage by stingless bees to avocado, citrus, macadamia, mango and passion flower, virtually all of which is associated with harvesting plant fibres for the construction of nests. Curiously, such activity is not always, nor necessarily, damaging to pollination. Greco *et al.* (2011) noted damage to stigmas of greenhouse peppers by *Austroplebeia australis* in Australia, yet the bee's activity still effected pollen transfer and pollen tube growth.

Stingless bees species also vary in their intensity of nest defence: the most vigorous individuals will fly into the mouth, nose and ears of human intruders, delivering painful bites to sensitive skin, an altogether frightening experience and an unacceptable risk to farm workers.

With some notable exceptions, industrial scale meliponiculture is underdeveloped, inconsistently represented across the tribe's global range, and in some regions declining (see section 11.3).

Moving stingless bee colonies from their natural habitats into managed agricultural landscapes can be fraught with disappointment. If the new simplified environment contains few competitors or natural enemies, and if enough food and nesting sites are nearby, the experiment may result in stable colonies near the crop. However, meliponines do not do well when stocked at high-density field colonies of the kind typical for honeybees (Villanueva-Gutiérrez *et al.*, 2013). It appears that meliponines are more vulnerable to intra- than interspecific competition (Hubbell and Johnson, 1977; Roubik and Villanueva-Gutiérrez, 2009). Colonies in high-density meliponaries are prone to intense resource competition, fighting and nest usurpation (Roubik *et al.*, 2018). The fact that bee floral resources are more temporally and spatially patchy in the tropics and more evenly distributed in temperate latitudes (reviewed in Dornhaus and Chittka, 2004b), supports an hypothesis that intraspecific competition is more keen in the Meliponini than the temperate Apini.

This restraint against overstocking must be balanced against the risk of understocking and consequent inbreeding depression – a problem that applies to any pioneer moved outside of its natural range, but especially so in equatorial zones where social bee species tend to be abundant and their niche breadths narrow (Biesmeijer and Slaa, 2006);

---

narrowness being a condition that predicts small geographic ranges (Slatyer *et al.*, 2013). In other words, it is comparatively easy to move a stingless bee colony and inadvertently isolate it from its breeding population. On this note, Kerr (1985) estimated that a small population of stingless bees needs at least 44 colonies to remain genetically viable.

Stingless bee species vary in their foraging constancy at a crop. Species also vary in their foraging recruitment strategies: some marshal all foragers to the few richest rewards (similar to honeybees), whereas others have individuals who forage as free agents (similar to bumble bees). Such a range of behaviours suggests customizable species, each excelling in a particular crop species, breeding system or growing condition, whether greenhouses or open fields (Slaa *et al.*, 2006).

If this discussion has so far failed to arrive at general guidelines for incorporating stingless bees into field pollination situations, it may be because general guidelines are not forthcoming for a group as large, sprawling and varied as the Meliponini. Far better to engage in the hard work of matching local particularities of crop, growing system and available pollinator through practical experience and research.

Stingless bees have many properties that make them attractive as crop pollinators. As social bees, stingless bees share the common character of a long-lived colony that outlasts the bloom interval of any focal plant. As such, the bees are consummate generalists (Biesmeijer *et al.*, 2005), even though individuals among any colony may specialize on a single plant species for a time (Slaa *et al.*, 2003). The absence of a sting underscores the fact that many species are gentle and amenable to pollination in enclosures or in close proximity to human traffic. This is important in the neotropics where it is not possible to use colonies of highly defensive Africanized honeybees in enclosures. Stingless bees forage normally in enclosures and given adequate climate control can forage year-round. This opens up the possibility of using these bees in greenhouse agriculture, as shown in tomatoes in Mexico where stingless bees performed as well or better than mechanical vibrators as measured by fruit-set and kilograms of fruit per square metre (Cauich *et al.*, 2004).

Of 1330 cultivated tropical plants, nearly 70% are pollinated by biotic agents, including stingless bees (Roubik, 1995). A later estimate credited stingless bees as floral visitors in up to 90% of tropical crops, as important pollinators in nine crops and as contributing pollinators in 60 crops (Heard, 1999). Today, over 25 species of stingless bees are pollinators of field crops in at least 12 different plant families in 12 tropical countries (Ramírez *et al.*, 2018).

Some influential reviews, notably those of Heard (1999), Slaa *et al.* (2006) and Ramírez *et al.* (2018) have summarized the state of knowledge on crop pollination by stingless bees. A digest of stingless bee associations with Brazilian crop plants was compiled by Giannini *et al.* (2015a), but these authors were focused on bee and plant networks not pollination per se, and their list includes records for which there is little evidence of consummated pollination. Listed in Table 11.1 are the crops reviewed by Slaa *et al.* (2006) and considered 'effectively pollinated by stingless bees'. Some post-2006 literature is included that relates to stingless bee pollination performance.

Meliponiculture shares with *Apis* spp. beekeeping the desirable quality of diversified revenue streams for beekeepers. Stingless bee honey is produced in small quantities but fetches premium prices throughout South and Central America where it is used in traditional medicine. Similarly, propolis collected from stingless bees is easier to harvest than its counterpart in *Apis* spp. hives and is valued for its properties as an antibacterial agent and wound dressing, qualities which are catching the attention of western medicine and pharmacology (Bankova and Popova, 2007). These multiple income streams have a stabilizing effect on what is still a cottage industry, strengthening its prospects for sourcing stingless bees for crop pollination.

11.3. Meliponiculture

Unlike apiculture or bombiculture for which equipment and management are worked out and prescribed the world over, meliponiculture is a local activity – an expression of innovation, received traditional methods and keen familiarity with the local bee species. As such, its practices are nearly as diverse as the bees themselves. For more detailed coverage of stingless beekeeping and management for pollination, readers are alerted to the reviews of Cortopassi-Laurino *et al.* (2006), Roubik *et al.* (2018) and Ramírez *et al.* (2018).

Hive containers can be as simple as a section of natural log, gourds, bottles or clay pottery. The ends of logs or pots are sealed over with mud or

Table 11.1. Crops considered 'effectively pollinated by stingless bees' by Slaa et al., 2006, and references cited therein. Post-2006 records of effective crop pollination by stingless bees are also included.

Bee taxa	Crop(s)	Pollination results	Reference
Melipona melanoventer, Melipona fuliginosa	annatto	• Study conducted under field conditions	Slaa et al. (2006)
Melipona quadrifasciata anthidioides	apple	• Stingless bees acted synergistically with Apis spp. hives to increase yield	Viana et al. (2014)
Trigona nigra, Nannotrigona perilampoides, Geotrigona acapulconis, Trigona nigerrima, Partamona bilineata, Nannotrigona perilampoides, Scaptotrigona pectoralis, Trigona nigra, Scaptotrigona mexicana, Trigona fulviventris, Plebeia frontalis	avocado	• Study conducted under field conditions	Slaa et al. (2006)
Meliponini	camu-camu	• Study conducted under field conditions	Slaa et al. (2006)
Trigona thoracica	carambola	• Study conducted under field conditions	Slaa et al. (2006)
Trigona corvina, Partamona cupira	chayote	• Study conducted under field conditions	Slaa et al. (2006)
Meliponini	coconut	• Study conducted under field conditions	Slaa et al. (2006)
Trigona (Lepidotrigona) terminata	coffee	• Study conducted under field conditions	Slaa et al. (2006)
Scaptotrigona aff. depilis, Nannotrigona testaceicornis	cucumber	• Study conducted under enclosed conditions	Slaa et al. (2006)
Tetragonisca angustula, Nannotrigona testaceicornis	cucumber	• Study conducted under enclosed conditions • T. angustula did not visit flowers • N. testaceicornis only collected nectar • Stingless bees did not dwindle in greenhouses • Fruit-set was 78% when bees excluded; with bees >97%	Nicodemo et al. (2013)
Trigona lurida Melipona fasciculata	cupuaçu aubergine	• Study conducted under field conditions • Study conducted under enclosed conditions • Bees foraged normally in greenhouses • M. fasciculata increased fruit-set by 29.5% over controls • M. fasciculata increased fruit weight by 96% over controls • Single visit by M. fasciculata sufficient to effect pollination	Slaa et al. (2006) Nunes-Silva et al. (2013b)
Trigona spp.	macadamia	• Study conducted under field conditions	Slaa et al. (2006)
Trigona spp.	mango	• Study conducted under field conditions	Slaa et al. (2006)
Meliponini	mapati	• Study conducted under field conditions	Slaa et al. (2006)

Species	Crop	Findings	Reference
Austroplebeia australis, Trigona carbonaria	pepper (green)	• Study conducted under enclosed conditions • T. carbonaria foraged more consistently on pepper flowers • Pollination performance of both species inconsistent across trials • Colonies thrived in greenhouse environment • A. australis caused damage to styles	Greco et al. (2011)
Melipona favosa, Melipona subnitida, Trigona carbonaria	pepper (sweet)	• Study conducted under enclosed conditions	Slaa et al. (2006)
Scaptotrigona mexicana, Tetragonisca angustula	rambutan	• Study conducted under enclosed conditions	Slaa et al. (2006)
Nannotrigona perilampoides, Tetragonisca angustula	sage (mealycup)	• Study conducted under enclosed conditions	Slaa et al. (2006)
Plebeia tobagoensis, Trigona minangkabau, Nannotrigona testaceicornis, Tetragonisca angustula	strawberry	• Study conducted under enclosed conditions	Slaa et al. (2006)
Melipona quadrifasciata, Nannotrigona perilampoides	tomato	• Study conducted under enclosed and open conditions	Slaa et al. (2006)
Melipona quadrifasciata	tomato	• Study conducted under enclosed conditions • Greenhouses stocked with M. quadrifasciata produced more and heavier fruit than greenhouse stocked with Apis spp.	Dos Santos et al. (2009)
Melipona subnitida, Scaptotrigona spp. nov.	watermelon	• Study conducted under enclosed conditions • M. subnitida did not forage on crop • Scaptotrigona spp. nov. actively foraged on crop • Scaptotrigona spp. nov. visited staminate and pistillate flowers from seeded and seedless varieties	Bomfim et al. (2015)

clay except for a small entryway. Ultimately, these bees are accommodating to a variety of cavity types. It is important that hives are made of rot-resistant, well insulated materials, are portable, screened or otherwise ventilated, afford minimally invasive means of dividing colonies, have separate compartments for brood and honey, have a bottom chamber for receiving colony waste that can be periodically cleaned out, and are affordable (Roubik *et al.*, 2018). Purpose-built compartmentalized hives are increasing in use, especially in Australia and Brazil.

It is possible to begin new colonies by putting out empty hives into the habitat as bait for reproductive swarms (Inoue *et al.*, 1993). Colonies of *Melipona* spp. can be divided by taking combs of emerging brood cells and honey pots from one hive and placing them in an empty hive. The newly emerged workers accept the space, and one of their sister queens emerges, mates and assumes egg laying (Cortopassi-Laurino *et al.*, 2006). It was pointed out by Roubik *et al.* (2018) that colony divisions are most successful during generous natural nectar flows. The transferred bees settle down quickly, even if the beekeeper denies them honey pots from the parent colony. In fact, the absence of stored honey stimulates the bees to rapidly stock their own nest with provisions. This practice also minimizes the odour of spilt honey and open pollen – strong attractors to nest enemies such as the sap beetle *Aethina tumida* and phorid flies. Vinegar traps are used in Brazil, both in-hive and out-, to control phorids (Cortopassi-Laurino *et al.*, 2006).

Today meliponiculture is practised in Mexico, Costa Rica, Brazil and Australia, and to a lesser extent in Tanzania, Angola, India, Malaysia, Thailand and the Philippines (Cortopassi-Laurino *et al.*, 2006; Chuttong *et al.*, 2014). The keeping of stingless bees reached its ancient high-water mark in the Maya civilization of present day Mexico and Guatemala, where today *Melipona beecheii* is favoured by keepers of stingless bees for its large body size and high honey yields. The industry, however, is in a period of decline driven by environmental changes, overharvest, and mismanagement (Villanueva-Gutiérrez *et al.*, 2005). Brazil is a leader in research on stingless bee biology and pollination, and its stingless beekeeping industry appears to be growing. Australia is another centre of innovation for meliponiculture, an outcome of public interest in conservation and a revival of indigenous farming practices. Honey and cerumen are in demand, and a market is emerging for crop pollination services with stingless bees (Cortopassi-Laurino *et al.*, 2006).

References

Ackerman, J.D. (2000) Abiotic pollen and pollination: ecological, functional, and evolutionary perspectives. In: Dafni, A., Hesse, M. and Pacini, E. (eds) *Pollen and Pollination.* Springer Wein, New York, 167–185.

Adams, S. and Senft, D. (1994) The busiest of bees: pollen bees outwork honey bees as crop pollinators. *Agricultural Research,* US Dept Agriculture, February 1994.

Aebi, A., Vaissière, B.E., Delaplane, K.S., Roubik, D.W. and Neumann, P. (2012) Back to the future: *Apis* versus non-*Apis* pollination—a response to Ollerton *et al. Trends in Ecology & Evolution* 27, 142–143.

Aguilar, I., Fonseca, A. and Biesmeijer, J.C. (2005) Recruitment and communication of food source location in three species of stingless bees (Hymenoptera, Apidae, Meliponini). *Apidologie* 36, 313–324.

Aizen, M.A. and Harder, L.D. (2009) The global stock of domesticated honey bees is growing slower than agricultural demand for pollination. *Current Biology* 19, 915–918.

Aizen, M.A., Garibaldi, L.A., Cunningham, S.A. and Klein, A.M. (2008) Long-term global trends in crop yield and production reveal no current pollination shortage but increasing pollinator dependency. *Current Biology* 18, 1572–1575.

Aizen, M.A., Garibaldi, L.A., Cunningham, S.A. and Klein, A.M. (2009) How much does agriculture depend on pollinators? Lessons from long-term trends in crop production. *Annals of Botany* 103, 1579–1588.

Alaux, C., Ducloz, F., Crauser, D. and Le Conte, Y. (2010) Diet effects on honeybee immunocompetence. *Biology Letters* 6, 562–565.

Albrecht, M., Kleijn, D., Williams, N.M., Tschumi, M., Blaauw, B.R., Bommarco, R., Campbell, A.J., Dainese, M., Drummond, F.A. and Entling, M.H., *et al.* (2020) The effectiveness of flower strips and hedgerows on pest control, pollination services and crop yield: a quantitative synthesis. *Ecology Letters.* 23(10), 1488–1498.

Alexandersson, R. and Johnson, S.D. (2002) Pollinator-mediated selection on flower-tube length in a hawk-moth-pollinated Gladiolus (Iridaceae). *Proceedings of the Royal Society of London: Series B: Biological Sciences* 269, 631–636.

Alger, S.A., Burnham, P.A., Lamas, Z.S., Brody, A.K. and Richardson, L.L. (2018) Home sick: impacts of migratory beekeeping on honey bee (*Apis mellif-era*) pests, pathogens, and colony size. *PeerJ* 6, e5812.

Allsopp, M.H., De Lange, W.J. and Veldtman, R. (2008) Valuing insect pollination services with cost of replacement. *PloS One* 3(9), e3128. https://doi.org/10.1371/journal.pone.0003128.

Alpatov, V. (1984) Bee races and red clover pollination. *Bee World* 29, 61–63.

Amador, G.J., Matherne, M., Waller, D.A., Mathews, M., Gorb, S.N. and Hu, D.L. (2017) Honey bee hairs and pollenkitt are essential for pollen capture and removal. *Bioinspiration & Biomimetics* 12, 026015.

Amand, P.S., Skinner, D. and Peaden, R. (2000) Risk of alfalfa transgene dissemination and scale-dependent effects. *Theoretical and Applied Genetics* 101, 107–114.

An, J.-D. and Chen, W.-F. (2011) Economic value of insect pollination for fruits and vegetables in China. *Acta Entomologica Sinica* 54, 443–450.

Anderson, D.L. and Giacon, H. (1992) Reduced pollen collection by honey bee (Hymenoptera: Apidae) colonies infected with *Nosema apis* and sacbrood virus. *Journal of Economic Entomology* 85, 47–51.

Arroyo, M. and Uslar, P. (1993) Breeding systems in a temperate mediterranean-type climate montane sclerophyllous forest in central Chile. *Botanical Journal of the Linnean Society* 111, 83–102.

Asase, A., Wade, S., Ofori-Frimpong, K., Hadley, P. and Norris, K. (2008) Carbon storage and the health of cocoa agroforestry ecosystems in south-eastern Ghana. In: Bombelli, A. and Valentini, R., (eds) *Africa and the Carbon Cycle, 2008, Accra, Ghana.* FAO, Rome, 131–143.

Asher, J. and Pickering, J. (2013) *Discover Life Bee Species Guide and World Checklist (Hymenoptera: Apoidea: Anthophila)* [Online]. Available at: http://www.discoverlife.org/mp/20q?guide=Apoidea_species (accessed 6 January 2021).

Ashman, T.-L., Knight, T.M., Steets, J.A., Amarasekare, P., Burd, M., Campbell, D.R., Dudash, M.R., Johnston, M.O., Mazer, S.J. and Mitchell, R.J. (2004) Pollen limitation of plant reproduction: ecological and evolutionary causes and consequences. *Ecology* 85, 2408–2421.

Ashworth, L., Aguilar, R., Martén-Rodríguez, S., Lopezaraiza-Mikel, M., Avila-Sakar, G., Rosas-Guerrero, V. and Quesada, M. (2015) Pollination syndromes: a global pattern of convergent evolution driven by the most effective pollinator. In: Pontarotti, P. (ed.) *Evolutionary Biology: Biodiversification From Genotype to Phenotype.* DOI 10.1007/978-3-319-19932-0_11. Springer, New York, 203–224.

Atmowidi, T., Buchori, D., Manuwoto, S., Suryobroto, B. and Hidayat, P. (2007) Diversity of pollinator insects in relation to seed set of mustard (*Brassica rapa* L.: Cruciferae). *HAYATI Journal of Biosciences* 14, 155–161.

Ayala, R. (1999) Revisión de las abejas sin aguijón de México (Himenoptera: Apidae: Meliponini). *Folia Entomologica Mexicana* 106, 1–123.

Ayers, G.S., Hoopingarner, R.A. and Howitt, A.J. (1987) Testing potential bee forage for attractiveness to bees. *American Bee Journal* 127, 91–98.

Bacon, O., Burton, V., McSwain, J., Marble, V., Stanger, W. and Thorp, R. (1965) *Pollinating Alfalfa with Leafcutting Bees (AXT 160).* University of California Agricultural Extension Service, Berkeley, CA.

Baird, C.R. and Bitner, R.M. (1991) *Loose Cell Management of Leafcutting Bees in Idaho.* University of Idaho College of Agriculture, *Current Information Series* 588.

Balfour, N.J., Garbuzov, M. and Ratnieks, F.L. (2013) Longer tongues and swifter handling: why do more bumble bees (*Bombus* spp.) than honey bees (*Apis mellifera*) forage on lavender (*Lavandula* spp.)? *Ecological Entomology* 38, 323–329.

Banaszak, J. (1992) Strategy for conservation of wild bees in an agricultural landscape. *Agriculture, Ecosystems & Environment* 40, 179–192.

Bankova, V. and Popova, M. (2007) Propolis of stingless bees: a promising source of biologically active compounds. *Pharmacognosy Reviews* 1(1), 88–92.

Barclay, J.S. and Moffett, J.O. (1984) The pollination value of honey bees to wildlife. *American Bee Journal* 124, 497–498, 551.

Barfield, A.S., Bergstrom, J.C., Ferreira, S., Covich, A.P. and Delaplane, K.S. (2015) An economic valuation of biotic pollination services in Georgia. *Journal of Economic Entomology* 108, 388–398.

Barrett, S.C. (2010) Darwin's legacy: the forms, function and sexual diversity of flowers. *Philosophical Transactions of the Royal Society B: Biological Sciences* 365, 351–368.

Barron, M., Wratten, S. and Donovan, B. (2000) A four-year investigation into the efficacy of domiciles for enhancement of bumble bee populations. *Agricultural and Forest Entomology* 2, 141–146.

Basualdo, M., Bedascarrasbure, E. and De Jong, D. (2000) Africanized honey bees (Hymenoptera: Apidae) have a greater fidelity to sunflowers than European bees. *Journal of Economic Entomology* 93, 304–307.

Batra, S.W. (1970) Behavior of the alkali bee, *Nomia melanderi*, within the nest (Hymenoptera: Halictidae). *Annals of the Entomological Society of America* 63, 400–406.

Batra, S.W. (1976) Comparative efficiency of alfalfa pollination by *Nomia melanderi*, *Megachile rotundata*, *Anthidium florentinum* and *Pithitis smaragdula* (Hymenoptera: Apoidea). *Journal of the Kansas Entomological Society* 49(1), 18–22.

Batra, S. (1982) The hornfaced bee for efficient pollination of small farm orchards [Online]. Available at: digitalcommons.usu.edu/cgi/viewcontent.cgi?article=1092&context=bee_lab_ba (accessed 6 January 2021). US Dept Agriculture, *ARS Miscellaneous Publication* 1422.

Batra, S.W. (1984) Solitary bees. *Scientific American* 250, 120–127.

Batra, S. (1989) Japanese hornfaced bees, gentle and efficient new pollinators. *Pomona* 22, 3–5.

Batra, S.W. and Bohart, G.E. (1969) Alkali bees: response of adults to pathogenic fungi in brood cells. *Science* 165, 607–607.

Bauder, J.A., Lieskonig, N.R. and Krenn, H.W. (2011) The extremely long-tongued Neotropical butterfly *Eurybia lycisca* (Riodinidae): proboscis morphology and flower handling. *Arthropod Structure & Development* 40, 122–127.

Bauer, D.M. and Wing, I.S. (2010) Economic consequences of pollinator declines: a synthesis. *Agricultural and Resource Economics Review* 39, 368–383.

Bauer, D.M. and Wing, I.S. (2016) The macroeconomic cost of catastrophic pollinator declines. *Ecological Economics* 126, 1–13.

Becher, M.A., Grimm, V., Thorbek, P., Horn, J., Kennedy, P.J. and Osborne, J.L. (2014) BEEHAVE: a systems model of honeybee colony dynamics and foraging to explore multifactorial causes of colony failure. *Journal of Applied Ecology* 51, 470–482.

Becher, M.A., Twiston-Davies, G., Penny, T.D., Goulson, D., Rotheray, E.L. and Osborne, J.L. (2018) Bumble-BEEHAVE: a systems model for exploring multifactorial causes of bumblebee decline at individual, colony, population and community level. *Journal of Applied Ecology* 55, 2790–2801.

Beekman, M. and Ratnieks, F. (2000) Long-range foraging by the honey-bee, *Apis mellifera* L. *Functional Ecology* 14, 490–496.

Beentje, H. (2016) *The Kew Plant Glossary, An Illustrated Dictionary of Plant Terms (2nd Edition).* Kew Publishing, Royal Botanic Gardens, Kew, pp. 192.

Bell, C.D., Soltis, D.E. and Soltis, P.S. (2010) The age and diversification of the angiosperms re-revisited. *American Journal of Botany* 97, 1296–1303.

Berger, L.A., Vaissiére, B.E., Moffett, J.O. and Merritt, S.J. (1988) *Bombus* spp. (Hymenoptera: Apidae) as pollinators of male-sterile upland cotton on the texas high plains. *Environmental Entomology* 17, 789–794.

Berry, J. (2011) Africanized honey bees in Georgia. *Bee Culture* 139, 53–55.

Biella, P., Bogliani, G., Cornalba, M., Manino, A., Neumayer, J., Porporato, M., Rasmont, P. and

Milanesi, P. (2017) Distribution patterns of the cold adapted bumblebee *Bombus alpinus* in the Alps and hints of an uphill shift (Insecta: Hymenoptera: Apidae). *Journal of Insect Conservation* 21, 357–366.

Biella, P., Ćetković, A., Gogala, A., Neumayer, J., Sárospataki, M., Šima, P. and Smetana, V. (2020) Northwestward range expansion of the bumblebee *Bombus haematurus* into Central Europe is associated with warmer winters and niche conservatism. *Insect Science* 00, 1–12. https://doi.org/10.1111/1744-7917.12800.

Biesmeijer, J.C. and Slaa, E.J. (2006) The structure of eusocial bee assemblages in Brazil. *Apidologie* 37, 240–258.

Biesmeijer, J.C., Slaa, E.J., Castro, M.S.D., Viana, B.F., Kleinert, A.D.M. and Imperatriz-Fonseca, V.L. (2005) Connectance of Brazilian social bee: food plant networks is influenced by habitat, but not by latitude, altitude or network size. *Biota Neotropica* 5, 85–93.

Biesmeijer, J.C., Roberts, S.P., Reemer, M., Ohlemüller, R., Edwards, M., Peeters, T., Schaffers, A., Potts, S.G., Kleukers, R. and Thomas, C. (2006) Parallel declines in pollinators and insect-pollinated plants in Britain and the Netherlands. *Science* 313, 351–354.

Birmingham, A.L. and Winston, M.L. (2004) Orientation and drifting behaviour of bumblebees (Hymenoptera: Apidae) in commercial tomato greenhouses. *Canadian Journal of Zoology* 82, 52–59.

Blaauw, B.R. and Isaacs, R. (2014) Flower plantings increase wild bee abundance and the pollination services provided to a pollination-dependent crop. *Journal of Applied Ecology* 51, 890–898.

Blitzer, E.J., Gibbs, J., Park, M.G. and Danforth, B.N. (2016) Pollination services for apple are dependent on diverse wild bee communities. *Agriculture, Ecosystems & Environment* 221, 1–7.

Bohart, G.E. (1955) Time relationships in the nest construction and life cycle of the alkali bee. *Annals of the Entomological Society of America* 48, 403–406.

Bohart, G.E. (1957) Pollination of alfalfa and red clover. *Annual Review of Entomology* 2, 355–380.

Bohart, G.E. (1972) Management of wild bees for the pollination of crops. *Annual Review of Entomology* 17, 287–312.

Bomfim, I.G.A., De Melo Bezerra, A.D., Nunes, A.C., De Aragão, F.A.S. and Freitas, B.M. (2015) Adaptive and foraging behavior of two stingless bee species in greenhouse mini watermelon pollination. *Sociobiology* 61, 502–509.

Bommarco, R., Marini, L. and Vaissière, B.E. (2012) Insect pollination enhances seed yield, quality, and market value in oilseed rape. *Oecologia* 169, 1025–1032.

Bosch, J. (1994a) Improvement of field management of *Osmia cornuta* (Latreille) (Hymenoptera, Megachilidae) to pollinate almond. *Apidologie* 25, 71–83.

Bosch, J. (1994b) The nesting behaviour of the mason bee *Osmia cornuta* (Latr) with special reference to its pollinating potential (Hymenoptera, Megachilidae). *Apidologie* 25, 84–93.

Bosch, J. (1994c) *Osmia cornuta* Latr. (Hym., Megachilidae) as a potential pollinator in almond orchards: releasing methods and nest-hole length. *Journal of Applied Entomology* 117, 151–157.

Bosch, J. (1995) Comparison of nesting materials for the orchard pollinator *Osmia cornuta* (Hymenoptera: Megachilidae). *Entomologia Generalis* 19, 285–289.

Bosch, J. and Kemp, W.P. (1999) Exceptional cherry production in an orchard pollinated with blue orchard bees. *Bee World* 80, 163–173.

Bosch, J. and Kemp, W. (2001) *How to manage the blue orchard bee, Osmia lignaria, as an orchard pollinator.* Sustainable Agriculture Network, National Agricultural Library, Beltsville, MD.

Bosch, J. and Kemp, W. (2002) Developing and establishing bee species as crop pollinators: the example of *Osmia* spp. (Hymenoptera: Megachilidae) and fruit trees. *Bulletin of Entomological Research* 92, 3–16.

Bosch, J. and Vicens, N. (2006) Relationship between body size, provisioning rate, longevity and reproductive success in females of the solitary bee *Osmia cornuta*. *Behavioral Ecology and Sociobiology* 60, 26–33.

Bosch, J., Kemp, W.P. and Trostle, G.E. (2006) Bee population returns and cherry yields in an orchard pollinated with *Osmia lignaria* (Hymenoptera: Megachilidae). *Journal of Economic Entomology* 99, 408–413.

Bourke, A.F. (2011) *Principles of social evolution.* Oxford Series in Ecology and Evolution, Oxford University Press, New York, pp. 267.

Bowe, L.M., Coat, G. and Depamphilis, C.W. (2000) Phylogeny of seed plants based on all three genomic compartments: extant gymnosperms are monophyletic and Gnetales' closest relatives are conifers. *Proceedings of the National Academy of Sciences* 97, 4092–4097.

Bowers, M.A. (1986) Resource availability and timing of reproduction in bumble bee colonies (Hymenoptera: Apidae). *Environmental Entomology* 15, 750–755.

Brading, P., El-Gabbas, A., Zalat, S. and Gilbert, F. (2009) Biodiversity economics: the value of pollination services to Egypt. *Egyptian Journal of Biology* 11, 46–51.

Brady, S.G., Sipes, S., Pearson, A. and Danforth, B.N. (2006) Recent and simultaneous origins of eusociality in halictid bees. *Proceedings of the Royal Society B: Biological Sciences* 273, 1643–1649.

Breeze, T., Bailey, A., Potts, S. and Balcombe, K. (2015) A stated preference valuation of the non-market benefits of pollination services in the UK. *Ecological Economics* 111, 76–85.

Brittain, C., Williams, N., Kremen, C. and Klein, A.-M. (2013) Synergistic effects of non-*Apis* bees and honey bees for pollination services. *Proceedings of the Royal Society B: Biological Sciences* 280, 20122767.

Brittain, C., Kremen, C., Garber, A. and Klein, A.-M. (2014) Pollination and plant resources change the nutritional quality of almonds for human health. *PloS One* 9, e90082.

Brosi, B.J. (2016) Pollinator specialization: from the individual to the community. *New Phytologist* 210, 1190–1194.

Brosi, B.J., Delaplane, K.S., Boots, M. and De Roode, J.C. (2017) Ecological and evolutionary approaches to managing honeybee disease. *Nature Ecology & Evolution* 1, 1250.

Bruckner, S., Steinhauer, N., Engelsma, J., Fauvel, A.M., Kulhanek, K., Malcolm, E., Meredith, A., Milbrath, M., Niño, E.L., Rangel, J., Rennich, K., Reynolds, D., Sagili, R., Tsuruda, J., Vanengelsdorp, D., Aurell, S.D., Wilson, M. and Williams, G. (2020) *2019–2020 Honey Bee Colony Losses in the United States: Preliminary Results* [Online]. Available at: www.beeinformed.org/wp-content/uploads/2020/06/BIP_2019_2020_Losses_Abstract.pdf (accessed 6 January 2021). Bee Informed Partnership.

Brunet, J. and Stewart, C.M. (2010) Impact of bee species and plant density on alfalfa pollination and potential for gene flow. *Psyche: A Journal of Entomology* 2010 Article ID 201858, pp. 7. https://doi.org/10.1155/2010/201858.

Brunet, J., Zhao, Y. and Clayton, M.K. (2019) Linking the foraging behavior of three bee species to pollen dispersal and gene flow. *PloS One* 14, e0212561.

Buchanan, G.A. (2016) *Feeding the world: agricultural research in the twenty-first century*. Texas A&M University Press, College Station, TX, USA.

Buchmann, S.L. (1983) Buzz pollination in angiosperms. In: Little, R.J. and Jones, C.E. (eds) *Handbook of Experimental Pollination Biology*. Van Nostrand Reinhold Company, New York, 63–113.

Buchmann, S.L. and Nabhan, G.P. (2012) *The forgotten pollinators*. Washington, DC, Island Press/Shearwater Books.

Butler, C. (1965) Sex attraction in *Andrena flavipes* Panzer (Hymenoptera: Apidae), with some observations on nest-site restriction. *Proceedings of the Royal Entomological Society of London* 40, 77–80.

Butler, C. and Fairey, E.M. (1964) Pheromones of the honeybee: biological studies of the mandibular gland secretion of the queen. *Journal of Apicultural Research* 3, 65–76.

Butz Huryn, V. (1997) Ecological impacts of introduced honey bees. *The Quarterly Review of Biology* 72, 275–297.

Byarugaba, D. (2004) Stingless bees (Hymenoptera: Apidae) of Bwindi impenetrable forest, Uganda and Abayanda indigenous knowledge. *International Journal of Tropical Insect Science* 24, 117–121.

Byng, J., Chase, M.W., Christenhusz, M., Fay, M., Judd, W.S., Mabberley, D., Sennikov, A., Soltis, D., Soltis, P.S. and Stevens, P.F. (2016) An update of the Angiosperm Phylogeny Group classification for the orders and families of flowering plants: APG IV. *Botanical Journal of the Linnean Society* 181, 1–20.

Calderone, N.W. (2012) Insect pollinated crops, insect pollinators and US agriculture: trend analysis of aggregate data for the period 1992–2009. *PloS One* 7, e37235.

Camargo, J.M. (1970) Ninhos e biologia de algumas especies de Meliponideos (Hymenoptera: Apidae) de regiao de Porto Velho, Territorio de Rondonia, Brasil. *Revista de Biología Tropical* 16, 207–239.

Camargo, J. and Roubik, D. (1991) Systematics and bionomics of the apoid obligate necrophages: the *Trigona hypogea* group (Hymenoptera: Apidae; Meliponinae). *Biological Journal of the Linnean Society* 44, 13–39.

Cameron, S.A., Lozier, J.D., Strange, J.P., Koch, J.B., Cordes, N., Solter, L.F. and Griswold, T.L. (2011) Patterns of widespread decline in North American bumble bees. *Proceedings of the National Academy of Sciences* 108, 662–667.

Cane, J.H. (2002) Pollinating bees (Hymenoptera: Apiformes) of US alfalfa compared for rates of pod and seed set. *Journal of Economic Entomology* 95, 22–27.

Cane, J.H. (2006) An evaluation of pollination mechanisms for purple prairie-clover, *Dalea purpurea* (Fabaceae: Amorpheae). *The American Midland Naturalist* 156, 193–197.

Cane, J.H. (2008) A native ground-nesting bee (*Nomia melanderi*) sustainably managed to pollinate alfalfa across an intensively agricultural landscape. *Apidologie* 39, 315–323.

Cane, J. and Payne, J. (1990) Native bee pollinates rabbiteye blueberry. Highlights of Agricultural Research, *Alabama Agricultural Experiment Station* 37(4), 4.

Cane, J.H. and Payne, J.A. (1993) Regional, annual, and seasonal variation in pollinator guilds: intrinsic traits of bees (Hymenoptera: Apoidea) underlie their patterns of abundance at *Vaccinium ashei* (Ericaceae). *Annals of the Entomological Society of America* 86, 577–588.

Cane, J.H. and Sipes, S. (2006) Characterizing floral specialization by bees: analytical methods and a revised lexicon for oligolecty. In: Waser, N.M. and Ollerton, J. (eds) *Plant-Pollinator Interactions: From Specialization to Generalization*. University of Chicago Press, Chicago, pp. 99–122.

Cane, J.H., Eickwort, G.C., Wesley, F.R. and Spielholz, J. (1985) Pollination ecology of *Vaccinium stamineum* (Ericaceae: Vaccinioideae). *American Journal of Botany* 72, 135–142.

Cane, J.H., Griswold, T. and Parker, F.D. (2007) Substrates and materials used for nesting by North American *Osmia* bees (Hymenoptera: Apiformes: Megachilidae). *Annals of the Entomological Society of America* 100, 350–358.

Cane, J.H., Dobson, H.E. and Boyer, B. (2017) Timing and size of daily pollen meals eaten by adult females of a solitary bee (*Nomia melanderi*)(Apiformes: Halictidae). *Apidologie* 48, 17–30.

Cardinal, S. and Danforth, B.N. (2011) The antiquity and evolutionary history of social behavior in bees. *PLOS One* 6, e21086.

Cardinal, S. and Danforth, B.N. (2013) Bees diversified in the age of eudicots. *Proceedings of the Royal Society B: Biological Sciences* 280(1755), 20122686.

Cardinal, S., Buchmann, S.L. and Russell, A.L. (2018) The evolution of floral sonication, a pollen foraging behavior used by bees (Anthophila). *Evolution* 72, 590–600.

Cariveau, D.P. and Winfree, R. (2015) Causes of variation in wild bee responses to anthropogenic drivers. *Current Opinion in Insect Science* 10, 104–109.

Carlson, K.M., Curran, L.M., Ratnasari, D., Pittman, A.M., Soares-Filho, B.S., Asner, G.P., Trigg, S.N., Gaveau, D.A., Lawrence, D. and Rodrigues, H.O. (2012) Committed carbon emissions, deforestation, and community land conversion from oil palm plantation expansion in West Kalimantan, Indonesia. *Proceedings of the National Academy of Sciences* 109, 7559–7564.

Carnell, J.D., Page, S., Goulson, D. and Hughes, W.O. (2019) Trialling techniques for rearing long-tongued bumblebees under laboratory conditions. *Apidologie* 51, 254–266.

Carreck, N. and Williams, I. (1998) The economic value of bees in the UK. *Bee World* 79, 115–123.

Carreck, N. and Williams, I.H. (2002) Food for insect pollinators on farmland: insect visits to flowers of annual seed mixtures. *Journal of Insect Conservation* 6, 13–23.

Cartar, R.V. and Dill, L.M. (1991) Costs of energy short-fall for bumble bee colonies: predation, social parasitism, and brood development. *The Canadian Entomologist* 123, 283–293.

Carvalheiro, L.G., Veldtman, R., Shenkute, A.G., Tesfay, G.B., Pirk, C.W.W., Donaldson, J.S. and Nicolson, S.W. (2011) Natural and within-farmland biodiversity enhances crop productivity. *Ecology Letters* 14, 251–259.

Carvell, C., Meek, W.R., Pywell, R.F., Goulson, D. and Nowakowski, M. (2007) Comparing the efficacy of agri-environment schemes to enhance bumble bee abundance and diversity on arable field margins. *Journal of Applied Ecology* 44, 29–40.

Cauich, O., Quezada-Euán, J.J.G., Macías-Macías, J.O., Reyes-Oregel, V., Medina-Peralta, S. and Parra-Tabla, V. (2004) Behavior and pollination efficiency of *Nannotrigona perilampoides* (Hymenoptera: Meliponini) on greenhouse tomatoes (*Lycopersicon esculentum*) in subtropical Mexico. *Journal of Economic Entomology* 97, 475–481.

Cavigli, I., Daughenbaugh, K.F., Martin, M., Lerch, M., Banner, K., Garcia, E., Brutscher, L.M. and Flenniken, M.L. (2016) Pathogen prevalence and abundance in honey bee colonies involved in almond pollination. *Apidologie* 47, 251–266.

Certal, A.C., Almeida, R.B., Bošković´, R., Oliveira, M.M. and Feijó, J.A. (2002) Structural and molecular analysis of self-incompatibility in almond (*Prunus dulcis*). *Sexual Plant Reproduction* 15, 13–20.

Cestaro, L.G., Alves, M.L.T.M.F., Silva, M.V.G.B. and Teixeira, É.W. (2017) Honey bee (*Apis mellifera*) health in stationary and migratory apiaries. *Sociobiology* 64, 42–49.

Chagnon, M., Gingras, J. and Deoliveira, D. (1993) Complementary aspects of strawberry pollination by honey and indigenous bees (Hymenoptera). *Journal of Economic Entomology* 86, 416–420.

Chameron, S., Schatz, B., Pastergue-Ruiz, I., Beugnon, G. and Collett, T.S. (1998) The learning of a sequence of visual patterns by the ant *Cataglyphis cursor*. *Proceedings of the Royal Society of London: Series B: Biological Sciences* 265, 2309–2313.

Chanderbali, A.S., Kim, S., Buzgo, M., Zheng, Z., Oppenheimer, D.G., Soltis, D.E. and Soltis, P.S. (2006) Genetic footprints of stamen ancestors guide perianth evolution in *Persea* (Lauraceae). *International Journal of Plant Sciences* 167, 1075–1089.

Chaplin-Kramer, R., Dombeck, E., Gerber, J., Knuth, K.A., Mueller, N.D., Mueller, M., Ziv, G. and Klein, A.-M. (2014) Global malnutrition overlaps with pollinator-dependent micronutrient production. *Proceedings of the Royal Society B: Biological Sciences* 281, 20141799.

Charman, T.G., Sears, J., Green, R.E. and Bourke, A.F. (2010) Conservation genetics, foraging distance and nest density of the scarce Great Yellow Bumblebee (*Bombus distinguendus*). *Molecular Ecology* 19, 2661–2674.

Chaudhary, O. and Chand, R. (2017) Economic benefits of animal pollination to Indian agriculture. *Indian Journal of Agricultural Sciences* 87, 1117–1138.

CHC (2020) *Industry Overview* [Online]. Canadian Honey Council. Available at: https://honeycouncil.ca/archive/honey_industry_overview.php (accessed 2020).

Chittka, L., Thomson, J.D. and Waser, N.M. (1999) Flower constancy, insect psychology, and plant evolution. *Naturwissenschaften* 86, 361–377.

Chopra, S.S., Bakshi, B.R. and Khanna, V. (2015) Economic dependence of US industrial sectors on animal-mediated pollination service. *Environmental Science & Technology* 49, 14441–14451.

Chuttong, B., Chanbang, Y. and Burgett, M. (2014) Meliponiculture: stingless bee beekeeping in Thailand. *Bee World* 91, 41–45.

Ciati, R.A. and Ruini, L.A. (2011) Double pyramid: healthy food for people and sustainable for the planet. In: Burlingame, B. and Dernini, S. (eds) *Sustainable Diets and Biodiversity: Directions and Solutions for Policy, Research and Action.*. Food and Agriculture Organization, Rome, pp. 280–293.

Ciotek, L., Giorgis, P., Benitez-Vieyra, S. and Cocucci, A.A. (2006) First confirmed case of pseudocopulation in terrestrial orchids of South America: pollination of *Geoblasta pennicillata* (Orchidaceae) by *Campsomeris bistrimacula* (Hymenoptera, Scoliidae). *Flora-Morphology, Distribution, Functional Ecology of Plants* 201, 365–369.

Clifford, P. (1973) Increasing bumble bee densities in red clover seed production areas. *New Zealand Journal of Experimental Agriculture* 1, 377–379.

Coiffard, C., Gomez, B., Daviero-Gomez, V. and Dilcher, D.L. (2012) Rise to dominance of angiosperm pioneers in European Cretaceous environments. *Proceedings of the National Academy of Sciences* 109, 20955–20959.

Colla, S.R. and Packer, L. (2008) Evidence for decline in eastern North American bumblebees (Hymenoptera: Apidae), with special focus on *Bombus affinis* Cresson. *Biodiversity and Conservation* 17, 1379.

Colla, S.R., Otterstatter, M.C., Gegear, R.J. and Thomson, J.D. (2006) Plight of the bumble bee: pathogen spillover from commercial to wild populations. *Biological Conservation* 129, 461–467.

Colla, S.R., Gadallah, F., Richardson, L., Wagner, D. and Gall, L. (2012) Assessing declines of North American bumble bees (*Bombus* spp.) using museum specimens. *Biodiversity and Conservation* 21, 3585–3595.

Collett, T., Fry, S. and Wehner, R. (1993) Sequence learning by honeybees. *Journal of Comparative Physiology A* 172, 693–706.

Connelly, H., Poveda, K. and Loeb, G. (2015) Landscape simplification decreases wild bee pollination services to strawberry. *Agriculture, Ecosystems & Environment* 211, 51–56.

Cook, D.C., Thomas, M.B., Cunningham, S.A., Anderson, D.L. and De Barro, P.J. (2007) Predicting the economic impact of an invasive species on an ecosystem service. *Ecological Applications* 17, 1832–1840.

Corbet, S.A., Fussell, M., Ake, R., Fraser, A., Gunson, C., Savage, A. and Smith, K. (1993) Temperature and the pollinating activity of social bees. *Ecological Entomology* 18, 17–30.

Corbet, S.A., Saville, N., Fussell, M., Prŷs-Jones, O. and Unwin, D. (1995) The competition box: a graphical aid to forecasting pollinator performance. *Journal of Applied Ecology* 32, 707–719.

Cortopassi-Laurino, M., Imperatriz-Fonseca, V.L., Roubik, D.W., Dollin, A., Heard, T., Aguilar, I., Venturieri, G.C., Eardley, C. and Nogueira-Neto, P. (2006) Global meliponiculture: challenges and opportunities. *Apidologie* 37, 275–292.

Couvillon, M.J. and Dornhaus, A. (2009) Location, location, location: larvae position inside the nest is correlated with adult body size in worker bumble-bees (*Bombus impatiens*). *Proceedings of the Royal Society B: Biological Sciences* 276, 2411–2418.

Couvillon, M.J., Fitzpatrick, G. and Dornhaus, A. (2010) Ambient air temperature does not predict whether small or large workers forage in bumble bees (*Bombus impatiens*). *Psyche: A Journal of Entomology* 2010, 536430, pp. 8. https://doi.org/10.1155/2010/536430.

Crane, E. (1999) *The world history of beekeeping and honey hunting*. Routledge, New York.

Crane, E., Walker, P. and Day, R. (1984) *Directory of important world honey sources*. International Bee Research Association, London, pp. 384.

Crepet, W.L. (1979) Insect pollination: a paleontological perspective. *BioScience* 29, 102–108.

Crepet, W.L. (2008) The fossil record of angiosperms: requiem or renaissance? *Annals of the Missouri Botanical Garden* 95, 3–33.

Crepet, W.L. and Niklas, K.J. (2009) Darwin's second "abominable mystery": why are there so many angiosperm species? *American Journal of Botany* 96, 366–381.

Cresswell, J., Bassom, A., Bell, S., Collins, S. and Kelly, T. (1995) Predicted pollen dispersal by honey-bees and three species of bumble-bees foraging on oilseed rape: a comparison of three models. *Functional Ecology* 9, 829–841.

Cresswell, J.E., Osborne, J.L. and Goulson, D. (2000) An economic model of the limits to foraging range in central place foragers with numerical solutions for bumblebees. *Ecological Entomology* 25, 249–255.

Cridland, J.M., Tsutsui, N.D. and Ramírez, S.R. (2017) The complex demographic history and evolutionary origin of the western honey bee, *Apis mellifera*. *Genome Biology and Evolution* 9, 457–472.

Cunningham, S.A. and Le Feuvre, D. (2013) Significant yield benefits from honeybee pollination of faba bean (*Vicia faba*) assessed at field scale. *Field Crops Research* 149, 269–275.

Cunningham, S.A., Fournier, A., Neave, M.J. and Le Feuvre, D. (2016) Improving spatial arrangement of honeybee colonies to avoid pollination shortfall and depressed fruit set. *Journal of Applied Ecology* 53, 350–359.

Currie, R., Winston, M. and Slessor, K. (1992a) Effect of synthetic queen mandibular pheromone sprays on honey bee (Hymenoptera: Apidae) pollination of berry crops. *Journal of Economic Entomology* 85, 1300–1306.

Currie, R., Winston, M., Slessor, K. and Mayer, D. (1992b) Effect of synthetic queen mandibular pheromone sprays on pollination of fruit crops by honey bees (Hymenoptera: Apidae). *Journal of Economic Entomology* 85, 1293–1299.

Daberkow, S., Korb, P. and Hoff, F. (2009) Structure of the US beekeeping industry: 1982–2002. *Journal of Economic Entomology* 102, 868–886.

Dafni, A. and Firmage, D. (2000) Pollen viability and longevity: practical, ecological and evolutionary implications. In: Dafni, A., Hesse, M. and Pacini, E. (eds) *Pollen and Pollination*. Springer, Vienna. https://doi.org/10.1007/978-3-7091-6306-1_6.

Dafni, A., Kevan, P., Gross, C.L. and Goka, K. (2010) *Bombus terrestris*, pollinator, invasive and pest: an assessment of problems associated with its widespread introductions for commercial purposes. *Applied Entomology and Zoology* 45, 101–113.

Dag, A. (2008) Bee pollination of crop plants under environmental conditions unique to enclosures. *Journal of Apicultural Research* 47, 162–165.

Dag, A., Weinbaum, S., Thorp, R. and Eisikowitch, D. (2000) Pollen dispensers (inserts) increase fruit set and yield in almonds under some commercial conditions. *Journal of Apicultural Research* 39, 117–123.

Danforth, B.N. (1991) Female foraging and intranest behavior of a communal bee, *Perdita portalis* (Hymenoptera: Andrenidae). *Annals of the Entomological Society of America* 84, 537–548.

Danka, R. and Rinderer, T. (1986) Africanized bees and pollination. *American Bee Journal* 126, 680–682.

Danka, R.G., Rinderer, T.E., Hellmich, R.L. and Collins, A.M. (1986a) Comparative toxicities of four topically applied insecticides to Africanized and European honey bees (Hymenoptera: Apidae). *Journal of Economic Entomology* 79, 18–21.

Danka, R.G., Rinderer, T.E., Hellmich, R.L. and Collins, A.M. (1986b) Foraging population sizes of Africanized and European honey bee (*Apis mellifera* L.) colonies. *Apidologie* 17, 193–202.

Danka, R.G., Rinderer, T.E., Collins, A.M. and Hellmich, R.L. (1987) Responses of Africanized honey bees (Hymenoptera: Apidae) to pollination-management stress. *Journal of Economic Entomology* 80, 621–624.

Danka, R.G., Hellmich, R.L., Collins, A.M., Rinderer, T.E. and Wright, V.L. (1990) Flight characteristics of foraging Africanized and European honey bees (Hymenoptera: Apidae). *Annals of the Entomological Society of America* 83, 855–859.

Danka, R., Villa, J.D. and Gary, N.E. (1993) Comparative foraging distances of Africanized, European and hybrid honey bees (*Apis mellifera* L) during pollination of cantaloupe. *BeeScience* 3, 16–21.

Darvill, B., Knight, M.E. and Goulson, D. (2004) Use of genetic markers to quantify bumblebee foraging range and nest density. *Oikos* 107, 471–478.

Darwin, C. (1859) *On the Origin of Species by Means of Natural Selection, or Preservation of Favoured Races in the Struggle for Life*. John Murray, London.

Darwin, C. (1862) *On the Various Contrivances by Which British and Foreign Orchids are Fertilized by Insects, and On the Good Effects of Intercrossing*. John Murray, London.

Darwin, F. and Seward, A. (1903) *More Letters of Charles Darwin: A Record of His Work in a Series of Hitherto Unpublished Letters*. John Murray, London.

Daszak, P., Cunningham, A.A. and Hyatt, A.D. (2000) Emerging infectious diseases of wildlife–threats to biodiversity and human health. *Science* 287, 443–449.

Davidson, M., Butler, R. and Howlett, B. (2010) *Apis mellifera* and *Megachile rotundata*: a comparison of behaviour and seed yield in a hybrid carrot seed crop. *New Zealand Journal of Crop and Horticultural Science* 38, 113–117.

Dawkins, R. (1989) *The Selfish Gene*, 2nd edn. Oxford University Press, Oxford, UK.

De Luca, P.A. and Vallejo-Marín, M. (2013) What's the 'buzz' about? The ecology and evolutionary significance of buzz-pollination. *Current Opinion in Plant Biology* 16, 429–435.

Debevec, A.H., Cardinal, S. and Danforth, B.N. (2012) Identifying the sister group to the bees: a molecular phylogeny of Aculeata with an emphasis on the superfamily Apoidea. *Zoologica Scripta* 41, 527–535.

Dedej, S. and Delaplane, K.S. (2003) Honey bee (Hymenoptera: Apidae) pollination of rabbiteye blueberry *Vaccinium ashei* var. 'Climax' is pollinator density-dependent. *Journal of Economic Entomology* 96, 1215–1220.

Dedej, S. and Delaplane, K.S. (2004) Nectar-robbing carpenter bees reduce seed-setting capability of honey bees (Hymenoptera: Apidae) in rabbiteye blueberry, *Vaccinium ashei*, 'Climax'. *Environmental Entomology* 33, 100–106.

Dedej, S. and Delaplane, K.S. (2005) Net energetic advantage drives honey bees (*Apis mellifera* L) to nectar larceny in *Vaccinium ashei* Reade. *Behavioral Ecology and Sociobiology* 57, 398–403.

Dedej, S., Delaplane, K.S. and Gocaj, E. (2000) A technical and economic evaluation of beekeeping in Albania. *Bee World* 81, 87–97.

Dedej, S., Delaplane, K.S. and Scherm, H. (2004) Effectiveness of honey bees in delivering the biocontrol agent *Bacillus subtilis* to blueberry flowers to suppress mummy berry disease. *Biological Control* 31, 422–427.

DeGrandi-Hoffman, G., Hoopingarner, R. and Baker, K. (1984) Identification and distribution of cross-pollinating honey bees (Hymenoptera: Apidae) in apple orchards. *Environmental Entomology* 13, 757–764.

Delaplane, K.S., Ellis, J.D. and Hood, W.M. (2010) A test for interactions between *Varroa destructor* (Acari: Varroidae) and *Aethina tumida* (Coleoptera: Nitidulidae) in colonies of honey bees (Hymenoptera: Apidae). *Annals of the Entomological Society of America* 103, 711–715.

Delaplane, K.S., Dag, A., Danka, R.G., Freitas, B.M., Garibaldi, L.A., Goodwin, R.M. and Hormaza, J.I.

(2013) Standard methods for pollination research with *Apis mellifera*. *Journal of Apicultural Research* 52, 1–28.

Di Pasquale, G., Salignon, M., Le Conte, Y., Belzunces, L.P., Decourtye, A., Kretzschmar, A., Suchail, S., Brunet, J.-L. and Alaux, C. (2013) Influence of pollen nutrition on honey bee health: do pollen quality and diversity matter? *PloS One* 8, e72016.

Di Prisco, G., Annoscia, D., Margiotta, M., Ferrara, R., Varricchio, P., Zanni, V., Caprio, E., Nazzi, F. and Pennacchio, F. (2016) A mutualistic symbiosis between a parasitic mite and a pathogenic virus undermines honey bee immunity and health. *Proceedings of the National Academy of Sciences* 113, 3203–3208.

Dohzono, I. and Yokoyama, J. (2010) Impacts of alien bees on native plant-pollinator relationships: a review with special emphasis on plant reproduction. *Applied Entomology and Zoology* 45, 37–47.

Dolezal, A.G., Carrillo-Tripp, J., Judd, T.M., Allen Miller, W., Bonning, B.C. and Toth, A.L. (2019) Interacting stressors matter: diet quality and virus infection in honeybee health. *Royal Society Open Science* 6, 181803.

Donovan, B. (1980) Interactions between native and introduced bees in New Zealand. *New Zealand Journal of Ecology* 3, 104–116.

Dornhaus, A. and Chittka, L. (1999) Evolutionary origins of bee dances. *Nature* 401, 38–38.

Dornhaus, A. and Chittka, L. (2001) Food alert in bumblebees (*Bombus terrestris*): possible mechanisms and evolutionary implications. *Behavioral Ecology and Sociobiology* 50, 570–576.

Dornhaus, A. and Chittka, L. (2004a) Information flow and regulation of foraging activity in bumble bees (*Bombus* spp.). *Apidologie* 35, 183–192.

Dornhaus, A. and Chittka, L. (2004b) Why do honey bees dance? *Behavioral Ecology and Sociobiology* 55, 395–401.

Dornhaus, A. and Chittka, L. (2005) Bumble bees (*Bombus terrestris*) store both food and information in honeypots. *Behavioral Ecology* 16, 661–666.

Dornhaus, A., Brockmann, A. and Chittka, L. (2003) Bumble bees alert to food with pheromone from tergal gland. *Journal of Comparative Physiology A* 189, 47–51.

Dos Santos, S.B., Roselino, A., Hrncir, M. and Bego, L. (2009) Pollination of tomatoes by the stingless bee *Melipona quadrifasciata* and the honey bee *Apis mellifera* (Hymenoptera, Apidae). *Genetics and Molecular Research* 8, 751–757.

Dramstad, W. and Fry, G. (1995) Foraging activity of bumblebees (*Bombus*) in relation to flower resources on arable land. *Agriculture, Ecosystems & Environment* 53, 123–135.

Dramstad, W.E., Fry, G.L. and Schaffer, M.J. (2003) Bumblebee foraging—is closer really better? *Agriculture, Ecosystems & Environment* 95, 349–357.

Drewnowski, A. and Popkin, B.M. (1997) The nutrition transition: new trends in the global diet. *Nutrition Reviews* 55, 31–43.

Driscoll, C.A., Macdonald, D.W. and O'Brien, S.J. (2009) From wild animals to domestic pets, an evolutionary view of domestication. *Proceedings of the National Academy of Sciences* 106, 9971–9978.

Drummond, F.A. (2016) Behavior of bees associated with the wild blueberry agro-ecosystem in the USA. *International Journal Entomology Nematology* 2, 21–26.

Dyer, F.C. (2002) When it pays to waggle. *Nature* 419, 885–886.

Eames, A.J. (1961) *Morphology of the angiosperms*. McGraw-Hill, New York.

Eckert, J. (1956) Honeybees increase asparagus seed. *American Bee Journal* 96, 153–154.

Eeraerts, M., Smagghe, G. and Meeus, I. (2019) Pollinator diversity, floral resources and semi-natural habitat, instead of honey bees and intensive agriculture, enhance pollination service to sweet cherry. *Agriculture, Ecosystems & Environment* 284, 106586.

Eeraerts, M., Smagghe, G. and Meeus, I. (2020) Bumble bee abundance and richness improves honey bee pollination behaviour in sweet cherry. *Basic and Applied Ecology* 43, 27–33.

Ehrlen, J. and Eriksson, O. (1995) Pollen limitation and population growth in a herbaceous perennial legume. *Ecology* 76, 652–656.

Eilers, E.J., Kremen, C., Greenleaf, S.S., Garber, A.K. and Klein, A.-M. (2011) Contribution of pollinator-mediated crops to nutrients in the human food supply. *PLoS One* 6(6), e21363. https://doi.org/10.1371/journal.pone.0021363.

Eischen, F. and Underwood, B. (1991) Cantaloupe pollination trials in the lower Rio Grande Valley. *The American Bee Journal* 131, 775.

Ellis, A. and Delaplane, K.S. (2008) Effects of nest invaders on honey bee (*Apis mellifera*) pollination efficacy. *Agriculture, Ecosystems & Environment* 127, 201–206.

Ellis, A. and Delaplane, K.S. (2009) An evaluation of Fruit-Boost™ as an aid for honey bee pollination under conditions of competing bloom. *Journal of Apicultural Research* 48, 15–18.

Ellis, J., Spiewok, S., Delaplane, K., Buchholz, S., Neumann, P. and Tedders, W. (2010) Susceptibility of *Aethina tumida* (Coleoptera: Nitidulidae) larvae and pupae to entomopathogenic nematodes. *Journal of Economic Entomology* 103, 1–9.

Ellis, A.M., Myers, S.S. and Ricketts, T.H. (2015) Do pollinators contribute to nutritional health? *PLoS One* 10(1), e114805. https://doi.org/10.1371/journal.pone.0114805.

Eltz, T., Brühl, C.A., Imiyabir, Z. and Linsenmair, K.E. (2003) Nesting and nest trees of stingless bees (Apidae: Meliponini) in lowland dipterocarp forests in

Sabah, Malaysia, with implications for forest management. *Forest Ecology and Management* 172, 301–313.

Endress, P.K. (2008) Perianth biology in the basal grade of extant angiosperms. *International Journal of Plant Sciences* 169, 844–862.

Endress, P.K. and Doyle, J.A. (2009) Reconstructing the ancestral angiosperm flower and its initial specializations. *American Journal of Botany* 96, 22–66.

Erbar, C. (2003) Pollen tube transmitting tissue: place of competition of male gametophytes. *International Journal of Plant Sciences* 164, S265–S277.

Eriksson, O. (2008) Evolution of seed size and biotic seed dispersal in angiosperms: paleoecological and neoecological evidence. *International Journal of Plant Sciences* 169, 863–870.

Evans, E., Burns, I. and Spivak, M. (2017) *Befriending Bumble Bees: A practical guide to raising local bumble bees*. University of Minnesota Extension Service, St. Paul, MN, USA.

Faegri, K. and Van der Pijl, L. (1979) *The principles of pollination ecology*. Oxford, Pergamon.

Fairey, D.T. and Lefkovitch, L.P. (1994) Collection of leaf pieces by *Megachile rotundata*: proportion used in nesting. *BeeScience* 3, 79–85.

Fairey, D., Lefkovitch, L. and Lieverse, J. (1989) The leafcutting bee, *Megachile rotundata* (F.): a potential pollinator for red clover. *Journal of Applied Entomology* 107, 52–57.

Fairon-Demaret, M. (1996) *Dorinnotheca streelii* Fairon-Demaret, gen. et sp. nov., a new early seed plant from the upper Famennian of Belgium. *Review of Palaeobotany and Palynology* 93, 217–233.

FAO (2003) *World agriculture: towards 2015/2030* [Online]. Rome. Available at: http://www.fao.org/3/y4252e/y4252e06.htm (accessed 6 January 2021).

Faustino, C., Silva-Matos, E., Mateus, S. and Zucchi, R. (2002) First record of emergency queen rearing in stingless bees (Hymenoptera, Apinae, Meliponini). *Insectes Sociaux* 49, 111–113.

Fenster, C.B. (1991) Selection on floral morphology by hummingbirds. *Biotropica* 23, 98–101.

Fiedler, A.K., Landis, D.A. and Wratten, S.D. (2008) Maximizing ecosystem services from conservation biological control: the role of habitat management. *Biological Control* 45, 254–271.

Fijen, T.P., Scheper, J.A., Boom, T.M., Janssen, N., Raemakers, I. and Kleijn, D. (2018) Insect pollination is at least as important for marketable crop yield as plant quality in a seed crop. *Ecology Letters* 21, 1704–1713.

Finnamore, A. and Neary, M. (1978) Blueberry pollinators of Nova Scotia, with a checklist of the blueberry pollinators in eastern Canada and northeastern United States. *Annales Societe Entomologique du Quebec* 23, 168–181.

Flanders, R., Wehling, W. and Craghead, A. (2003) Laws and regulations on the import, movement, and release of bees in the United States. In: Strickler, K. and Cane, J.H. (eds) *For Nonnative Crops, Whence Pollinators of the Future?* Entomological Society of America, Lanham, MD, pp. 99–111.

Fleming, T.H., Geiselman, C. and Kress, W.J. (2009) The evolution of bat pollination: a phylogenetic perspective. *Annals of Botany* 104, 1017–1043.

Francis, R.M., Nielsen, S.L. and Kryger, P. (2013) Varroa-virus interaction in collapsing honey bee colonies. *PloS One* 8, e57540.

Free, J. (1963) The flower constancy of honeybees. *The Journal of Animal Ecology* 32, 119–131.

Free, J.B. (1993) *Insect pollination of crops*. Academic press, London.

Frison, T.H. (1926) Experiments in attracting queen bumblebees to artificial domiciles. *Journal of Economic Entomology* 19, 149–155.

Fussell, M. and Corbet, S.A. (1991) Forage for bumble bees and honey bees in farmland: a case study. *Journal of Apicultural Research* 30, 87–97.

Fussell, M. and Corbet, S.A. (1992) Flower usage by bumble-bees: a basis for forage plant management. *Journal of Applied Ecology* 29, 451–465.

Gallai, N., Salles, J.-M., Settele, J. and Vaissière, B.E. (2009) Economic valuation of the vulnerability of world agriculture confronted with pollinator decline. *Ecological Economics* 68, 810–821.

Gamboa, V., Ravoet, J., Brunain, M., Smagghe, G., Meeus, I., Figueroa, J., Riaño, D. and De Graaf, D.C. (2015) Bee pathogens found in *Bombus atratus* from Colombia: a case study. *Journal of Invertebrate Pathology* 129, 36–39.

Garbuzov, M., Samuelson, E.E. and Ratnieks, F.L. (2015) Survey of insect visitation of ornamental flowers in Southover Grange garden, Lewes, UK. *Insect Science* 22, 700–705.

Garibaldi, L.A., Steffan-Dewenter, I., Kremen, C., Morales, J.M., Bommarco, R., Cunningham, S.A., Carvalheiro, L.G., Chacoff, N.P., Dudenhöffer, J.H. and Greenleaf, S.S. (2011) Stability of pollination services decreases with isolation from natural areas despite honey bee visits. *Ecology Letters* 14, 1062–1072.

Garibaldi, L.A., Steffan-Dewenter, I., Winfree, R., Aizen, M.A., Bommarco, R., Cunningham, S.A., Kremen, C., Carvalheiro, L.G., Harder, L.D. and Afik, O. (2013) Wild pollinators enhance fruit set of crops regardless of honey bee abundance. *Science* 339, 1608–1611.

Garratt, M.P., Breeze, T.D., Jenner, N., Polce, C., Biesmeijer, J.C. and Potts, S.G. (2014a) Avoiding a bad apple: insect pollination enhances fruit quality and economic value. *Agriculture, Ecosystems & Environment* 184, 34–40.

Garratt, M.P.D., Truslove, C., Coston, D., Evans, R., Moss, E., Dodson, C., Jenner, N., Biesmeijer, J. and Potts, S. (2014b) Pollination deficits in UK apple orchards. *Journal of Pollination Ecology* 12, 9–14.

Garratt, M.P., Bishop, J., Degani, E., Potts, S.G., Shaw, R.F., Shi, A. and Roy, S. (2018) Insect pollination as an agronomic input: strategies for oilseed rape production. *Journal of Applied Ecology* 55, 2834–2842.

Gary, N.E. and Witherell, P.C. (1977) Distribution of foraging bees of three honey bee stocks located near onion and safflower fields. *Environmental Entomology* 6, 785–788.

Gathmann, A. and Tscharntke, T. (2002) Foraging ranges of solitary bees. *Journal of Animal Ecology* 71, 757–764.

Geib, J.C., Strange, J.P. and Galen, C. (2015) Bumble bee nest abundance, foraging distance, and host-plant reproduction: implications for management and conservation. *Ecological Applications* 25, 768–778.

Genersch, E., Von Der Ohe, W., Kaatz, H., Schroeder, A., Otten, C., Büchler, R., Berg, S., Ritter, W., Mühlen, W. and Gisder, S. (2010) The German bee monitoring project: a long term study to understand periodically high winter losses of honey bee colonies. *Apidologie* 41, 332–352.

Genise, J.F., Sciutto, J.C., Laza, J.H., González, M.G. and Bellosi, E.S. (2002) Fossil bee nests, coleopteran pupal chambers and tuffaceous paleosols from the late cretaceous laguna palacios formation, Central Patagonia (Argentina). *Palaeogeography, Palaeoclimatology, Palaeoecology* 177, 215–235.

Ghazoul, J. (2005) Buzziness as usual? Questioning the global pollination crisis. *Trends in Ecology & Evolution* 20, 367–373.

Giannini, T., Boff, S., Cordeiro, G., Cartolano, E., Veiga, A., Imperatriz-Fonseca, V. and Saraiva, A. (2015a) Crop pollinators in Brazil: a review of reported interactions. *Apidologie* 46, 209–223.

Giannini, T.C., Cordeiro, G.D., Freitas, B.M., Saraiva, A.M. and Imperatriz-Fonseca, V.L. (2015b) The dependence of crops for pollinators and the economic value of pollination in Brazil. *Journal of Economic Entomology* 108, 849–857.

Giannini, T.C., Garibaldi, L.A., Acosta, A.L., Silva, J.S., Maia, K.P., Saraiva, A.M., Guimarães Jr, P.R. and Kleinert, A.M. (2015c) Native and non-native supergeneralist bee species have different effects on plant-bee networks. *PloS One* 10, e0137198.

Giles, V. and Ascher, J.S. (2006) A survey of the bees of the black rock forest preserve, New York (Hymenoptera: Apoidea). *Journal of Hymenoptera Research* 15, 208–231.

Giray, T., Kence, M., Oskay, D., Döke, M.A. and Kence, A. (2010) Scientific note: colony losses survey in Turkey and causes of bee deaths. *Apidologie* 41, 451–453.

Godini, A., De Palma, L. and Palasciano, M. (1992) Role of self-pollination and reciprocal stigma/anthers position on fruit set of eight self-compatible almonds. *HortScience* 27, 887–889.

Golick, D.A. and Ellis, M.D. (2006) An update on the distribution and diversity of *Bombus* in Nebraska (Hymenoptera: Apidae). *Journal of the Kansas Entomological Society* 79, 341–347.

Gonzalez, A., Rowe, C., Weeks, P., Whittle, D., Gilbert, F. and Barnard, C. (1995) Flower choice by honey bees (*Apis mellifera* L.): sex-phase of flowers and preferences among nectar and pollen foragers. *Oecologia* 101, 258–264.

González-Varo, J.P., Biesmeijer, J.C., Bommarco, R., Potts, S.G., Schweiger, O., Smith, H.G., Steffan-Dewenter, I., Szentgyörgyi, H., Woyciechowski, M. and Vilà, M. (2013) Combined effects of global change pressures on animal-mediated pollination. *Trends in Ecology & Evolution* 28, 524–530.

Goodman, R. (1974) Rate of brood rearing in the effect of pollen trapping on honeybee colonies. *Australasian Beekeeper* 76, 39–41.

Goodrich, B. and Goodhue, R. (2016) Honey bee colony strength in the California almond pollination market. *University of California Giannini Foundation of Agricultural Economics ARE Update* 19, 5–8.

Gordon, D.M., Barthell, J.F., Page Jr, R.E., Kim Fondrk, M. and Thorp, R.W. (1995) Colony performance of selected honey bee (Hymenoptera: Apidae) strains used for alfalfa pollination. *Journal of Economic Entomology* 88, 51–57.

Goulson, D. (2010) Impacts of non-native bumblebees in Western Europe and North America. *Applied Entomology and Zoology* 45, 7–12.

Goulson, D. and Darvill, B. (2004) Niche overlap and diet breadth in bumblebees; are rare species more specialized in their choice of flowers? *Apidologie* 35, 55–63.

Goulson, D., Hanley, M.E., Darvill, B., Ellis, J. and Knight, M.E. (2005) Causes of rarity in bumblebees. *Biological Conservation* 122, 1–8.

Goulson, D., Lye, G.C. and Darvill, B. (2008) Diet breadth, coexistence and rarity in bumblebees. *Biodiversity and Conservation* 17, 3269–3288.

Goulson, D., Nicholls, E., Botías, C. and Rotheray, E.L. (2015) Bee declines driven by combined stress from parasites, pesticides, and lack of flowers. *Science* 347, 1255957.

Goulson, D., O'Connor, S. and Park, K.J. (2018) The impacts of predators and parasites on wild bumblebee colonies. *Ecological Entomology* 43, 168–181.

Graystock, P., Goulson, D. and Hughes, W.O. (2014) The relationship between managed bees and the prevalence of parasites in bumblebees. *PeerJ* 2, e522.

Graystock, P., Blane, E.J., McFrederick, Q.S., Goulson, D. and Hughes, W.O. (2016a) Do managed bees drive parasite spread and emergence in wild bees? *International Journal for Parasitology: Parasites and Wildlife* 5, 64–75.

Graystock, P., Meeus, I., Smagghe, G., Goulson, D. and Hughes, W.O. (2016b) The effects of single and mixed infections of *Apicystis bombi* and deformed wing virus in *Bombus terrestris. Parasitology* 143, 358–365.

Greco, M.K., Spooner-Hart, R.N., Beattie, A.G., Barchia, I. and Holford, P. (2011) Australian stingless bees improve greenhouse capsicum production. *Journal of Apicultural Research* 50, 102–115.

Greenleaf, S.S. and Kremen, C. (2006a) Wild bee species increase tomato production and respond differently to surrounding land use in Northern California. *Biological Conservation* 133, 81–87.

Greenleaf, S.S. and Kremen, C. (2006b) Wild bees enhance honey bees' pollination of hybrid sunflower. *Proceedings of the National Academy of Sciences* 103, 13890–13895.

Greenleaf, S.S., Williams, N.M., Winfree, R. and Kremen, C. (2007) Bee foraging ranges and their relationship to body size. *Oecologia* 153, 589–596.

Griffin, R., Macfarlane, R. and Van den Ende, H. (1990) Rearing and domestication of long tongued bumble bees in New Zealand. *VI International Symposium on Pollination* 288, 149–153.

Grimaldi, D. (1999) The co-radiations of pollinating insects and angiosperms in the Cretaceous. *Annals of the Missouri Botanical Garden* 86, 373–406.

Grimaldi, D., Engel, M.S. and Engel, M.S. (2005) *Evolution of the insects*. Cambridge University Press, New York, pp. 755.

Grixti, J.C., Wong, L.T., Cameron, S.A. and Favret, C. (2009) Decline of bumble bees (*Bombus*) in the North American Midwest. *Biological Conservation* 142, 75–84.

Gross, H.R., Hamm, J.J. and Carpenter, J.E. (1994) Design and application of a hive-mounted device that uses honey bees (Hymenoptera: Apidae) to disseminate *Heliothis* nuclear polyhedrosis virus. *Environmental Entomology* 23, 492–501.

Gruber, B., Eckel, K., Everaars, J. and Dormann, C.F. (2011) On managing the red mason bee (*Osmia bicornis*) in apple orchards. *Apidologie* 42, 564.

Guedot, C., Bosch, J. and Kemp, W.P. (2009) Relationship between body size and homing ability in the genus *Osmia* (Hymenoptera; Megachilidae). *Ecological Entomology* 34, 158–161.

Guimarães-Cestaro, L., Martins, M.F., Martínez, L.C., Alves, M.L.T.M.F., Guidugli-Lazzarini, K.R., Nocelli, R.C.F., Malaspina, O., Serrão, J.E. and Teixeira, É.W. (2020) Occurrence of virus, microsporidia, and pesticide residues in three species of stingless bees (Apidae: Meliponini) in the field. *The Science of Nature* 107, 1–14.

Gurr, G.M., Lu, Z., Zheng, X., Xu, H., Zhu, P., Chen, G., Yao, X., Cheng, J., Zhu, Z. and Catindig, J.L. (2016) Multi-country evidence that crop diversification promotes ecological intensification of agriculture. *Nature Plants* 2, 1–4.

Hagen, M., Wikelski, M. and Kissling, W.D. (2011) Space use of bumblebees (*Bombus* spp.) revealed by radio-tracking. *PloS One* 6, e19997.

Haider, M., Dorn, S., Sedivy, C. and Müller, A. (2014) Phylogeny and floral hosts of a predominantly pollen generalist group of mason bees (Megachilidae: Osmiini). *Biological Journal of the Linnean Society* 111, 78–91.

Hallett, A.C., Mitchell, R.J., Chamberlain, E.R. and Karron, J.D. (2017) Pollination success following loss of a frequent pollinator: the role of compensatory visitation by other effective pollinators. *AoB Plants* 9(3), May 2017, plx020, https://doi.org/10.1093/aobpla/plx020.

Hammel, B., Vollet-Neto, A., Menezes, C., Nascimento, F.S., Engels, W. and Grüter, C. (2016) Soldiers in a stingless bee: work rate and task repertoire suggest they are an elite force. *The American Naturalist* 187, 120–129.

Hanley, M.E. and Goulson, D. (2003) Introduced weeds pollinated by introduced bees: cause or effect? *Weed Biology and Management* 3, 204–212.

Hansen, P. (1967) 14C-studies on apple trees. I. The effect of the fruit on the translocation and distribution of photosynthates. *Physiologia Plantarum* 20, 382–391.

Hartfelder, K. and Engels, W. (1989) The composition of larval food in stingless bees: evaluating nutritional balance by chemosystematic methods. *Insectes Sociaux* 36, 1–14.

Hatcher, M.J. and Dunn, A.M. (2011) *Parasites in ecological communities: from interactions to ecosystems*, Cambridge University Press, New York, pp. 464.

Hatjina, F. (1998) Hive-entrance fittings as a simple and cost-effective way to increase cross-pollination by honey bees. *Bee World* 79, 71–80.

Hawkes, C. (2006) Uneven dietary development: linking the policies and processes of globalization with the nutrition transition, obesity and diet-related chronic diseases. *Globalization and Health* 2, 4 (2006). https://doi.org/10.1186/1744-8603-2-4.

Heard, T.A. (1999) The role of stingless bees in crop pollination. *Annual Review of Entomology* 44, 183–206.

Heard, M., Carvell, C., Carreck, N., Rothery, P., Osborne, J. and Bourke, A. (2007) Landscape context not patch size determines bumble-bee density on flower mixtures sown for agri-environment schemes. *Biology Letters* 3, 638–641.

Heinrich, B. (1976) Resource partitioning among some eusocial insects: bumblebees. *Ecology* 57, 874–889.

Heinrich, B. (1979) *Bumblebee economics*. Harvard University Press, Cambridge, MA.

Heinrich, B. (1993) *The Hot-blooded insects: Strategies and Mechanisms of Thermoregulation*. Springer-Verlag, Berlin.

Heithaus, E.R. (1979) Flower-feeding specialization in wild bee and wasp communities in seasonal neotropical habitats. *Oecologia* 42, 179–194.

Heller, S., Joshi, N.K., Leslie, T., Rajotte, E.G. and Biddinger, D.J. (2019) Diversified floral resource plantings support bee communities after apple bloom in commercial orchards. *Scientific Reports* 9, 1–13.

Hellmich, R.L., Kulincevic, J.M. and Rothenbuhler, W.C. (1985) Selection for high and low pollenhoarding honey bees. *Journal of Heredity* 76, 155–158.

Henry, M., Becher, M.A., Osborne, J.L., Kennedy, P.J., Aupinel, P., Bretagnolle, V., Brun, F., Grimm, V., Horn, J. and Requier, F. (2017) Predictive systems models can help elucidate bee declines driven by multiple combined stressors. *Apidologie* 48, 328–339.

Henselek, Y., Eilers, E.J., Kremen, C., Hendrix, S.D. and Klein, A.-M. (2018) Pollination requirements of almond (*Prunus dulcis*): combining laboratory and field experiments. *Journal of Economic Entomology* 111, 1006–1013.

Herrera, J. (1988) Pollination relationships in southern Spanish Mediterranean shrublands. *Journal of Ecology* 76, 274–287.

Higo, H., Winston, M. and Slessor, K. (1995) Mechanisms by which honey bee (Hymenoptera: Apidae) queen pheromone sprays enhance pollination. *Annals of the Entomological Society of America* 88, 366–373.

Hobbs, G. (1956) Ecology of the leaf-cutter bee *Megachile perihirta* Ckll. (Hymenoptera: Megachilidae) in relation to production of alfalfa seed. *The Canadian Entomologist* 88, 625–631.

Hobbs, G. (1967) Obtaining and protecting red-clover pollinating species of *Bombus* (Hymenoptera: Apidae). *The Canadian Entomologist* 99, 943–951.

Hobbs, G., Virostek, J. and Nummi, W. (1960) Establishment of *Bombus* spp. (Hymenoptera: Apidae) in artificial domiciles in southern Alberta. *The Canadian Entomologist* 92, 868–872.

Hoehn, P., Tscharntke, T., Tylianakis, J.M. and Steffan-Dewenter, I. (2008) Functional group diversity of bee pollinators increases crop yield. *Proceedings of the Royal Society B: Biological Sciences* 275, 2283–2291.

Hoff, F.L. and Willett, L.S. (1994) The US beekeeping industry. US Dept Agriculture, Economic Research Service, *Agriculture Economic Report* 680, pp. 82.

Holzschuh, A., Dudenhöffer, J.-H. and Tscharntke, T. (2012) Landscapes with wild bee habitats enhance pollination, fruit set and yield of sweet cherry. *Biological Conservation* 153, 101–107.

Horskins, K. and Turner, V. (1999) Resource use and foraging patterns of honeybees, *Apis mellifera*, and native insects on flowers of *Eucalyptus costata*. *Australian Journal of Ecology* 24, 221–227.

Houlahan, J.E., Findlay, C.S., Schmidt, B.R., Meyer, A.H. and Kuzmin, S.L. (2000) Quantitative evidence for global amphibian population declines. *Nature* 404, 752–755.

Howlett, B. (2012) Hybrid carrot seed crop pollination by the fly *Calliphora vicina* (Diptera: Calliphoridae). *Journal of Applied Entomology* 136, 421–430.

Hu, S., Dilcher, D.L., Jarzen, D.M. and Taylor, D.W. (2008) Early steps of angiosperm–pollinator coevolution. *Proceedings of the National Academy of Sciences* 105, 240–245.

Hubbell, S.P. and Johnson, L.K. (1977) Competition and nest spacing in a tropical stingless bee community. *Ecology* 58, 949–963.

Ingersoll, T.E., Sewall, B.J. and Amelon, S.K. (2013) Improved analysis of long-term monitoring data demonstrates marked regional declines of bat populations in the eastern United States. *PLoS One* 8, e65907.

Inoue, T., Nakamura, K., Salmah, S. and Abbas, I. (1993) Population dynamics of animals in unpredictably-changing tropical environments. *Journal of Biosciences* 18, 425–455.

Inoue, M.N., Yokoyama, J. and Washitani, I. (2008) Displacement of Japanese native bumblebees by the recently introduced *Bombus terrestris* (L.) (Hymenoptera: Apidae). *Journal of Insect Conservation* 12, 135–146.

Inouye, D.W. (1980) The effect of proboscis and corolla tube lengths on patterns and rates of flower visitation by bumblebees. *Oecologia* 45, 197–201.

Irshad, M. and Stephen, E. (2013) Value of insect pollinators to agriculture of Pakistan. *International Journal of Agronomy and Agricultural Research* 3, 14–21.

Isaacs, R. and Kirk, A.K. (2010) Pollination services provided to small and large highbush blueberry fields by wild and managed bees. *Journal of Applied Ecology* 47, 841–849.

Ishii, H.S., Kadoya, T., Kikuchi, R., Suda, S.-I. and Washitani, I. (2008) Habitat and flower resource partitioning by an exotic and three native bumble bees in central Hokkaido, Japan. *Biological Conservation* 141, 2597–2607.

Ito, T., Konno, I., Kubota, S., Ochiai, T., Sonoda, T., Hayashi, Y., Fukuda, T., Yokoyama, J., Nakayama, H. and Kameya, T. (2011) Production and characterization of interspecific hybrids between *Asparagus kiusianus* Makino and *A. officinalis* L. *Euphytica* 182, 285.

Jacobson, M.M., Tucker, E.M., Mathiasson, M.E. and Rehan, S.M. (2018) Decline of bumble bees in northeastern North America, with special focus on *Bombus terricola*. *Biological Conservation* 217, 437–445.

Jalali-Khanabadi, B.-A., Mozaffari-Khosravi, H. and Parsaeyan, N. (2010) Effects of almond dietary supplementation on coronary heart disease lipid risk factors and serum lipid oxidation parameters in men with mild hyperlipidemia. *The Journal of Alternative and Complementary Medicine* 16, 1279–1283.

James, R. (2005) Impact of disinfecting nesting boards on chalkbrood control in the alfalfa leafcutting bee. *Journal of Economic Entomology* 98, 1094–1100.

James, R. (2011) Chalkbrood transmission in the alfalfa leafcutting bee: the impact of disinfecting bee cocoons in loose cell management systems. *Environmental Entomology* 40, 782–787.

James, R. and Pitts-Singer, T. (2005) *Ascosphaera aggregata* contamination on alfalfa leafcutting bees in a loose cell incubation system. *Journal of Invertebrate Pathology* 89, 176–178.

Jarvis, C.E. and Linné, C.V. (2007) *Order out of chaos: Linnean plant names and their types*. Linnean Society of London in association with the Natural History Museum, London.

Javorek, S., Mackenzie, K. and Vander Kloet, S. (2002) Comparative pollination effectiveness among bees (Hymenoptera: Apoidea) on lowbush blueberry (Ericaceae: *Vaccinium angustifolium*). *Annals of the Entomological Society of America* 95, 345–351.

Jean, R.P. (2010) *Studies of bee diversity in Indiana: the influence of collection methods on species captures, and a state checklist based on museum collections*. Ph.D., Indiana State University, pp. 235.

Jin, B., Zhang, L., Lu, Y., Wang, D., Jiang, X.X., Zhang, M. and Wang, L. (2012) The mechanism of pollination drop withdrawal in *Ginkgo biloba* L. *BMC Plant Biology* 12, 59.

Johansen, C. and Eves, J. (1971) Management of alkali bees for alfalfa seed production. Washington State University Cooperative Extension Service, Pullman, WA.

Johansen, C.A., Mayer, D. and Eves, J. (1978) Biology and management of the alkali bee, *Nomia melanderi* Cockrell (Hymenoptera: Halictidae). *Melanderia*. 28, 25–46.

Johnson, S.D. and Anderson, B. (2010) Coevolution between food-rewarding flowers and their pollinators. *Evolution: Education and Outreach* 3, 32.

Johnson, S.A., Tompkins, M.M., Tompkins, H. and Colla, S.R. (2019) Artificial domicile use by bumble bees (*Bombus*; Hymenoptera: Apidae) in Ontario, Canada. *Journal of Insect Science* 19, 7.

Jung, C. (2008) Economic value of honeybee pollination on major fruit and vegetable crops in Korea. *Korean Journal of Apiculture* 23, 147–152.

Kacelnik, A., Houston, A.I. and Schmid-Hempel, P. (1986) Central-place foraging in honey bees: the effect of travel time and nectar flow on crop filling. *Behavioral Ecology and Sociobiology* 19, 19–24.

Kaminski, L.-A., Slessor, K.N., Winston, M.L., Hay, N.W. and Borden, J.H. (1990) Honeybee response to queen mandibular pheromone in laboratory bioassays. *Journal of Chemical Ecology* 16, 841–850.

Kapheim, K.M., Pan, H., Li, C., Blatti, C., Harpur, B.A., Ioannidis, P., Jones, B.M., Kent, C.F., Ruzzante, L. and Sloofman, L. (2019) Draft genome assembly and population genetics of an agricultural pollinator, the solitary alkali bee (Halictidae: *Nomia melanderi*). *G3: Genes, Genomes, Genetics* 9, 625–634.

Kasina, J., Mburu, J., Kraemer, M. and Holm-Mueller, K. (2009) Economic benefit of crop pollination by bees: a case of Kakamega small-holder farming in western Kenya. *Journal of Economic Entomology* 102, 467–473.

Kawakita, A., Ascher, J.S., Sota, T., Kato, M. and Roubik, D.W. (2008) Phylogenetic analysis of the corbiculate bee tribes based on 12 nuclear protein-coding genes (Hymenoptera: Apoidea: Apidae). *Apidologie* 39, 163–175.

Kerr, W. (1985) Número máximo e mínimo de colônias de Meliponideos que devem ser colocados em um local. *Boletim Capel* 40, 7–8.

King, C., Ballantyne, G. and Willmer, P.G. (2013) Why flower visitation is a poor proxy for pollination: measuring single-visit pollen deposition, with implications for pollination networks and conservation. *Methods in Ecology and Evolution* 4, 811–818.

Kissinger, C.N., Cameron, S.A., Thorp, R.W., White, B. and Solter, L.F. (2011) Survey of bumble bee (*Bombus*) pathogens and parasites in Illinois and selected areas of northern California and southern Oregon. *Journal of Invertebrate Pathology* 107, 220–224.

Klatt, B.K., Holzschuh, A., Westphal, C., Clough, Y., Smit, I., Pawelzik, E. and Tscharntke, T. (2014) Bee pollination improves crop quality, shelf life and commercial value. *Proceedings of the Royal Society B: Biological Sciences* 281, 20132440.

Kleijn, D., Winfree, R., Bartomeus, I., Carvalheiro, L.G., Henry, M., Isaacs, R., Klein, A.-M., Kremen, C., M'Gonigle, L.K. and Rader, R. (2015) Delivery of crop pollination services is an insufficient argument for wild pollinator conservation. *Nature Communications* 6, 1–9.

Klein, A.M., Steffan–Dewenter, I. and Tscharntke, T. (2003) Fruit set of highland coffee increases with the diversity of pollinating bees. *Proceedings of the Royal Society of London. Series B: Biological Sciences* 270, 955–961.

Klein, A.-M., Vaissiere, B.E., Cane, J.H., Steffan-Dewenter, I., Cunningham, S.A., Kremen, C. and Tscharntke, T. (2007) Importance of pollinators in changing landscapes for world crops. *Proceedings of the Royal Society B: Biological Sciences* 274, 303–313.

Klein, A.M., Brittain, C., Hendrix, S.D., Thorp, R., Williams, N. and Kremen, C. (2012) Wild pollination services to California almond rely on semi-natural habitat. *Journal of Applied Ecology* 49, 723–732.

Klein, S., Cabirol, A., Devaud, J.-M., Barron, A.B. and Lihoreau, M. (2017) Why bees are so vulnerable to environmental stressors. *Trends in Ecology & Evolution* 32, 268–278.

Klinkhamer, P.G. and De Jong, T.J. (2002) Sex allocation in hermaphrodite plants. In: *Sex ratios: concepts and research methods* (Hardy, I.C.W., ed.). Cambridge University Press, Cambridge, UK, 333–348.

Klostermeyer, E. and Gerber, H.S. (1969) Nesting behavior of *Megachile rotundata* (Hymenoptera: Megachilidae) monitored with an event recorder. *Annals of the Entomological Society of America* 62, 1321–1325.

Knight, M.E., Martin, A.P., Bishop, S., Osborne, J.L., Hale, R.J., Sanderson, R.A. and Goulson, D. (2005) An interspecific comparison of foraging range and nest density of four bumblebee (*Bombus*) species. *Molecular Ecology* 14, 1811–1820.

Knight, T.M., Steets, J.A. and Ashman, T.L. (2006) A quantitative synthesis of pollen supplementation experiments highlights the contribution of resource reallocation to estimates of pollen limitation. *American Journal of Botany* 93, 271–277.

Koch, L., Lunau, K. and Wester, P. (2017) To be on the safe site–Ungroomed spots on the bee's body and their importance for pollination. *PloS One* 12, e0182522.

Kodad, O. and Socias I Company, R. (2008) Variability of oil content and of major fatty acid composition in almond (*Prunus amygdalus* Batsch) and its relationship with kernel quality. *Journal of Agricultural and Food Chemistry* 56, 4096–4101.

Koh, I., Lonsdorf, E.V., Williams, N.M., Brittain, C., Isaacs, R., Gibbs, J. and Ricketts, T.H. (2016) Modeling the status, trends, and impacts of wild bee abundance in the United States. *Proceedings of the National Academy of Sciences* 113, 140–145.

Kojima, Y., Toki, T., Morimoto, T., Yoshiyama, M., Kimura, K. and Kadowaki, T. (2011) Infestation of Japanese native honey bees by tracheal mite and virus from non-native European honey bees in Japan. *Microbial Ecology* 62, 895–906.

Kono, Y. and Kohn, J.R. (2015) Range and frequency of Africanized honey bees in California (USA). *PLoS One* 10, e0137407.

Kosior, A., Celary, W., Olejniczak, P., Fijał, J., Król, W., Solarz, W. and Płonka, P. (2007) The decline of the bumble bees and cuckoo bees (Hymenoptera: Apidae: Bombini) of Western and Central Europe. *Oryx* 41, 79–88.

Kotthoff, U., Wappler, T. and Engel, M.S. (2013) Greater past disparity and diversity hints at ancient migrations of European honey bee lineages into Africa and Asia. *Journal of Biogeography* 40, 1832–1838.

Kovach, J., Petzoldt, R. and Harman, G.E. (2000) Use of honey bees and bumble bees to disseminate *Trichoderma harzianum* 1295–22 to strawberries for *Botrytis* control. *Biological Control* 18, 235–242.

Kremen, C., Bugg, R.L., Nicola, N., Smith, S.A., Thorp, R.W. and Williams, N.M. (2002) Native bees, native plants and crop pollination in California. *Fremontia* 30, 41–49.

Kremen, C., Williams, N.M., Bugg, R.L., Fay, J.P. and Thorp, R.W. (2004) The area requirements of an ecosystem service: crop pollination by native bee communities in California. *Ecology Letters* 7, 1109–1119.

Kremen, C., M'Gonigle, L.K. and Ponisio, L.C. (2018) Pollinator community assembly tracks changes in floral resources as restored hedgerows mature in agricultural landscapes. *Frontiers in Ecology and Evolution* 6, 170.

Kreyer, D., Oed, A., Walther-Hellwig, K. and Frankl, R. (2004) Are forests potential landscape barriers for foraging bumblebees? Landscape scale experiments with *Bombus terrestris* agg. and *Bombus pascuorum* (Hymenoptera, Apidae). *Biological Conservation* 116, 111–118.

Kritsky, G. (2015) *The tears of Re: beekeeping in ancient Egypt*. Oxford University Press, New York.

Kuhn, E.D. and Ambrose, J.T. (1984) Pollination of 'Delicious' apple by megachilid bees of the genus *Osmia* (Hymenoptera: Megachilidae). *Journal of the Kansas Entomological Society* 57, 169–180.

Kumar, J., Mishra, R. and Gupta, J. (1985) The effect of mode of pollination on *Allium* species with observation on insects as pollinators. *Journal of Apicultural Research* 24, 62–66.

Lach, L., Kratz, M. and Baer, B. (2015) Parasitized honey bees are less likely to forage and carry less pollen. *Journal of Invertebrate Pathology* 130, 64–71.

Ladurner, E., Recla, L., Wolf, M., Zelger, R. and Burgio, G. (2004) *Osmia cornuta* (Hymenoptera Megachilidae) densities required for apple pollination: a cage study. *Journal of Apicultural Research* 43, 118–122.

Langridge, D. and Goodman, R. (1982) Honeybee pollination of oilseed rape, cultivar Midas. *Australian Journal of Experimental Agriculture* 22, 124–126.

Larsen, O., Gleffe, G. and Tengo, J. (1986) Vibration and sound communication in solitary bees and wasps. *Physiological Entomology* 11, 287–296.

Lautenbach, S., Seppelt, R., Liebscher, J. and Dormann, C.F. (2012) Spatial and temporal trends of global pollination benefit. *PLoS One* 7(4): e35954. doi:10.1371/journal.pone.0035954.

Leonard, W.R. (2002) Dietary change was a driving force in human evolution. *Scientific American* 287, 106–116.

Leonhardt, S.D., Gallai, N., Garibaldi, L.A., Kuhlmann, M. and Klein, A.-M. (2013) Economic gain, stability of pollination and bee diversity decrease from southern to northern Europe. *Basic and Applied Ecology* 14, 461–471.

Lerer, H., Bailey, W., Mills, P. and Pankiw, P. (1982) Pollination activity of *Megachile rotundata* (Hymenoptera: Apoidae). *Environmental Entomology* 11, 997–1000.

Leslie, A.B., Beaulieu, J.M. and Mathews, S. (2017) Variation in seed size is structured by dispersal syndrome and cone morphology in conifers and other nonflowering seed plants. *New Phytologist* 216, 429–437.

Levin, M.D. (1986) Using honey bees to pollinate crops. US Dept Agriculture, leaflet 549.

Lewis, H. (1973) The origin of diploid neospecies in *Clarkia*. *The American Naturalist* 107, 161–170.

Li, J.-K. and Huang, S.-Q. (2009) Effective pollinators of Asian sacred lotus (*Nelumbo nucifera*): contemporary pollinators may not reflect the historical pollination syndrome. *Annals of Botany* 104, 845–851.

Linsley, E. and McSwain, J. (1947) Factors influencing the effectiveness of insect pollinators of alfalfa in California. *Journal of Economic Entomology* 40, 349–357.

Litman, J.R., Griswold, T. and Danforth, B.N. (2016) Phylogenetic systematics and a revised generic classification of anthidiine bees (Hymenoptera: Megachilidae). *Molecular Phylogenetics and Evolution* 100, 183–198.

Lloyd, D.G. (1965) Evolution of self-compatibility and racial differentiation in *Leavenworthia* (Cruciferae). *Contributions from the Gray Herbarium of Harvard University* 195, 3–134.

Lloyd, D.G. and Wells, M.S. (1992) Reproductive biology of a primitive angiosperm, *Pseudowintera colorata* (Winteraceae), and the evolution of pollination systems in the Anthophyta. *Plant Systematics and Evolution* 181, 77–95.

Loper, G. and Danka, R. (1991) Pollination tests with Africanized honey bees in Southern Mexico, 1986–88. *American Bee Journal* 131, 191–193.

Losey, J.E. and Vaughan, M. (2006) The economic value of ecological services provided by insects. *Bioscience* 56, 311–323.

Lotter, J.D.V. (1960) Recent developments in the pollination technique of deciduous fruit trees. *Deciduous Fruit Grower* 10, 182–190, 212–224, 304–311.

Luck, G.W., Spooner, P.G., Watson, D.M., Watson, S.J. and Saunders, M.E. (2014) Interactions between almond plantations and native ecosystems: lessons learned from north-western Victoria. *Ecological Management & Restoration* 15, 4–15.

Lye, G.C., Park, K.J., Holland, J.M. and Goulson, D. (2011) Assessing the efficacy of artificial domiciles for bumblebees. *Journal for Nature Conservation* 19, 154–160.

Maccagnani, B., Ladurner, E., Santi, F. and Burgio, G. (2003) *Osmia cornuta* (Hymenoptera, Megachilidae) as a pollinator of pear (*Pyrus communis*): fruit- and seed-set. *Apidologie* 34, 207–216.

Maccagnani, B., Pisman, M. and Smagghe, G. (2020) Dispensers for entomovectoring: for every bee a different type? In: Smagghe, G., Boecking, O., Maccagnani, B., M. and M., Kevan, P.G. (eds) *Entomovectoring for Precision Biocontrol and Enhanced Pollination of Crops*. Springer, Switzerland, 95–122.

Macfarlane, R., Griffin, R. and Read, P. (1983) Bumble bee management options to improve 'Grasslands Pawera' red clover seed yields. *Proceedings of the New Zealand Grassland Association* 44, 47–53.

Macías-Macías, O., Chuc, J., Ancona-Xiu, P., Cauich, O. and Quezada-Euán, J. (2009) Contribution of native bees and Africanized honey bees (Hymenoptera: Apoidea) to Solanaceae crop pollination in tropical

México. *Journal of Applied Entomology* 133, 456–465.

Macías-Macías, J.O. and Quezada-Euán, J.J.G. (2015) Stingless bees in a temperate climate: oviposition behavior and duration of ontogenic development stages in *Melipona colimana* (Hymenoptera: Meliponini). *Journal of Apicultural Research* 54, 255–259.

Maeta, Y. (1990) Utilization of wild bees. *Farming Japan* 24, 13–22.

Maeta, Y. and Adachi, K. (2005) Nesting behaviors of the alfalfa leaf-cutting bee, *Megachile* (*Eutricharaea*) *rotundata* (Fabricius) (Hymenoptera, Megachilidae). *Chugoku Kontyu* 18, 5–21.

Maeta, Y. and Kitamura, T. (2005) On the number of eggs laid by one individual of females in the alfalfa leaf-cutting bee, *Megachile* (*Eutricharaea*) *rotundata* (Fabricius) (Hymenoptera, Megachilidae). *Chugoku Kontyu* 19, 39–43.

Majewski, J. (2014) Economic value of pollination of major crops in Poland. *Economic Science for Rural Development* 34, 14–21.

Makino, T.T. and Sakai, S. (2004) Findings on spatial foraging patterns of bumblebees (*Bombus ignitus*) from a bee-tracking experiment in a net cage. *Behavioral Ecology and Sociobiology* 56, 155–163.

Mallinger, R.E. and Gratton, C. (2015) Species richness of wild bees, but not the use of managed honeybees, increases fruit set of a pollinator-dependent crop. *Journal of Applied Ecology* 52, 323–330.

Maloof, J.E. and Inouye, D.W. (2000) Are nectar robbers cheaters or mutualists? *Ecology* 81, 2651–2661.

Mandelik, Y., Winfree, R., Neeson, T. and Kremen, C. (2012) Complementary habitat use by wild bees in agro-natural landscapes. *Ecological Applications* 22, 1535–1546.

Manning, A. (1956) Some aspects of the foraging behaviour of bumble-bees. *Behaviour* 9, 164–200.

Manning, R. and Wallis, I.R. (2005) Seed yields in canola (*Brassica napus* cv. Karoo) depend on the distance of plants from honeybee apiaries. *Australian Journal of Experimental Agriculture* 45, 1307–1313.

Martinet, B., Rasmont, P., Cederberg, B., Evrard, D., Ødegaard, F., Paukkunen, J. and Lecocq, T. (2015) Forward to the north: two Euro-Mediterranean bumblebee species now cross the Arctic Circle. *Annales de la Société Entomologique de France (NS)* 51, 303–309.

Martins, A.C., Gonçalves, R.B. and Melo, G.A. (2013) Changes in wild bee fauna of a grassland in Brazil reveal negative effects associated with growing urbanization during the last 40 years. *Zoologia (Curitiba)* 30, 157–176.

Mateus, S. and Noll, F.B. (2004) Predatory behavior in a necrophagous bee *Trigona hypogea* (Hymenoptera; Apidae, Meliponini). *Naturwissenschaften* 91, 94–96.

Mateus, S., Ferreira-Caliman, M., Menezes, C. and Grüter, C. (2019) Beyond temporal-polyethism: division of labor in the eusocial bee *Melipona marginata*. *Insectes Sociaux* 66, 317–328.

Matheson, A. (1993) World bee health report. *Bee World* 74, 176–212.

Matheson, A. (1995) World bee health update. *Bee World* 76, 31–39.

Mathewson, J.A. (1968) Nest construction and life history of the eastern cucurbit bee, *Peponapis pruinosa* (Hymenoptera: Apoidea). *Journal of the Kansas Entomological Society* 41, 255–261.

Matsumoto, S., Abe, A. and Maejima, T. (2009) Foraging behavior of *Osmia cornifrons* in an apple orchard. *Scientia Horticulturae* 121, 73–79.

Mayer, D. and Johansen, C. (1988) WSU research examines bee hive pollen dispensers. *Goodfruit Grower April* 39, 32–33.

Mayer, D. and Lunden, J. (1991) Honey bee foraging on dandelion and apple in apple orchards. *Journal of the Entomological Society of British Columbia* 88, 15–17.

Mayer, D. and Lunden, J. (1996) A comparison of commercially managed bumblebees and honey bees (Hymenoptera: Apidae) for pollination of pears. *VII International Symposium on Pollination* 437, 283–288.

Mayer, D. and Miliczky, E. (1998) Emergence, male behavior, and mating in the alkali bee, *Nomia melanderi* Cockerell (Hymenoptera: Halictidae). *Journal of the Kansas Entomological Society* 71, 61–68.

Mayer, D., Lunden, J. and Kious, C. (1988) Effects of dipping alfalfa leaf-cutting bee nesting materials on chalkbrood disease. *Applied Agricultural Research* 3, 167–169.

Mburu, J., Hein, L.G., Gemmill, B. and Collette, L. (2006) *Economic valuation of pollination services: review of methods*. United Nations FAO, Rome.

McCall, C. and Primack, R.B. (1992) Influence of flower characteristics, weather, time of day, and season on insect visitation rates in three plant communities. *American Journal of Botany* 79, 434–442.

McCorquodale, D. and Owen, R. (1994) Laying sequence, diploid males, and nest usurpation in the leafcutter bee, *Megachile rotundata* (Hymenoptera: Megachilidae). *Journal of Insect Behavior* 7, 731.

McGregor, S.E. (1976) *Insect pollination of cultivated crop plants*. Agricultural Research Service, US Dept Agriculture, *Agriculture Handbook* 496, pp. 411.

McKinney, M.I. and Park, Y.-L. (2012) Nesting activity and behavior of *Osmia cornifrons* (Hymenoptera: Megachilidae) elucidated using videography. *Psyche: A Journal of Entomology* 2012, Article ID 814097, 7 pages, 2012. https://doi.org/10.1155/2012/814097.

Meeus, I., Brown, M.J., De Graaf, D.C. and Smagghe, G. (2011) Effects of invasive parasites on bumble bee declines. *Conservation Biology* 25, 662–671.

Meeus, I., Pisman, M., Smagghe, G. and Piot, N. (2018) Interaction effects of different drivers of wild bee decline and their influence on host–pathogen dynamics. *Current Opinion in Insect Science* 26, 136–141.

Melathopoulos, A.P., Cutler, G.C. and Tyedmers, P. (2015) Where is the value in valuing pollination ecosystem services to agriculture? *Ecological Economics* 109, 59–70.

Mena Granero, A., Guerra Sanz, J.M., Egea Gonzalez, F.J., Martinez Vidal, J.L., Dornhaus, A., Ghani, J., Roldán Serrano, A. and Chittka, L. (2005) Chemical compounds of the foraging recruitment pheromone in bumblebees. *Naturwissenschaften* 92, 371–374.

Merino, N., Aronson, H.S., Bojanova, D.P., Feyhl-Buska, J., Wong, M.L., Zhang, S. and Giovannelli, D. (2019) Living at the extremes: extremophiles and the limits of life in a planetary context. *Frontiers in Microbiology* 10, 780.

Meynhardt, J.T. and Malan, A.H. (1963) Translocation of sugars in doublestem grape-vines. *South African Journal of Agricultural Science* 6, 337–338.

Michener, C.D. (1974) *The Social Behavior of the Bees: A Comparative Study*. Harvard University Press, Cambridge, MA.

Michener, C.D. (2000) *The Bees of the World*. Johns Hopkins University Press, Baltimore, MD.

Michener, C.D. (2013) The Meliponini. In: Vit, P., Pedro, S.R. and Roubik, D.W. (eds) *Pot-honey: a legacy of stingless bees*. Springer, New York, pp. 3–17.

Michener, C.D., McGinley, R.J. and Danforth, B.N. (1994) *The bee genera of North and Central America (Hymenoptera: Apoidea)*. Smithsonian Institution Press, Washington, DC.

Mishra, A., Afik, O., Cabrera, M.L., Delaplane, K.S. and Mowrer, J.E. (2013) Inorganic nitrogen derived from foraging honey bees could have adaptive benefits for the plants they visit. *PLoS One* 8, e70591.

Molet, M., Chittka, L. and Raine, N.E. (2009) Potential application of the bumblebee foraging recruitment pheromone for commercial greenhouse pollination. *Apidologie* 40, 608–616.

Monzón, V.H., Bosch, J. and Retana, J. (2004) Foraging behavior and pollinating effectiveness of *Osmia cornuta* (Hymenoptera: Megachilidae) and *Apis mellifera* (Hymenoptera: Apidae) on "Comice" pear. *Apidologie* 35, 575–585.

Morandin, L.A. and Kremen, C. (2013) Bee preference for native versus exotic plants in restored agricultural hedgerows. *Restoration Ecology* 21, 26–32.

Morandin, L.A. and Winston, M.L. (2005) Wild bee abundance and seed production in conventional, organic, and genetically modified canola. *Ecological Applications* 15, 871–881.

Morandin, L.A. and Winston, M.L. (2006) Pollinators provide economic incentive to preserve natural land in agroecosystems. *Agriculture, Ecosystems & Environment* 116, 289–292.

Morandin, L., Laverty, T. and Kevan, P. (2001) Bumble bee (Hymenoptera: Apidae) activity and pollination levels in commercial tomato greenhouses. *Journal of Economic Entomology* 94, 462–467.

Morandin, L., Long, R. and Kremen, C. (2016) Pest control and pollination cost–benefit analysis of hedgerow restoration in a simplified agricultural landscape. *Journal of Economic Entomology* 109, 1020–1027.

Morris, W.F., Vázquez, D.P. and Chacoff, N.P. (2010) Benefit and cost curves for typical pollination mutualisms. *Ecology* 91, 1276–1285.

Morrissette, R., Francoeur, A. and Perron, J.-M. (1985) Importance des abeilles sauvages (Apoidea) dans la pollinisation des bleuetiers nains (*Vaccinium* spp.) en Sagamie, Québec. *Revue d'entomologie du Québec* 30, 44–53.

Morse, R.A. (1991) Honeybees forever. *Trends in Ecology & Evolution* 6, 337.

Mwebaze, P., Marris, G.C., Budge, G.E., Brown, M., Potts, S.G., Breeze, T.D. and Macleod, A. (2010) Quantifying the value of ecosystem services: a case study of honeybee pollination in the UK. *Contributed Paper for the 12th Annual BIOECON Conference.*

Naug, D. (2009) Nutritional stress due to habitat loss may explain recent honeybee colony collapses. *Biological Conservation* 142, 2369–2372.

Naumann, K., Winston, M.L., Slessor, K.N. and Smirle, M.J. (1994) Synthetic honey bee (Hymenoptera: Apidae) queen mandibular gland pheromone applications affect pear and sweet cherry pollination. *Journal of Economic Entomology* 87, 1595–1599.

Ng, M., Fleming, T., Robinson, M., Thomson, B., Graetz, N., Margono, C., Mullany, E.C., Biryukov, S., Abbafati, C. and Abera, S.F. (2014) Global, regional, and national prevalence of overweight and obesity in children and adults during 1980–2013: a systematic analysis for the Global Burden of Disease Study 2013. *The Lancet* 384, 766–781.

Nicodemo, D., Malheiros, E.B., De Jong, D. and Nogueira Couto, R.H. (2013) Enhanced production of parthenocarpic cucumbers pollinated with stingless bees and Africanized honey bees in greenhouses. *Semina: Ciências Agrárias* 34, 3625–3634.

Nieh, J.C. (2004) Recruitment communication in stingless bees (Hymenoptera, Apidae, Meliponini). *Apidologie* 35, 159–182.

Nieto, A., Roberts, S.P., Kemp, J., Rasmont, P., Kuhlmann, M., Criado, M.G., Biesmeijer, J.C., Bogusch, P., Dathe, H.H. and De La Rúa, P. (2017) European Red List of Bees, Publications Office of the European Union, Luxembourg.

Nunes-Silva, P., Hnrcir, M., Shipp, L., Imperatriz-Fonseca, V.L. and Kevan, P.G. (2013a) The behaviour of *Bombus impatiens* (Apidae, Bombini) on tomato (*Lycopersicon esculentum* Mill., Solanaceae) flowers: pollination and reward perception. *Journal of Pollination Ecology* 11, 33–40.

Nunes-Silva, P., Hrncir, M., Da Silva, C.I., Roldão, Y.S. and Imperatriz-Fonseca, V.L. (2013b) Stingless bees, *Melipona fasciculata*, as efficient pollinators of eggplant (*Solanum melongena*) in greenhouses. *Apidologie* 44, 537–546.

Nye, W.P. and Mackensen, O. (1970) Selective breeding of honeybees for alfalfa pollen collection: with tests in high and low alfalfa pollen collection regions. *Journal of Apicultural Research* 9, 61–64.

O'Grady, J.H. (1987) *Market Failure in the Provision of Honeybee Pollination: A Heuristic Investigation.* MSc Thesis, University of Vermont.

Ohashi, K., Thomson, J.D. and D'Souza, D. (2006) Trapline foraging by bumble bees: IV. Optimization of route geometry in the absence of competition. *Behavioral Ecology* 18, 1–11.

Ollerton, J. and Coulthard, E. (2009) Evolution of animal pollination. *Science* 326, 808–809.

Ollerton, J., Alarcón, R., Waser, N.M., Price, M.V., Watts, S., Cranmer, L., Hingston, A., Peter, C.I. and Rotenberry, J. (2009) A global test of the pollination syndrome hypothesis. *Annals of Botany* 103, 1471–1480.

Ollerton, J., Winfree, R. and Tarrant, S. (2011) How many flowering plants are pollinated by animals? *Oikos* 120, 321–326.

Ollerton, J., Price, V., Armbruster, W.S., Memmott, J., Watts, S., Waser, N.M., Totland, Ø., Goulson, D., Alarcón, R. and Stout, J.C. (2012) Overplaying the role of honey bees as pollinators: a comment on Aebi and Neumann (2011). *Trends in Ecology and Evolution* 27, 141.

Olsson, O., Bolin, A., Smith, H.G. and Lonsdorf, E.V. (2015) Modeling pollinating bee visitation rates in heterogeneous landscapes from foraging theory. *Ecological Modelling* 316, 133–143.

Osborne, J.L., Williams, I.H. and Corbet, S.A. (1991) Bees, pollination and habitat change in the European community. *Bee World* 72, 99–116.

Osborne, J.L., Martin, A.P., Carreck, N.L., Swain, J.L., Knight, M.E., Goulson, D., Hale, R.J. and Sanderson, R.A. (2008) Bumblebee flight distances in relation to the forage landscape. *Journal of Animal Ecology* 77, 406–415.

Osorio-Beristain, M., Domínguez, C., Eguiarte, L. and Benrey, B. (1997) Pollination efficiency of native and invading Africanized bees in the tropical dry forest annual plant, *Kallstroemia grandiflora* Torr ex Gray. *Apidologie* 28, 11–16.

Oster, G.F. and Wilson, E.O. (1979) *Caste and Ecology in the Social Insects.* Princeton University Press, Princeton, NJ.

Otterstatter, M.C. and Thomson, J.D. (2008) Does pathogen spillover from commercially reared bumble bees threaten wild pollinators? *PloS One* 3, e2771.

Pacini, E. (2000) From anther and pollen ripening to pollen presentation. In: Dafni, A., Hesse, M. and Pacini,

E. (eds) *Pollen and pollination*. Springer, Vienna, https://doi.org/10.1007/978-3-7091-6306-1_6.

Paini, D. (2004) Impact of the introduced honey bee (*Apis mellifera*) (Hymenoptera: Apidae) on native bees: a review. *Austral Ecology* 29, 399–407.

Parker, F.D., Batra, S. and Tepedino, V.J. (1987) New pollinators for our crops. *Agricultural Zoology Reviews* 2, 279–304

Parker, A.J., Tran, J.L., Ison, J.L., Bai, J.D.K., Weis, A.E. and Thomson, J.D. (2015) Pollen packing affects the function of pollen on corbiculate bees but not non-corbiculate bees. *Arthropod-Plant Interactions* 9, 197–203.

Parmentier, L., Smagghe, G., De Graaf, D.C. and Meeus, I. (2016) *Varroa destructor* Macula-like virus, Lake Sinai virus and other new RNA viruses in wild bumblebee hosts (*Bombus pascuorum*, *Bombus lapidarius* and *Bombus pratorum*). *Journal of Invertebrate Pathology* 134, 6–11.

Parsche, S., Fründ, J. and Tscharntke, T. (2011) Experimental environmental change and mutualistic vs. antagonistic plant flower–visitor interactions. *Perspectives in Plant Ecology, Evolution and Systematics* 13, 27–35.

Patten, K.D., Shanks, C.H. and Mayer, D.F. (1993) Evaluation of herbaceous plants for attractiveness to bumble bees for use near cranberry farms. *Journal of Apicultural Research* 32, 73–79.

Pearson, W. and Braiden, V. (1990) Seasonal pollen collection by honeybees from grass/shrub highlands in Canterbury, New Zealand. *Journal of Apicultural Research* 29, 206–213.

Pellett, F.C. (1976) *American honey plants*. Dadant and Sons, Hamilton, IL.

Pelto, G.H. and Pelto, P.J. (1983) Diet and delocalization: dietary changes since 1750. *The Journal of Interdisciplinary History* 14, 507–528.

Pendrel, B. and Plowright, R. (1981) Larval feeding by adult bumble bee workers (Hymenoptera: Apidae). *Behavioral Ecology and Sociobiology* 8, 71–76.

Pereboom, J., Velthuis, H. and Duchateau, M. (2003) The organisation of larval feeding in bumblebees (Hymenoptera, Apidae) and its significance to caste differentiation. *Insectes Sociaux* 50, 127–133.

Perry, C.J., Søvik, E., Myerscough, M.R. and Barron, A.B. (2015) Rapid behavioral maturation accelerates failure of stressed honey bee colonies. *Proceedings of the National Academy of Sciences* 112, 3427–3432.

Petanidou, T. and Smets, E. (1995) The potential of marginal lands for bees and apiculture: nectar secretion in Mediterranean shrublands. *Apidologie* 26, 39–52.

Petersen, J., Huseth, A. and Nault, B. (2014) Evaluating pollination deficits in pumpkin production in New York. *Environmental Entomology* 43, 1247–1253.

Peterson, J. and Roitberg, B. (2006) Impact of resource levels on sex ratio and resource allocation in the solitary bee, *Megachile rotundata*. *Environmental Entomology* 35, 1404–1410.

Peterson, A., Furgala, B. and Holdaway, F. (1960) Pollination of red clover in Minnesota. *Journal of Economic Entomology* 53, 546–550.

Pettis, J.S. and Delaplane, K.S. (2010) Coordinated responses to honey bee decline in the USA. *Apidologie* 41, 256–263.

Phillips, B.B., Williams, A., Osborne, J.L. and Shaw, R.F. (2018) Shared traits make flies and bees effective pollinators of oilseed rape (*Brassica napus* L.). *Basic and Applied Ecology* 32, 66–76.

Pitts-Singer, T.L. (2007) Olfactory response of megachilid bees, *Osmia lignaria*, *Megachile rotundata*, and *M. pugnata*, to individual cues from old nest cavities. *Environmental Entomology* 36, 402–408.

Pitts-Singer, T.L. and Bosch, J. (2010) Nest establishment, pollination efficiency, and reproductive success of *Megachile rotundata* (Hymenoptera: Megachilidae) in relation to resource availability in field enclosures. *Environmental Entomology* 39, 149–158.

Pitts-Singer, T.L. and Cane, J.H. (2011) The alfalfa leafcutting bee, *Megachile rotundata*: the world's most intensively managed solitary bee. *Annual Review of Entomology* 56, 221–237.

Pitts-Singer, T.L. and James, R.R. (2008) Do weather conditions correlate with findings in failed, provision-filled nest cells of *Megachile rotundata* (Hymenoptera: Megachilidae) in western North America? *Journal of Economic Entomology* 101, 674–685.

Plowright, R. and Jay, S. (1966) Rearing bumble bee colonies in captivity. *Journal of Apicultural Research* 5, 155–165.

Plowright, R. and Laverty, T. (1987) Bumble bees and crop pollination in Ontario. *Proceedings of the Entomological Society of Ontario* 118, 155–160.

Plowright, C. and Plowright, R. (1997) The advantage of short tongues in bumble bees (*Bombus*)—analyses of species distributions according to flower corolla depth, and of working speeds on white clover. *The Canadian Entomologist* 129, 51–59.

Pomeroy, N. and Plowright, R. (1980) Maintenance of bumble bee colonies in observation hives (Hymenoptera: Apidae). *The Canadian Entomologist* 112, 321–326.

Popkin, B.M., Adair, L.S. and Ng, S.W. (2012) Global nutrition transition and the pandemic of obesity in developing countries. *Nutrition Reviews* 70, 3–21.

Potts, S.G., Roberts, S.P., Dean, R., Marris, G., Brown, M.A., Jones, R., Neumann, P. and Settele, J. (2010) Declines of managed honey bees and beekeepers in Europe. *Journal of Apicultural Research* 49, 15–22.

Purkiss, T. and Lach, L. (2019) Pathogen spillover from *Apis mellifera* to a stingless bee. *Proceedings of the Royal Society B* 286, 20191071.

Pyke, G.H. (1978) Optimal foraging: movement patterns of bumblebees between inflorescences. *Theoretical Population Biology* 13, 72–98.

Pyke, G.H., Pulliam, H.R. and Charnov, E.L. (1977) Optimal foraging: a selective review of theory and tests. *The Quarterly Review of Biology* 52, 137–154.

Pyke, G.H., Thomson, J.D., Inouye, D.W. and Miller, T.J. (2016) Effects of climate change on phenologies and distributions of bumble bees and the plants they visit. *Ecosphere* 7, e01267.

Quezada-Euán, J.J.G., Nates-Parra, G., Maués, M.M., Roubik, D.W. and Imperatriz-Fonseca, V.L. (2018) The economic and cultural values of stingless bees (Hymenoptera: Meliponini) among ethnic groups of tropical America. *Sociobiology* 65, 534–557.

Ramankutty, N., Mehrabi, Z., Waha, K., Jarvis, L., Kremen, C., Herrero, M. and Rieseberg, L.H. (2018) Trends in global agricultural land use: implications for environmental health and food security. *Annual Review of Plant Biology* 69, 789–815.

Ramírez, V.M., Ayala, R. and González, H.D. (2018) Crop pollination by stingless bees. In: Vit, P., Pedro, S.R. and Roubik, D.W. (eds) *Pot-Pollen in Stingless Bee Melittology.* Springer, Switzerland. DOI 10.1007/978-3-319-61839-5, pp. 139–153.

Rank, G. and Goerzen, D. (1982) Effect of incubation temperatures on emergence of *Megachile rotundata* (Hymenoptera: Megachilidae). *Journal of Economic Entomology* 75, 467–471.

Rank, G., Rank, F. and Watts, R. (1990) Chalkbrood (*Ascosphaera aggregata* Skou.) resistance of a univoltine strain of the alfalfa leafcutting bee, *Megachile rotundata* F. *Journal of Applied Entomology* 109, 524–527.

Ransome, H.M. (1937) *The sacred bee in ancient times and folklore.* George Allen & Unwin, London.

Ranta, E. and Tiainen, M. (1982) Structure in seven bumblebee communities in eastern Finland in relation to resource availability. *Holarctic Ecology* 5, 48–54.

Rao, S. and Strange, J.P. (2012) Bumble bee (Hymenoptera: Apidae) foraging distance and colony density associated with a late-season mass flowering crop. *Environmental Entomology* 41, 905–915.

Rasmussen, C., Orihuela-Pasquel, P. and Sánchez-Bocanegra, V.H. (2009) *Trigona* Jurine, 1807 bees (Hymenoptera: Apidae) as pests of physic nut (Euphorbiaceae: Jatropha curcas) in Peru. *Entomotropica* 24, 31–32.

Ratnieks, F.L. and Helanterä, H. (2009) The evolution of extreme altruism and inequality in insect societies. *Philosophical Transactions of the Royal Society of London B: Biological Sciences* 364, 3169–3179.

Redhead, J.W., Dreier, S., Bourke, A.F., Heard, M.S., Jordan, W.C., Sumner, S., Wang, J. and Carvell, C. (2016) Effects of habitat composition and landscape structure on worker foraging distances of five bumble bee species. *Ecological Applications* 26, 726–739.

Reilly, J., Artz, D., Biddinger, D., Bobiwash, K. and Boyle, N. (2020) Crop production in the USA is frequently limited by a lack of pollinators. *Proceedings of the Royal Society B* 287, 20200922.

Ren, D., Labandeira, C.C., Santiago-Blay, J.A., Rasnitsyn, A., Shih, C., Bashkuev, A., Logan, M.A.V., Hotton, C.L. and Dilcher, D. (2009) A probable pollination mode before angiosperms: Eurasian, long-proboscid scorpionflies. *Science* 326, 840–847.

Renner, S.S. and Ricklefs, R.E. (1995) Dioecy and its correlates in the flowering plants. *American Journal of Botany* 82, 596–606.

Reyes-Centeno, H., Ghirotto, S., Détroit, F., Grimaud-Hervé, D., Barbujani, G. and Harvati, K. (2014) Genomic and cranial phenotype data support multiple modern human dispersals from Africa and a southern route into Asia. *Proceedings of the National Academy of Sciences* 111, 7248–7253.

Reynaldi, F.J., Sguazza, G.H., Albicoro, F.J., Pecoraro, M.R. and Galosi, C.M. (2013) First molecular detection of co-infection of honey bee viruses in asymptomatic *Bombus atratus* in South America. *Brazilian Journal of Biology* 73, 797–800.

Ribeiro, M., Duchateau, M. and Velthuis, H. (1996) Comparison of the effects of two kinds of commercially available pollen on colony development and queen production in the bumble bee *Bombus terrestris* L (Hymenoptera, Apidae). *Apidologie* 27, 133–144.

Rice, A.L., West Jr, K.P. and Black, R.E. (2004) Vitamin A deficiency. In: Ezzati, M., Lopez, A., Rodgers, A. and Murray, C. (eds) *Comparative quantification of health risks: global and regional burden of disease attributes to selected major risk factors.* World Health Organization, Geneva, 211–256.

Richards, K.W. (1984) *Alfalfa leafcutter bee management in western Canada.* Agriculture Canada, Ottawa.

Richards, K. (1990) Effectiveness of the alfalfa leafcutter bee as a pollinator of legume forage crops. *VI International Symposium on Pollination* 288, 180–184.

Richards, K.W. (1993) Non-*Apis* bees as crop pollinators. *Revue Suisse de Zoologie* 100, 807–822.

Richards, K. (1996) Effect of environment and equipment on productivity of alfalfa leafcutter bees (Hymenoptera: Megachilidae) in southern Alberta, Canada. *The Canadian Entomologist* 128, 47–56.

Richards, K.W. (2020) Effectiveness of the alfalfa leafcutter bee *Megachile rotundata* Fab. to pollinate four perennial legumes. *Journal of Apicultural Research* 59, 69–76.

Ricketts, T.H., Daily, G.C., Ehrlich, P.R. and Michener, C.D. (2004) Economic value of tropical forest to coffee production. *Proceedings of the National Academy of Sciences* 101, 12579–12582.

Ricketts, T.H., Regetz, J., Steffan-Dewenter, I., Cunningham, S.A., Kremen, C., Bogdanski, A., Gemmill-Herren, B., Greenleaf, S.S., Klein, A.M. and Mayfield, M.M. (2008) Landscape effects on crop pollination services: are there general patterns? *Ecology Letters* 11, 499–515.

Rodríguez-Gironés, M.A. and Santamaría, L. (2004) Why are so many bird flowers red? *PLoS Biology* 2, e350.

Rodríguez-Gironés, M.A. and Santamaría, L. (2005) Resource partitioning among flower visitors and evolution of nectar concealment in multi-species communities. *Proceedings of the Royal Society B: Biological Sciences* 272, 187–192.

Rodríguez-Gironés, M.A. and Santamaría, L. (2006) Models of optimal foraging and resource partitioning: deep corollas for long tongues. *Behavioral Ecology* 17, 905–910.

Rodríguez-Gironés, M.A. and Santamaría, L. (2007) Resource competition, character displacement, and the evolution of deep corolla tubes. *The American Naturalist* 170, 455–464.

Rollin, O. and Garibaldi, L.A. (2019) Impacts of honeybee density on crop yield: a meta-analysis. *Journal of Applied Ecology* 56, 1152–1163.

Rosas-Guerrero, V., Aguilar, R., Martén-Rodríguez, S., Ashworth, L., Lopezaraiza-Mikel, M., Bastida, J.M. and Quesada, M. (2014) A quantitative review of pollination syndromes: do floral traits predict effective pollinators? *Ecology Letters* 17, 388–400.

Röseler, P.-F. (1985) A technique for year-round rearing of *Bombus terrestris* (Apidae, Bombini) colonies in captivity. *Apidologie* 16, 165–170.

Rosenberg, K.V., Dokter, A.M., Blancher, P.J., Sauer, J.R., Smith, A.C., Smith, P.A., Stanton, J.C., Panjabi, A., Helft, L. and Parr, M. (2019) Decline of the North American avifauna. *Science* 366, 120–124.

Rossi, B.H., Nonacs, P. and Pitts-Singer, T.L. (2010) Sexual harassment by males reduces female fecundity in the alfalfa leafcutting bee, *Megachile rotundata*. *Animal Behaviour* 79, 165–171.

Roubik, D.W. (1978) Competitive interactions between neotropical pollinators and Africanized honey bees. *Science* 201, 1030–1032.

Roubik, D.W. (1980) Foraging behavior of competing Africanized honeybees and stingless bees. *Ecology* 61, 836–845.

Roubik, D.W. (1983) Experimental community studies: time-series tests of competition between African and Neotropical bees. *Ecology* 64, 971–978.

Roubik, D.W. (1995) *Pollination of cultivated plants in the tropics*. FAO, Rome.

Roubik, D.W. (2002) The value of bees to the coffee harvest. *Nature* 417, 708–708.

Roubik, D.W. and Villanueva-Gutiérrez, R. (2009) Invasive Africanized honey bee impact on native solitary bees: a pollen resource and trap nest analysis. *Biological Journal of the Linnean Society* 98, 152–160.

Roubik, D.W. and Wolda, H. (2001) Do competing honey bees matter? Dynamics and abundance of native bees before and after honey bee invasion. *Population Ecology* 43, 53–62.

Roubik, D., Arturo, J., Segura, L. and Franco De Camargo, J.M. (1997) New stingless bee genus endemic to Central American cloudforests: phylogenetic and biogeographic implications (Hymenoptera: Apidae: Meliponini). *Systematic Entomology* 22, 67–80.

Roubik, D., Heard, T. and Kwapong, P. (2018) Stingless bee colonies and pollination. In: Roubik, D.W. (ed.) *The Pollination of cultivated plants: a compendium for practitioners*. FAO, Balboa.

Rowe, L., Gibson, D., Landis, D., Gibbs, J. and Isaacs, R. (2018) A comparison of drought-tolerant prairie plants to support managed and wild bees in conservation programs. *Environmental Entomology* 47, 1128–1142.

Rundlöf, M., Andersson, G.K., Bommarco, R., Fries, I., Hederström, V., Herbertsson, L., Jonsson, O., Klatt, B.K., Pedersen, T.R. and Yourstone, J. (2015) Seed coating with a neonicotinoid insecticide negatively affects wild bees. *Nature* 521, 77–80.

Russo, L. (2016) Positive and negative impacts of nonnative bee species around the world. *Insects* 7, 69.

Ryder, J.T., Cherrill, A., Prew, R., Shaw, J., Thorbek, P. and Walters, K.F. (2020) Impact of enhanced *Osmia bicornis* (Hymenoptera: Megachilidae) populations on pollination and fruit quality in commercial sweet cherry (*Prunus avium* L.) orchards. *Journal of Apicultural Research* 59, 77–87.

Sagili, R.R. and Burgett, D. (2011) Evaluating honey bee colonies for pollination: a guide for commercial growers and beekeepers. *Pacific Northwest Extension Publication* PNW 623.

Sampson, B.J., Stringer, S.J. and Marshall, D.A. (2013) Blueberry floral attributes and their effect on the pollination efficiency of an oligolectic bee, *Osmia ribifloris* Cockerell (Megachilidae: Apoidea). *HortScience* 48, 136–142.

Samuelson, A.E., Gill, R.J., Brown, M.J. and Leadbeater, E. (2018) Lower bumblebee colony reproductive success in agricultural compared with urban environments. *Proceedings of the Royal Society B: Biological Sciences* 285, 20180807.

Sánchez, D. and Vandame, R. (2013) Stingless bee food location communication: from the flowers to the honey pots. In: Vit, P., Pedro, S.R. and Roubik, D.W. (eds) *Pot-honey: a legacy of stingless bees*. Springer, New York, pp. 187–199.

Sanjerehei, M.M. (2014) The economic value of bees as pollinators of crops in Iran. *Annual Research & Review in Biology* 4, 2957–2964.

Sapir, G., Baras, Z., Azmon, G., Goldway, M., Shafir, S., Allouche, A., Stern, E. and Stern, R. (2017) Synergistic effects between bumblebees and honey bees in apple orchards increase cross pollination, seed number and fruit size. *Scientia Horticulturae* 219, 107–117.

Sárospataki, M., Novák, J. and Molnár, V. (2005) Assessing the threatened status of bumble bee species (Hymenoptera: Apidae) in Hungary, Central Europe. *Biodiversity & Conservation* 14, 2437–2446.

Schemske, D.W. (1978) Sexual reproduction in an Illinois population of *Sanguinaria canadensis* L. *American Midland Naturalist* 100, 261–268.

Schemske, D.W. and Lande, R. (1985) The evolution of self-fertilization and inbreeding depression in plants. II. Empirical observations. *Evolution* 39, 41–52.

Scheper, J., Reemer, M., Van Kats, R., Ozinga, W.A., Van der Linden, G.T., Schaminée, J.H., Siepel, H. and Kleijn, D. (2014) Museum specimens reveal loss of pollen host plants as key factor driving wild bee decline in The Netherlands. *Proceedings of the National Academy of Sciences* 111, 17552–17557.

Schiestl, F.P. (2010) Pollination: sexual mimicry abounds. *Current Biology* 20, R1020–R1022.

Schindler, M. and Peters, B. (2011) Mason bees *Osmia bicornis* and *Osmia cornuta* as suitable orchard pollinators? *Erwerbsobstbau* 52, 111–116.

Schmid-Hempel, P. and Durrer, S. (1991) Parasites, floral resources and reproduction in natural populations of bumblebees. *Oikos* 62, 342–350.

Schoener, T.W. (1979) Generality of the size-distance relation in models of optimal feeding. *The American Naturalist* 114, 902–914.

Scott-Dupree, C.D. and Winston, M.L. (1987) Wild bee pollinator diversity and abundance in orchard and uncultivated habitats in the Okanagan Valley, British Columbia. *The Canadian Entomologist* 119, 735–745.

Sedivy, C. and Dorn, S. (2014) Towards a sustainable management of bees of the subgenus *Osmia* (Megachilidae; *Osmia*) as fruit tree pollinators. *Apidologie* 45, 88–105.

Seeley, T.D. (1985) Honeybee ecology: a study of adapatation in social life. Princeton University Press, Princeton, NJ, 201 pp.

Seidelmann, K. (2006) Open-cell parasitism shapes maternal investment patterns in the Red Mason bee *Osmia rufa*. *Behavioral Ecology* 17, 839–848.

Seidelmann, K., Ulbrich, K. and Mielenz, N. (2010) Conditional sex allocation in the Red Mason bee, *Osmia rufa*. *Behavioral Ecology and Sociobiology* 64, 337–347.

Sekita, N. (2001) Managing *Osmia cornifrons* to pollinate apples in Aomori Prefecture, Japan. *Acta Horticulturae* 561, 303–307.

Serrano, A.R. and Guerra-Sanz, J.M. (2006) Quality fruit improvement in sweet pepper culture by bumblebee pollination. *Scientia Horticulturae* 110, 160–166.

Sgolastra, F., Kemp, W.P., Maini, S. and Bosch, J. (2012) Duration of prepupal summer dormancy regulates synchronization of adult diapause with winter temperatures in bees of the genus *Osmia*. *Journal of Insect Physiology* 58, 924–933.

Sheffield, C.S. (2008) Summer bees for spring crops? Potential problems with *Megachile rotundata* (Fab.)

(Hymenoptera: Megachilidae) as a pollinator of low-bush blueberry (Ericaceae). *Journal of the Kansas Entomological Society* 81, 276–287.

Silva, E. and Dean, B.B. (2000) Effect of nectar composition and nectar concentration on honey bee (Hymenoptera: Apidae) visitations to hybrid onion flowers. *Journal of Economic Entomology* 93, 1216–1221.

Simmons, N.B., Seymour, K.L., Habersetzer, J. and Gunnell, G.F. (2008) Primitive Early Eocene bat from Wyoming and the evolution of flight and echolocation. *Nature* 451, 818.

Singh, R., Levitt, A.L., Rajotte, E.G., Holmes, E.C., Ostiguy, N., Lipkin, W.I., Depamphilis, C.W., Toth, A.L. and Cox-Foster, D.L. (2010) RNA viruses in hymenopteran pollinators: evidence of inter-taxa virus transmission via pollen and potential impact on non-Apis hymenopteran species. *Plos One* 5, e14357.

Slaa, E.J. (2006) Population dynamics of a stingless bee community in the seasonal dry lowlands of Costa Rica. *Insectes Sociaux* 53, 70–79.

Slaa, E.J., Tack, A.J. and Sommeijer, M.J. (2003) The effect of intrinsic and extrinsic factors on flower constancy in stingless bees. *Apidologie* 34, 457–468.

Slaa, E.J., Chaves, L.A.S., Malagodi-Braga, K.S. and Hofstede, F.E. (2006) Stingless bees in applied pollination: practice and perspectives. *Apidologie* 37, 293–315.

Sladen, F.W.L. (1912) *The humble-bee: its life history and how to domesticate it*. Macmillan, London.

Slatyer, R.A., Hirst, M. and Sexton, J.P. (2013) Niche breadth predicts geographical range size: a general ecological pattern. *Ecology Letters* 16, 1104–1114.

Slessor, K.N., Kaminski, L.-A., King, G., Borden, J.H. and Winston, M.L. (1988) Semiochemical basis of the retinue response to queen honey bees. *Nature* 332, 354–356.

Slessor, K.N., Kaminski, L.-A., King, G. and Winston, M.L. (1990) Semiochemicals of the honeybee queen mandibular glands. *Journal of Chemical Ecology* 16, 851–860.

Small, E., Brookes, B., Lefkovitch, L.P. and Fairey, D.T. (1997) A preliminary analysis of the floral preferences of the alfalfa leafcutting bee, *Megachile rotundata*. *Canadian Field Naturalist* 111, 445–453.

Smith, P., Bustamante, M., Ahammad, H., Clark, H., Dong, H., Elsiddig, E.A., Haberl, H., Harper, R., House, J. and Jafari, M. (2014) Agriculture, forestry and other land use (AFOLU). In: Edenhofer, O., Pichs-Madruga, R., Sokona, Y., Farahani, E., Kadner, S. and Al., E. (eds) *Climate change 2014: mitigation of climate change: contribution of Working Group III to the Fifth Assessment Report of the Intergovernmental Panel on Climate Change*. Cambridge University Press, Cambridge, UK, 813–912.

Smith, M.R., Singh, G.M., Mozaffarian, D. and Myers, S.S. (2015) Effects of decreases of animal pollinators on human nutrition and global health: a modelling analysis. *The Lancet* 386, 1964–1972.

Socias, R., Alonso, J. and Aparisi, J.G. (2004) Fruit set and productivity in almond as related to self-compatibility, flower morphology and bud density. *The Journal of Horticultural Science and Biotechnology* 79, 754–758.

Soroka, J., Goerzen, D., Falk, K. and Bett, K. (2001) Alfalfa leafcutting bee (Hymenoptera: Megachilidae) pollination of oilseed rape (*Brassica napus* L.) under isolation tents for hybrid seed production. *Canadian Journal of Plant Science* 81, 199–204.

Southwick, E.E. and Southwick Jr, L. (1992) Estimating the economic value of honey bees (Hymenoptera: Apidae) as agricultural pollinators in the United States. *Journal of Economic Entomology* 85, 621–633.

Southwick, E.E., Loper, G.M. and Sadwick, S.E. (1981) Nectar production, composition, energetics and pollinator attractiveness in spring flowers of western New York. *American Journal of Botany* 68, 994–1002.

Spaethe, J., Tautz, J. and Chittka, L. (2001) Visual constraints in foraging bumblebees: flower size and color affect search time and flight behavior. *Proceedings of the National Academy of Sciences* 98, 3898–3903.

Spears Jr, E.E. (1983) A direct measure of pollinator effectiveness. *Oecologia* 57, 196–199.

Spiewok, S. and Neumann, P. (2006) Infestation of commercial bumblebee (*Bombus impatiens*) field colonies by small hive beetles (*Aethina tumida*). *Ecological Entomology* 31, 623–628.

Stang, M., Klinkhamer, P.G. and Van der Meijden, E. (2006) Size constraints and flower abundance determine the number of interactions in a plant–flower visitor web. *Oikos* 112, 111–121.

Stanghellini, M.S., Ambrose, J.T. and Schultheis, J.R. (1997) The effects of honey bee and bumble bee pollination on fruit set and abortion of cucumber and watermelon. *American Bee Journal* 137, 386–391.

Stanghellini, M.S., Ambrose, J.T. and Schultheis, J.R. (2002) Diurnal activity, floral visitation and pollen deposition by honey bees and bumble bees on field-grown cucumber and watermelon. *Journal of Apicultural Research* 41, 27–34.

Stanley, J., Sah, K., Subbanna, A.R., Preetha, G. and Gupta, J. (2017) How efficient is *Apis cerana* (Hymenoptera: Apidae) in pollinating cabbage, *Brassica oleracea* var. capitata? Pollination behavior, pollinator effectiveness, pollinator requirement, and impact of pollination. *Journal of Economic Entomology* 110, 826–834.

Stavert, J.R., Liñán-Cembrano, G., Beggs, J.R., Howlett, B.G., Pattemore, D.E. and Bartomeus, I. (2016) Hairiness: the missing link between pollinators and pollination. *PeerJ* 4, e2779.

Steffan, S.A., Dharampal, P.S., Danforth, B.N., Gaines-Day, H.R., Takizawa, Y. and Chikaraishi, Y. (2019) Omnivory in bees: elevated trophic positions among all major bee families. *The American Naturalist* 194, 414–421.

Steffan-Dewenter, I. and Kuhn, A. (2003) Honeybee foraging in differentially structured landscapes. *Proceedings of the Royal Society of London: Series B: Biological Sciences* 270, 569–575.

Steffan-Dewenter, I., Münzenberg, U., Bürger, C., Thies, C. and Tscharntke, T. (2002) Scale-dependent effects of landscape context on three pollinator guilds. *Ecology* 83, 1421–1432.

Steffan-Dewenter, I., Potts, S.G. and Packer, L. (2005) Pollinator diversity and crop pollination services are at risk. *Trends in Ecology & Evolution* 20, 651–652.

Stein, K., Coulibaly, D., Stenchly, K., Goetze, D., Porembski, S., Lindner, A., Konaté, S. and Linsenmair, E.K. (2017) Bee pollination increases yield quantity and quality of cash crops in Burkina Faso, West Africa. *Scientific Reports* 7, 1–10.

Stephen, W. (1981) The design and function of field domiciles and incubators for leafcutting bee management (*Megachile rotundata* [Fabricius]). Oregon State University, *Agricultural Experiment Station Bulletin* 654.

Stephen, W. (1982) Chalkbrood control in the leafcutting bee. *Proceedings of the 1st International Symposium of Alfalfa Leafcutting Bee Management*, Saskatoon, Saskatchewan, Canada, 98–107.

Stern, R., Eisikowitch, D. and Dag, A. (2001) Sequential introduction of honeybee colonies and doubling their density increases cross-pollination, fruit-set and yield in 'Red Delicious' apple. *The Journal of Horticultural Science and Biotechnology* 76, 17–23.

Stillman, R.A., Railsback, S.F., Giske, J., Berger, U. and Grimm, V. (2015) Making predictions in a changing world: the benefits of individual-based ecology. *BioScience* 65, 140–150.

Stork, N.E. (2018) How many species of insects and other terrestrial arthropods are there on Earth? *Annual Review of Entomology* 63, 31–45.

Stout, J.C., Nombre, I., De Bruijn, B., Delaney, A., Doke, D.A., Gyimah, T., Kamano, F., Kelly, R., Lovett, P. and marshall, E. (2018) Insect pollination improves yield of shea (*Vitellaria paradoxa* subsp. *paradoxa*) in the agroforestry parklands of West Africa. *Journal of Pollination Ecology* 22, 11–20.

Strange, J.P. (2009) Raising bumble bees at home: a guide to getting started. Available at: https://www.ars.usda.gov/ARSUserFiles/20800500/BumbleBeeRearingGuide.pdf.

Strange, J.P. (2010) Nest initiation in three North American bumble bees (*Bombus*): gyne number and presence of honey bee workers influence establishment success and colony size. *Journal of Insect Science* 10, 130.

Straub, L., Williams, G.R., Pettis, J., Fries, I. and Neumann, P. (2015) Superorganism resilience: eusociality and susceptibility of ecosystem service providing insects to stressors. *Current Opinion in Insect Science* 12, 109–112.

Strickler, K. (1996) Seed and bee yields as a function of forager populations: alfalfa pollination as a model system. *Journal of the Kansas Entomological Society* 69, 201–215.

Strickler, K. (1997) Flower production and pollination in *Medicago sativa* L. grown for seed: model and monitoring studies. *Acta Horticulturae* 437, 109–113.

Strickler, K. and Freitas, S. (1999) Interactions between floral resources and bees (Hymenoptera: Megachilidae) in commercial alfalfa seed fields. *Environmental Entomology* 28, 178–187.

Strickler, K. and Vinson, J.W. (2000) Simulation of the effect of pollinator movement on alfalfa seed set. *Environmental Entomology* 29, 907–918.

Stubbs, C.S. and Drummond, F.A. (1997) Management of the alfalfa leafcutting bee, *Megachile rotundata* (Hymenoptera: Megachilidae), for pollination of wild lowbush blueberry. *Journal of the Kansas Entomological Society* 70, 81–93.

Sundarasami, A., Sridhar, A. and Mani, K. (2019) Halophilic archaea as beacon for exobiology: recent advances and future challenges. *Advances in Biological Science Research.* Elsevier, 197–214.

Switzer, C.M., Hogendoorn, K., Ravi, S. and Combes, S.A. (2016) Shakers and head bangers: differences in sonication behavior between Australian *Amegilla murrayensis* (blue-banded bees) and North American *Bombus impatiens* (bumblebees). *Arthropod-Plant Interactions* 10, 1–8.

Syed, R.A. (1979) Studies on oil palm pollination by insects. *Bulletin of Entomological Research* 69, 213–224.

Szabo, N.D., Colla, S.R., Wagner, D.L., Gall, L.F. and Kerr, J.T. (2012) Do pathogen spillover, pesticide use, or habitat loss explain recent North American bumble-bee declines? *Conservation Letters* 5, 232–239.

Szalanski, A. and Tripodi, A. (2014) Assessing the utility of a PCR diagnostics marker for the identification of Africanized honey bee, *Apis mellifera* L.,(Hymenoptera: Apidae) in the United States. *Sociobiology* 61, 234–236.

Takhtajan, A. (1991) *Evolutionary trends in flowering plants.* Columbia University Press, New York.

Tasei, J. (1994) Effect of different narcosis procedures on initiating oviposition of prediapausing *Bombus terrestris* queens. *Entomologia Experimentalis et Applicata* 72, 273–279.

Tasei, J.-N. and Aupinel, P. (1994) Effect of photoperiodic regimes on the oviposition of artificially overwintered *Bombus terrestris* L. queens and the production of sexuals. *Journal of Apicultural Research* 33, 27–33.

Tautz, J. and Rostás, M. (2008) Honeybee buzz attenuates plant damage by caterpillars. *Current Biology* 18, R1125–R1126.

Teeling, E.C., Springer, M.S., Madsen, O., Bates, P., O'Brien, S.J. and Murphy, W.J. (2005) A molecular phylogeny for bats illuminates biogeography and the fossil record. *Science* 307, 580–584.

Teuber, L.R. and Thorp, R.W. (1987) The relationship of alfalfa nectar production to seed yield and honey bee visitation. *Proceedings of the Alfalfa Seed Production Symposium.* Davis, CA, 25–30.

Thien, L.B., Azuma, H. and Kawano, S. (2000) New perspectives on the pollination biology of basal angiosperms. *International Journal of Plant Sciences* 161, S225–S235.

Thogmartin, W.E., Wiederholt, R., Oberhauser, K., Drum, R.G., Diffendorfer, J.E., Altizer, S., Taylor, O.R., Pleasants, J., Semmens, D. and Semmens, B. (2017) Monarch butterfly population decline in North America: identifying the threatening processes. *Royal Society Open Science* 4, 170760.

Thomann, M., Imbert, E., Devaux, C. and Cheptou, P.-O. (2013) Flowering plants under global pollinator decline. *Trends in Plant Science* 18, 353–359.

Thompson, C.E., Biesmeijer, J.C., Allnutt, T.R., Pietravalle, S. and Budge, G.E. (2014) Parasite pressures on feral honey bees (*Apis mellifera* sp.). *PloS One* 9, e105164.

Thompson, J.N., Segraves, K.A. and Althoff, D.M. (2017) Coevolution and macroevolution. In: *Evolutionary developmental biology: a reference guide* (Nuño de la Rosa, Müller, eds.), Springer, Switzerland, 1–13.

Thomson, J.D. (1996) Trapline foraging by bumblebees: I. Persistence of flight-path geometry. *Behavioral Ecology* 7, 158–164.

Thomson, J.D. and Plowright, R. (1980) Pollen carryover, nectar rewards, and pollinator behavior with special reference to *Diervilla lonicera. Oecologia* 46, 68–74.

Thomson, S., Hansen, D., Flint, K. and Vandenberg, J. (1992) Dissemination of bacteria antagonistic to *Erwinia amylovora* by honey bees. *Plant Diseases* 76, 1052–1056.

Thorp, R.W. (1979) Structural, behavioral, and physiological adaptations of bees (Apoidea) for collecting pollen. *Annals of the Missouri Botanical Garden*, 788–812.

Thorp, R. (2005) Species profile: *Bombus franklini.* In: Shepherd, M., Vaughan, D. and Black, S. (eds) *Red list of pollinator insects of North America.* Xerces Society, Portland, Oregon.

Tiffney, B.H. and Mazer, S.J. (1995) Angiosperm growth habit, dispersal and diversification reconsidered. *Evolutionary Ecology* 9, 93–117.

Tilman, D. and Clark, M. (2014) Global diets link environmental sustainability and human health. *Nature* 515, 518–522.

Tilman, D., Isbell, F. and Cowles, J.M. (2014) Biodiversity and ecosystem functioning. *Annual Review of Ecology, Evolution, and Systematics* 45, 471–493.

Tokarev, Y.S., Huang, W.-F., Solter, L.F., Malysh, J.M., Becnel, J.J. and Vossbrinck, C.R. (2020) A formal redefinition of the genera *Nosema* and *Vairimorpha* (Microsporidia: Nosematidae) and reassignment of species based on molecular phylogenetics. *Journal of Invertebrate Pathology* 169, 107279.

Tong, Z.-Y. and Huang, S.-Q. (2018) Safe sites of pollen placement: a conflict of interest between plants and bees? *Oecologia* 186, 163–171.

Torchio, P.F. (1966) *A survey of alfalfa pollinators and polination in the San Joaquin Valley of California with emphasis on establishment of the alkali bee.* MS thesis, Oregon State University, pp. 108.

Torchio, P.F. (1976) Use of *Osmia lignaria* Say (Hymenoptera: Apoidea, Megachilidae) as a pollinator in an apple and prune orchard. *Journal of the Kansas Entomological Society* 49, 475–482.

Torchio, P.F. (1982) Field experiments with *Osmia lignaria propinqua* Cresson as a pollinator in almond orchards: III, 1977 studies (Hymenoptera: Megachilidae). *Journal of the Kansas Entomological Society* 55, 101–116.

Torchio, P.F. (1985) Field experiments with the pollinator species, *Osmia lignaria propinqua* Cresson, in apple orchards: V (1979–1980), methods of introducing bees, nesting success, seed counts, fruit yields (Hymenoptera: Megachilidae). *Journal of the Kansas Entomological Society* 58, 448–464.

Torchio, P.F. (1987) Use of non-honey bee species as pollinators of crops. *Proceedings of the Entomological Society of Ontario* 118, 111–124.

Torchio, P.F. (1990) *Osmia ribifloris*, a native bee species developed as a commercially managed pollinator of highbush blueberry (Hymenoptera: Megachilidae). *Journal of the Kansas Entomological Society* 63, 427–436.

Torchio, P.F. (1991) Bees as crop pollinators and the role of solitary species in changing environments. *Acta Horticulturae* 288, 49–61.

Torres-Ruiz, A. and Jones, R.W. (2012) Comparison of the efficiency of the bumble bees *Bombus impatiens* and *Bombus ephippiatus* (Hymenoptera: Apidae) as pollinators of tomato in greenhouses. *Journal of Economic Entomology* 105, 1871–1877.

Tosi, S., Nieh, J.C., Sgolastra, F., Cabbri, R. and Medrzycki, P. (2017) Neonicotinoid pesticides and nutritional stress synergistically reduce survival in honey bees. *Proceedings of the Royal Society B: Biological Sciences* 284, 20171711.

Traynor, J. (2017) A history of almond pollination in California. *Bee World* 94, 69–79.

Trillo, A., Montero-Castaño, A., González-Varo, J.P., González-Moreno, P., Ortiz-Sánchez, F.J. and Vilà, M. (2019) Contrasting occurrence patterns of man-

aged and native bumblebees in natural habitats across a greenhouse landscape gradient. *Agriculture, Ecosystems & Environment* 272, 230–236.

Trostle, G. and Torchio, P.F. (1994) Comparative nesting behavior and immature development of *Megachile rotundata* (Fabricius) and *Megachile apicalis* Spinola (Hymenoptera: Megachilidae). *Journal of the Kansas Entomological Society* 67, 53–72.

Tscharntke, T., Klein, A.M., Kruess, A., Steffan-Dewenter, I. and Thies, C. (2005) Landscape perspectives on agricultural intensification and biodiversity–ecosystem service management. *Ecology Letters* 8, 857–874.

Vaidya, C., Fisher, K. and Vandermeer, J. (2018) Colony development and reproductive success of bumblebees in an urban gradient. *Sustainability* 10, 1936.

Vaissière, B., Morison, N. and Subirana, M. (2001) Ineffectiveness of pollen dispensers to improve apricot pollination. *XII International Symposium on Apricot Culture and Decline* 701, 637–642.

Vaissière, B., Freitas, B.M. and Gemmill-Herren, B. (2011) *Protocol to detect and assess pollination deficits in crops: a handbook for its use.* FAO, Rome.

Van den Eijnde, J., De Ruijter, A. and Van der Steen, J. (1990) Method for rearing *Bombus terrestris* continuously and the production of bumblebee colonies for pollination purposes. *VI International Symposium on Pollination* 288, 154–158.

Van der Zee, R., Pisa, L., Andonov, S., Brodschneider, R., Charriere, J.-D., Chlebo, R., Coffey, M.F., Crailsheim, K., Dahle, B. and Gajda, A. (2012) Managed honey bee colony losses in Canada, China, Europe, Israel and Turkey, for the winters of 2008–9 and 2009–10. *Journal of Apicultural Research* 51, 100–114.

Van Ravestijn, W. and Van der Sande, J. (1990) Use of bumblebees for the pollination of glasshouse tomatoes. *VI International Symposium on Pollination* 288, 204–212.

Vandenberg, J.D., Fichter, B.L. and Stephen, W. (1980) Spore load of *Ascosphaera* species on emerging adults of the alfalfa leafcutting bee, *Megachile rotundata*. *Applied and Environmental Microbiology* 39, 650–655.

Vansell, G.H. and Todd, F.E. (1946) Alfalfa tripping by insects. *Journal of the American Society of Agronomy* 38, 470–488.

Varatharajan, R., Maisnam, S., Shimray, C.V. and Rachana, R. (2016) Pollination potential of thrips (Insecta: Thysanoptera)–an overview. *Zoo's Print* 31, 6–12.

Vaudo, A.D., Biddinger, D.J., Sickel, W., Keller, A. and López-Uribe, M.M. (2020) Introduced bees (*Osmia cornifrons*) collect pollen from both coevolved and novel host-plant species within their family-level phylogenetic preferences. *Royal Society Open Science* 7, 200225.

Vázquez, D.P. and Aizen, M.A. (2004) Asymmetric specialization: a pervasive feature of plant–pollinator interactions. *Ecology* 85, 1251–1257.

Vázquez, D.P., Morris, W.F. and Jordano, P. (2005) Interaction frequency as a surrogate for the total effect of animal mutualists on plants. *Ecology Letters* 8, 1088–1094.

Veddeler, D., Olschewski, R., Tscharntke, T. and Klein, A.-M. (2008) The contribution of non-managed social bees to coffee production: new economic insights based on farm-scale yield data. *Agroforestry Systems* 73, 109–114.

Velthuis, H.H. and Van Doorn, A. (2006) A century of advances in bumblebee domestication and the economic and environmental aspects of its commercialization for pollination. *Apidologie* 37, 421–451.

Viana, B.F., Da Encarnação Coutinho, J.G., Garibaldi, L.A., Bragança Gastagnino, G.L., Peres Gramacho, K. and Oliveira Da Silva, F. (2014) Stingless bees further improve apple pollination and production. *Journal of Pollination Ecology* 14, 261–269.

Vicens, N. and Bosch, J. (2000a) Nest site orientation and relocation of populations of the orchard pollinator *Osmia cornuta* (Hymenoptera: Megachilidae). *Environmental Entomology* 29, 69–75.

Vicens, N. and Bosch, J. (2000b) Pollinating efficacy of *Osmia cornuta* and *Apis mellifera* (Hymenoptera: Megachilidae, Apidae) on 'red Delicious' apple. *Environmental Entomology* 29, 235–240.

Villanueva-G, R. (1994) Nectar sources of European and Africanized honey bees (*Apis mellifera* L.) in the Yucatán peninsula, Mexico. *Journal of Apicultural Research* 33, 44–58.

Villanueva-G, R., Roubik, D.W. and Colli-Ucán, W. (2005) Extinction of *Melipona beecheii* and traditional beekeeping in the Yucatán peninsula. *Bee World* 86, 35–41.

Villanueva-G, R., Roubik, D.W., Colli-Ucán, W., Güemez-Ricalde, F.J. and Buchmann, S.L. (2013) A critical view of colony losses in managed Mayan honey-making bees (Apidae: Meliponini) in the heart of Zona Maya. *Journal of the Kansas Entomological Society* 86, 352–362.

Vinchesi, A. and Walsh, D. (2014) Quadrat method for assessing the population abundance of a commercially managed native soil-nesting bee, *Nomia melanderi* (Hymenoptera: Halictidae), in proximity to alfalfa seed production in the Western United States. *Journal of Economic Entomology* 107, 1695–1699.

Visscher, P., Vetter, R. and Baptista, F. (1997) Africanized bees, 1990–1995: initial rapid invasion has slowed in the US. *California Agriculture* 51, 22–24.

Von Frisch, K. and Chadwick, L.E. (1967) *The dance language and orientation of bees*. Belknap Press of Harvard University Press, Cambridge, MA.

Waddington, K.D. (1980) Flight patterns of foraging bees relative to density of artificial flowers and distribution of nectar. *Oecologia* 44, 199–204.

Wallace, H. and Lee, L. (1999) Pollen source, fruit set and xenia in mandarins. *The Journal of Horticultural Science and Biotechnology* 74, 82–86.

Wallberg, A., Han, F., Wellhagen, G., Dahle, B., Kawata, M., Haddad, N., Simões, Z.L.P., Allsopp, M.H., Kandemir, I. and De La Rúa, P. (2014) A worldwide survey of genome sequence variation provides insight into the evolutionary history of the honeybee *Apis mellifera*. *Nature Genetics* 46, 1081.

Walther-Hellwig, K. and Frankl, R. (2000) Foraging habitats and foraging distances of bumblebees, *Bombus* spp. (Hym., Apidae), in an agricultural landscape. *Journal of Applied Entomology* 124, 299–306.

Wang, D.-M., Meng, M.-C. and Guo, Y. (2016) Pollen organ *Telangiopsis* sp. of Late Devonian seed plant and associated vegetative frond. *PloS One* 11, e0147984.

Wardhaugh, C.W. (2015) How many species of arthropods visit flowers? *Arthropod-Plant Interactions* 9, 547–565.

Wardhaugh, C.W., Edwards, W. and Stork, N.E. (2013) Variation in beetle community structure across five microhabitats in Australian tropical rainforest trees. *Insect Conservation and Diversity* 6, 463–472.

Wardlaw, I.F. (1968) The control and pattern of movement of carbohydrates in plants. *The Botanical Review* 34, 79–105.

Webster, T., Thorp, R., Briggs, D., Skinner, J. and Parisian, T. (1985) Effects of pollen traps on honey bee (Hymenoptera: Apidae) foraging and brood rearing during almond and prune pollination. *Environmental Entomology* 14, 683–686.

West, T.P. and McCutcheon, T.W. (2009) Evaluating *Osmia cornifrons* as pollinators of highbush blueberry. *International Journal of Fruit Science* 9, 115–125.

Westerkamp, C. (1991) Honeybees are poor pollinators—why? *Plant Systematics and Evolution* 177, 71–75.

Westerkamp, C. (1996) Pollen in bee-flower relations some considerations on melittophily. *Botanica Acta* 109, 325–332.

Westphal, C., Steffan-Dewenter, I. and Tscharntke, T. (2003) Mass flowering crops enhance pollinator densities at a landscape scale. *Ecology Letters* 6, 961–965.

Westphal, C., Steffan-Dewenter, I. and Tscharntke, T. (2009) Mass flowering oilseed rape improves early colony growth but not sexual reproduction of bumblebees. *Journal of Applied Ecology* 46, 187–193.

Whittington, R., Winston, M.L., Tucker, C. and Parachnowitsch, A.L. (2004) Plant-species identity of pollen collected by bumblebees placed in greenhouses for tomato pollination. *Canadian Journal of Plant Science* 84, 599–602.

WHO (2009) *Global prevalence of Vitamin A deficiency in populations at risk 1995–2005: WHO global data-*

base on Vitamin A deficiency. World Health Organization, Geneva.

Wichelns, D., Weaver, T.F. and Brooks, P.M. (1992) Estimating the impact of alkali bees on the yield and acreage of alfalfa seed. *Journal of Production Agriculture* 5, 512–518.

Wietzke, A., Westphal, C., Gras, P., Kraft, M., Pfohl, K., Karlovsky, P., Pawelzik, E., Tscharntke, T. and Smit, I. (2018) Insect pollination as a key factor for strawberry physiology and marketable fruit quality. *Agriculture, Ecosystems & Environment* 258, 197–204.

Wilcock, C. and Neiland, R. (2002) Pollination failure in plants: why it happens and when it matters. *Trends in Plant Science* 7, 270–277.

Wilkaniec, Z. and Giejdasz, K. (2003) Suitability of nesting substrates for the cavity-nesting bee *Osmia rufa*. *Journal of Apicultural Research* 42, 29–31.

Wille, A. and Michener, C.D. (1973) The nest architecture of stingless bees with special reference to those of Costa Rica (Hymenoptera, Apidae). *Revista de Biologia Tropical* 21, 9–279

Williams, P.H. (1985) A preliminary cladistic investigation of relationships among the bumble bees (Hymenoptera, Apidae). *Systematic Entomology* 10, 239–255.

Williams, P.H. (1998) An annotated checklist of bumble bees with an analysis of patterns of description (Hymenoptera: Apidae, Bombini). *Bulletin of the British Museum (Natural History) Entomology* 67, 79–152.

Williams, P. (2005) Does specialization explain rarity and decline among British bumblebees? a response to Goulson *et al. Biological Conservation* 122, 33–43.

Williams, I.H., Carreck, N. and Little, D. (1993) Nectar sources for honey bees and the movement of honey bee colonies for crop pollination and honey production in England. *Bee World* 74, 160–175.

Williams, N.M., Regetz, J. and Kremen, C. (2012) Landscape-scale resources promote colony growth but not reproductive performance of bumble bees. *Ecology* 93, 1049–1058.

Williams, N.M., Ward, K.L., Pope, N., Isaacs, R., Wilson, J., May, E.A., Ellis, J., Daniels, J., Pence, A. and Ullmann, K. (2015) Native wildflower plantings support wild bee abundance and diversity in agricultural landscapes across the United States. *Ecological Applications* 25, 2119–2131.

Willig, R. (1976) Consumer's surplus without apology. *American Economic Review* 66, 589–597.

Willmer, P.G. and Finlayson, K. (2014) Big bees do a better job: intraspecific size variation influences pollination effectiveness. *Journal of Pollination Ecology* 14, 244–254.

Wilson, E.O. (1971) *The insect societies*. Belknap Press, Harvard Universtiy Press, Cambridge, MA.

Wilson, E.O. and Hölldobler, B. (2009) *The superorganism: the beauty, elegance, and strangeness of insect societies*. W. W Norton & Company, New York.

Wilson, E.O. and Kinne, O. (1990) *Success and dominance in ecosystems: the case of the social insects*. Ecology Institute, Oldendorf/Luhe, Germany.

Winfree, R. and Kremen, C. (2009) Are ecosystem services stabilized by differences among species? A test using crop pollination. *Proceedings of the Royal Society B: Biological Sciences* 276, 229–237.

Winfree, R., Williams, N.M., Dushoff, J. and Kremen, C. (2007) Native bees provide insurance against ongoing honey bee losses. *Ecology Letters* 10, 1105–1113.

Winfree, R., Williams, N.M., Gaines, H., Ascher, J.S. and Kremen, C. (2008) Wild bee pollinators provide the majority of crop visitation across land-use gradients in New Jersey and Pennsylvania, USA. *Journal of Applied Ecology* 45, 793–802.

Winfree, R., Aguilar, R., Vázquez, D.P., Lebuhn, G. and Aizen, M.A. (2009) A meta-analysis of bees' responses to anthropogenic disturbance. *Ecology* 90, 2068–2076.

Winfree, R., Gross, B.J. and Kremen, C. (2011) Valuing pollination services to agriculture. *Ecological Economics* 71, 80–88.

Winston, M. (1987) *The biology of the honey bee*. Harvard University Press, Cambridge, MA.

Winter, K., Adams, L., Thorp, R., Inouye, D., Day, L., Ascher, J. and Buchmann, S. (2006) *Importation of non-native bumble bees into North America: potential consequences of using* Bombus terrestris *and other non-native bumble bees for greenhouse crop pollination in Canada, Mexico, and the United States*. North American Pollinator Protection Campaign, San Francisco.

Wolf, S. and Moritz, R.F. (2008) Foraging distance in *Bombus terrestris* L.(Hymenoptera: Apidae). *Apidologie* 39, 419–427.

Wolfe, L.M. and Sowell, D.R. (2006) Do pollination syndromes partition the pollinator community? A test using four sympatric morning glory species. *International Journal of Plant Sciences* 167, 1169–1175.

Woodcock, B., Edwards, M., Redhead, J., Meek, W., Nuttall, P., Falk, S., Nowakowski, M. and Pywell, R. (2013) Crop flower visitation by honeybees, bumblebees and solitary bees: behavioural differences and diversity responses to landscape. *Agriculture, Ecosystems & Environment* 171, 1–8.

Woodcock, B., Savage, J., Bullock, J., Nowakowski, M., Orr, R., Tallowin, J. and Pywell, R. (2014) Enhancing floral resources for pollinators in productive agricultural grasslands. *Biological Conservation* 171, 44–51.

Yanega, D. (1988) Social plasticity and early-diapausing females in a primitively social bee. *Proceedings of the National Academy of Sciences* 85, 4374–4377.

Yang, D. (1999) The status of species diversity and conservation strategy of bumble bees, a pollination insect in Lancang River Basin of Yunnan, China. *Chinese Biodiversity* 7, 170–174.

Yang, G.-H. (2005) Harm of introducing the western honeybee *Apis mellifera* L. to the Chinese honeybee

Apis cerana F. and its ecological impact. *Acta Entomologica Sinica* 3, 015.

Yang, G.-H., Li, J.-P. and Li, M.-H. (2005) Studies on pollinating soybean male sterile plant in caged plots using bumble bee (*Bombus ignites*) and alfalfa leaf-cutting bee (*Megachile rotundata*). *Jilin Agricultural Sciences* 3, 21–22, 28.

Yoneda, M., Furuta, H., Tsuchida, K., Okabe, K. and Goka, K. (2008) Commercial colonies of *Bombus terrestris* (Hymenoptera: Apidae) are reservoirs of the tracheal mite *Locustacarus buchneri* (Acari: Podapolipidae). *Applied Entomology and Zoology* 43, 73–76.

Yucel, B. and Duman, I. (2005) Effects of foraging activity of honeybees (*Apis mellifera* L.) on onion (*Allium cepa*) seed production and quality. *Pakistan Journal of Biological Sciences* 8, 123–126.

Zayed, A. and Packer, L. (2005) Complementary sex determination substantially increases extinction proneness of haplodiploid populations. *Proceedings of the National Academy of Sciences* 102, 10742–10746.

Zheng, H.-Q., Gong, H.-R., Huang, S.-K., Sohr, A., Hu, F.-L. and Chen, Y.P. (2015) Evidence of the synergistic interaction of honey bee pathogens *Nosema ceranae* and deformed wing virus. *Veterinary Microbiology* 177, 1–6.

Zisovich, A., Goldway, M., Schneider, D., Steinberg, S., Stern, E. and Stern, R. (2012) Adding bumblebees (*Bombus terrestris* L., Hymenoptera: Apidae) to pear orchards increases seed number per fruit, fruit set, fruit size and yield. *The Journal of Horticultural Science and Biotechnology* 87, 353–359.

Zurbuchen, A., Landert, L., Klaiber, J., Müller, A., Hein, S. and Dorn, S. (2010) Maximum foraging ranges in solitary bees: only few individuals have the capability to cover long foraging distances. *Biological Conservation* 143, 669–676.

Index

Note: Page numbers in **bold** type refer to **figures.** Page numbers in *italic* type refer to *tables*